ROCK MECHANICS FOR NATURAL RESOURCES AND INFRASTRUCTURE DEVELOPMENT

Proceedings in Earth and geosciences series

The Proceedings in Earth and geosciences series contains proceedings of peer-reviewed international conferences dealing in earth and geosciences. The main topics covered by the series include: geotechnical engineering, underground construction, mining, rock mechanics, soil mechanics and hydrogeology.

Volume 5

ISSN: 2639-7749
eISSN: 2639-7757

PROCEEDINGS OF THE 14TH INTERNATIONAL CONGRESS ON ROCK MECHANICS AND ROCK ENGINEERING (ISRM 2019), FOZ DO IGUASSU, BRAZIL, 13-18 SEPTEMBER 2019

Rock Mechanics for Natural Resources and Infrastructure Dcvelopment

- Invited Lectures

Editors

Sergio A.B. da Fontoura

Department of Civil Engineering, Pontifical Catholic University of Rio de Janeiro, Rio de Janeiro, Brazil

Ricardo José Rocca

School of Civil Engineering, National University of Cordoba, Cordoba, Argentina

José Félix Pavón Mendoza

Jose Pavon & Associates, Asuncion, Paraguay

CRC Press
Taylor & Francis Group
Boca Raton London New York

CRC Press is an imprint of the
Taylor & Francis Group, an **informa** business
A BALKEMA BOOK

CRC Press/Balkema is an imprint of the Taylor & Francis Group, an informa business

Typeset by Integra Software Services Pvt. Ltd., Pondicherry, India

Library of Congress Cataloging-in-Publication Data

Applied for

Published by: CRC Press/Balkema
Schipholweg 107C, 2316 XC Leiden, The Netherlands

First issued in paperback 2023

ISBN: 978-1-03-257091-4 (pbk)
ISBN: 978-0-367-42285-1 (hbk)
ISBN: 978-0-367-82318-4 (ebk)

DOI: 10.1201/9780367823184

Proceedings in Earth and geosciences
Volume 5
Proceedings in Earth and geosciences (Print) ISSN 2639-7749
Proceedings in Earth and geosciences (Online) ISSN 2639-7757

Publisher's Note
The publisher has gone to great lengths to ensure the quality of this reprint but points out that some imperfections in the original copies may be apparent.

Rock Mechanics for Natural Resources and Infrastructure Development –
Fontoura, Rocca & Pavón Mendoza (Eds)
© 2020 ISRM, ISBN 978-0-367-42285-1

Table of contents

Rock Mechanics for Natural Resources and Infrastructure Development –
Fontoura, Rocca & Pavón Mendoza (Eds)
© 2020 ISRM, ISBN 978-0-367-42285-1

Preface

Every four years, the International Society for Rock Mechanics and Rock Engineering, ISRM, holds its Congress, a series that started in 1966 in Lisbon, Portugal, and that it is now in its 14th edition. ISRM Council awarded in 2014 the task of organizing its 2019 International Congress to the National Groups of Brazil, Argentina and Paraguay; it is with great honor that the Chairs of the Congress present this volume to the rock mechanics community. As in the previous occasions, the Congress mobilizes researchers, professors, engineers and students around contemporaneous themes relevant to rock mechanics and rock engineering. This volume contains the written version of 7 Keynote Lectures and 449 papers distributed over ten chapters, covering up-to-date studies in topics that cover fundamental research in rock mechanics, laboratory and experimental field studies, petroleum, mining and civil engineering applications.

The Proceedings also include the prestigious ISRM Award Lectures, namely, the Leopold Muller Award Lecture, this time around delivered by professor Peter K. Kaiser and the Manuel Rocha Award Lecture, and delivered by Dr. Quinghua Lei.

Rock Mechanics for Natural Resources and Infrastructure Development –
Fontoura, Rocca & Pavón Mendoza (Eds)
© 2020 ISRM, ISBN 978-0-367-42285-1

Committees

ADVISORY COMMITTEE

The advisory committee, according to the ISRM By-Laws, is made up of the members of the ISRM Board. For the 14th ISRM Congress, the ISRM Board 2015-2019 members are listed below.

Eda Quadros (*ISRM President*)
William Joughin (*Vice President for Africa*)
Charlie Chunlin Li (*Vice President for Europe*)
Doug Stead (*Vice President for North America*)
Seokwon Jeon (*Vice President for Asia*)
Sergio A.B. da Fontoura (*Vice President for South America*)
Stuart Read (*Vice President for Australasia*)
Manchao He (*Vice President at-Large*)
Petr Konicek (*Vice President at-Large*)
Norikazu Shimizu (*Vice President at-Large*)
Resat Ulusay (*ISRM President-elected 2019-2023*)
Luis Lamas (*ISRM General Secretary*)

EXECUTIVE ORGANIZING COMMITTEE

Sergio A.B. da Fontoura (*Chair*)
José Félix Pavón Mendoza (*Co-Chair*)
Ricardo José Rocca (*Co-Chair*)
Vivian Rodrigues Marchesi (*General Secretary*)
Carlos Emmanuel R. Lautenschlager (*General Secretary*)
Lineu A. Ayres da Silva (*Treasurer*)
Anna Luiza M. Ayres da Silva (*Treasurer*)
Bismarck Gomes Souza Junior (*Editorial Committee*)
Guilherme Lima Righetto (*Editorial Committee*)
Lilia Maria Cruz Metzger Raymundo (*Executive Secretary*)
MCI Group (*Executive Secretariat*)
Levitatur (*Travel Agency*)

PAPER REVIEW COMMITTEE

Abdelkareem Alzo'ubi
Alvaro Gonzalez
Anna Maria Ferrero
Antonio Samaniego
Ariel Bustamante
Bojana Grujic
Conrad Boley
Dominique J M Ngan-Tillard
Erika Prina Howald
Essaieb Hamdi
Eva Friedman
Fanny Descamps
Frederic Pellet
Gharouni Nik
Igor Pesevski
Iliya Garkov
Ito Takatoshi
José Muralha

José Pavon
Juha Antikainen
Ki-Bok Min
Kirsten Laackmann
Koichi Shin
Krishna Panthi
Laura Pyrak Nolte
M. Sharifzadeh
Manchao He
Martin Grenon
Mauro Menéndez
Michael Du Plessis
Michael Tsesarsky
Nikoletta Rozgonyi-Boissinot
Oldemar Bermudez
Patricio Gomez
Patrick Mushangwe
Petr Konicek
Pham Quoc Tuan
Phung Manh Dac
Predrag Miscevic
Prem Krishna K.C.
Resat Ulusay
Rhido K. Wattimena
Ricardo Rocca
Rini Asnida Abdullah
Salma Soussi
Sevda Dehkhoda
Stuart Read
T. Okada
Tai-Tien Wang
U. Kar Winn
Uday Chander
Valentin Castellano Pedroza
Vladimir Noskov
Wulf Schubert

PAPER REVIEWERS

Abbas Taheri
Aldo Farfan
Alvaro J. Gonzalez-Garcia
Ana Luiza M. Ayres Da Silva
Andrea Segalini
Angelo Zenobio
Anselmo Machado Borba
Ausama Giwelli
Aydin Bilgin
Bailin Wu
Baotang Shen
Bismarck Gomes Souza Junior
Bona Park
Carla Carrapatoso
Carlos Emmanuel R. Lautenschläger
Chaoshui Xu
Charlie Li
Christophe Auvray
Clóvis Gonzatti
Conrad Boley
Cristhian Bernardo M. Monsalve
Dashnor Hoxha
Deepak Adhikary
Didier Subrin
Eda Quadros
Eduardo César Sansone
Erast Gaziev
Ergun Tuncay
Erick Slis Raggio Santos
Erik Johansson
Erika Prina Howald
Ero Vinicius Silva
Eva Hrubesova
Federico Vagnon
Fredrik Johansson
Geoff Kilgour
Gessica Umili
Goh Thian Lai
Guilherme Lima Righetto
Gunzburger Yann
Hideaki Yasuhara
Hiroshi Morioka
Igor Fernandes Gomes
Isabel Reig
Ismet Canbulat
Iñaki García Mendive
Jacopo Abbruzzese
Jae-joon Song
Jaewon Lee
Jair Carlos Koppe
Jannie Maritz
Javier González-Gallego
Jean Sulem
Jean-Michel Pereira
Joachim Stahlmann
Jochen Fillibeck
Joel Sarout
John Henning
Jorge Lopez Molina
José Serón
João Armelin
Jung-Wook Par
Junichi Kodama
Ken Mills
Koji Uenishi
Konstantin Morozov

krishna Kanta Panthi
Kwang-yeom Kim
Larisa Nazarova
Lauri Uotinen
Leandro R. Alejano Monge
Leonardo Cabral Pereira
Les Gardner
Luc Beauchamp
Luis Arnaldo Mejía Camones
Maja Prskalo
Makoto Ishimaru
Manchao He
Manoj Verman
Marc Panet
Marc-Andre Brideau
Marcelo Heidemann
Maria Migliazza
Marina Pirulli
Martin Feinendegen
Martin Grenon
Masaji Kato
Mauro Muñiz Menéndez
Maxim Karasev
Mehdi Ghoreychi
Mehdi Serati
Michael Du Plessis
Michael Tsesarsky
Miguel Cano
Miguel Stanichevsky
Milan Broz
Milene Sabino Lana
Mohsen Nicksiar
Mostafa Sharifzadeh
Murat Karakus
Muriel Gasc
Nathalia Christina Passos
Nelson Inoue
Nicolas Gatelier
Nicolas Guy
Omberai Mandingaisa
Patricio Gomez
Paul Couto
Paulo Cesar De A. Maia
Pawan Kumar Shrestha
Pedro Alameda-Hernández
Petr Konicek
Phil Dight
Philippe Vaskou
Pierre Berest
Prem Krishna Kc
Qianbing Zhang
R.K. Goel
Raquel Quadros Velloso
Resat Ulusay
Reuber Cota
Ricardo J. Rocca
Roberto Juan Quevedo Quispe
Roberto M. Flores
Roberto Tomas
Sam Proskin
Sergio A.B. Da Fontoura
Sevda Dehkhoda
Siegfried Maiolino
Silvia Garcia
Sripad Ramachandra Naik
Stuart Read
Svetlana Melentijevic
Takashi Sasaoka
Talita Caroline Miranda
Teijiro Saito
Topias Siren
Toru Takahashi
Valentin Castellanos
Victor Rechitskiy
Vincent Maury
Vivian Rodrigues Marchesi
Vladimir Noskov
William Joughin
Wulf Schubert
Yoshitaka Nara
Youn-kyou Lee
Young Zoo Lee

Acknowledgements

To put together such a volume requires a great amount of work and this cannot be done without the contribution of many people or organizations. First and foremost, the editors would like to thank Dr. Guilherme Lima Righetto and M.S. Bismarck Gomes Souza Junior for their work managing the paper submission and review processes. Without their intense and focused work these proceedings would not exist. The efforts by all the reviewers are acknowledged. Thanks are extended to Dr. Vivian Rodrigues Marchesi and Dr. Carlos Emmanuel R. Lautenschlager who helped in many ways, especially during the final moments, the edition of this volume. Many thanks to the International Society for Rock Mechanics and Rock Engineering (ISRM) Board (2015-2019), Brazilian Association of Soil Mechanics and Geotechnical Engineering (ABMS) and to the Brazilian Rock Mechanics Committee (CBMR) for the decisive support offered to the Organizing Committee during the preparation of the Congress. Thanks are due to the Brazilian National Research Council (Conselho Nacional de Pesquisas e Desenvolvimento – CNPq) and to CAPES (Coordenação de Aperfeiçoamento de Pessoal de Nível Superior), Agency from the Ministry of Education of Brazil, the funds to publish these proceedings. Thanks are extended to the authors of the papers and to the publisher, Taylor & Francis Group.

Rock Mechanics for Natural Resources and Infrastructure Development – Fontoura, Rocca & Pavón Mendoza (Eds)
© 2020 ISRM, ISBN 978-0-367-42285-1

About the editors

Sergio A.B. da Fontoura graduated in Civil Engineering in 1972 and obtained a PhD degree in Geotechnical Engineering at the University of Alberta, Canada, in 1980. He has served as President of the Brazilian Tunneling Committee, Brazilian Rock Mechanics Committee and, at the time of this publication, serves as ISRM Vice-President for South America. He is a professor at the Civil Engineering Department of the Pontifical Catholic University of Rio de Janeiro, Brazil.

Ricardo J. Rocca graduated in Geological Engineering in 1979 (U.N. Cordoba) and earned a M. Eng. degree at the University of California, Berkeley (1986). He is the Secretary of the ISRM National Group of Argentina. He is Correspondent Member of the National Academy of Engineering of Argentina and Emeritus Professor of the National University of Cordoba, Argentina.

José F. Pavon graduated in Civil Engineering in 1991 and obtained an MSc degree in Geotechnical Engineering at the University of São Paulo, Brazil, in 1994. He is the present President of the Paraguayan Geotechnical Society since 2016. Nowadays he is a professor at the Civil Engineering Department of the Catholic University in Asuncion, Paraguay. Besides his academic achievements, he is also the founder and Director of a Consultant Engineering Company focusing on Geotechnical area.

Keynote Lectures

Rock Mechanics for Natural Resources and Infrastructure Development –
Fontoura, Rocca & Pavón Mendoza (Eds)
© 2020 Taylor & Francis Group, London, ISBN 978-0-367-42284-4

The role of rock mechanics in oil field development

E. Fjær
SINTEF Industry, Trondheim, Norway

ABSTRACT: Rock mechanics play a crucial role in oil field development. The hydrocarbon resources are resting in porous rocks, and most reservoirs can only be accessed by drilling thousands of meters through rocks of highly variable consistency. It has been estimated that 5-10% of the drilling costs in challenging areas are related to rock instability problems, offering a huge cost cutting potential for rock mechanical applications. Rock instability may also be a problem during production from sand and chalk reservoirs, while in some areas induced fracturing of the rock is crucial for economical production. Recently, it has been shown that rock mechanics may offer a huge cost saving potential during field abandonment, as the rock itself may restore the sealing capability of the cap rock given the right conditions. On a larger scale, reservoir compaction and surface subsidence also imply rock mechanical challenges.

1 THE BEGINNING

The history of the petroleum industry goes back to the 1850'ies, and over the following century it grew to become one of the most important industries worldwide. For a long time however, the industry had little interest in the mechanical aspects of the rock from where these valuable resources were harvested. Even in the late 1970'ies the petroleum engineers paid little attention to geomechanical analyses, and statements like "we just drill the damn thing" reflected a common attitude. At that time, things were about to change, however. Spurred by the rapid increase in the oil price during that decade (Figure 1), the desire to produce from deeper and less permeable reservoirs called for solutions where rock mechanics had something to offer. Rock mechanics emerged as a subject of interest for the industry.

2 FRACTURING

In the early 1980'ies, hydraulic fracturing became a hot topic in the industry. This stimulation technique, which had been used commercially in the industry since the early 1950'ies, implies that the fluid pressure within the well is increased until the formation around the well cracks up and a fracture is formed from the well deep into the formation. The free surface, through which hydrocarbons can flow into the well, is then multiplied by orders of magnitude. This enhances the productivity of the well, especially in low permeable reservoirs.

Rock stiffness affects the size and shape of the fracture. The minimum principal in situ stress has a key role in the fracturing process. It controls the opening of the fracture, and it also controls the direction where the stress grows. Stress gradients and stress contrasts control the extent of the fracture, hence it is important to have reasonable estimates of these parameters up front. Thus, contributions from rock mechanics is essential for a successful fracturing operation.

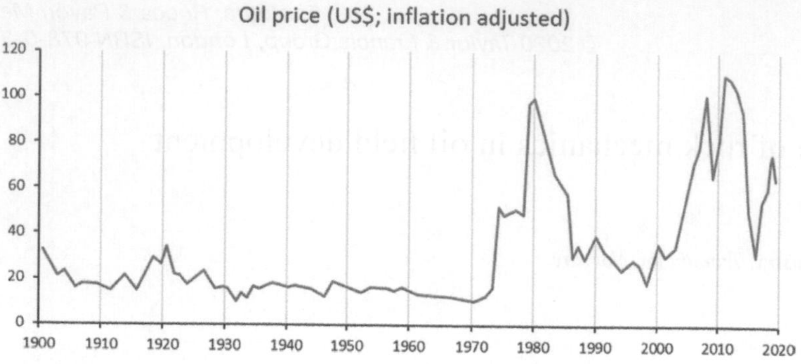

Figure 1. Development of the oil price.

The dramatic increase in the oil price during the years from 2004 to 2008 opened up a new arena for large scale hydrocarbon production: shale gas. This implies production from very low permeable rock, making hydraulic fracturing an absolute necessity for economic production. This activity spurred renewed interest in the well-known concept brittleness, as optimization the fracturing process requires identification of the brittle formations. Several definitions of a quantitative number describing brittleness – a brittleness index – already existed, and many more have appeared since, adding confusion as much as improved description. The main reason for this somewhat chaotic situation is that rock failure is a complex process that cannot be precisely described by a single parameter. The various definitions have emerged from different data sets combining different mechanical parameters with different measurable quantities.

Hydraulic fracturing – or fracking – is in many ways a brilliant method for enhancing productivity and making production feasible from otherwise unproductive fields. Environmental concerns have triggered strong public resistance, however. Fracture growth into aquifers and ground water supplies is a major concern. Moreover, the method requires huge amounts of water – in the order of 10,000 m^3 per well – which is extracted from local resources. The method also uses several harmful substances that may potentially contaminate the ground water. For these reasons, the method has been banned in some countries. Its rather poor reputation is not entirely justified though. It appears that most reported incidents of groundwater contamination associated with fracturing originate from leakage along the well rather than from the fractures themselves.

3 RESERVOIR GEOMECHANICS

In 1984 it was discovered that a platform on the Ekofisk field in the North Sea was slowly sinking, revealing subsidence of the seafloor above the reservoir 3 kilometers below. The oil production had reduced the pore pressure, leading to corresponding higher effective stress on the rock and compaction of the reservoir. When a reservoir compacts, the surrounding rock will move in as more space becomes available, leading to deformations and stress alterations in a much larger volume than the reservoir itself. To some extent, a stress arch will be established in the rock above the reservoir, making the adjacent rock taking over a part of the weight of the overlaying rock mass (Figure 2). This effect – essentially the same as the stress arch above an open tunnel – reduces further compaction of the reservoir, and prevents most of the rock movement from reaching the surface. The efficiency of the stress arch depends on size and depth of the reservoir, the reservoir geometry, and the mechanical properties of the rock within and around the reservoir. In most cases, it is sufficiently efficient so that the subsidence of the surface is hardly detectable and has no practical consequences. In later years, there has been a growing interest however, to map out surface subsidence by use of highly sensitive techniques, as a means to monitor the drainage patterns within the reservoir.

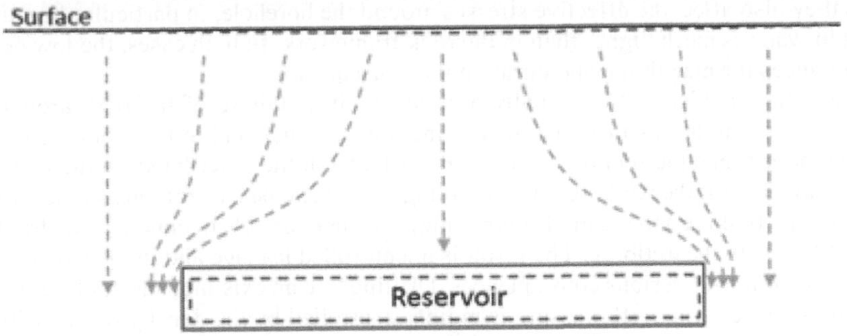

Figure 2. Stress arching above a shrinking reservoir.

Even though surface subsidence above reservoirs rarely leads to practical problems, the problems can be huge if they occur. Thus, it is important to be able to predict up front whether subsidence will happen, and plan accordingly. For the Ekofisk operator, a consequence of the unexpected subsidence was a jack up operation on the platforms, at a cost of 600 million USD. In addition, a massive water injection program was started in order to maintain the reservoir pressure. The reservoir compaction has not been entirely problematic, however. On the contrary, the high porosity chalk this reservoir consist of has been cracking up and to some extent collapsing as a result of the production, leading to enhanced productivity and an extension of the lifetime of the reservoir way beyond any expectations when production started nearly 50 years ago.

4 DRILLING

Towards the end of 1986 the oil price fell by almost 50% within a few weeks. Cutting costs suddenly became a top priority. This did not mean that the industry lost interest in rock mechanics. On the contrary, it was realized that rock mechanics might hold the key to significant cost reductions. Drilling costs are generally high – today the cost of a single well may reach 100 million dollars – and it was found that borehole instability problems were responsible for about 5–10% of the drilling costs. This was obviously a target for cost cutting where rock mechanical analyses could contribute. R & D budgets for studies on borehole stability increased considerably over the following years, and several oil companies established their own rock mechanics laboratory. Shales, which had attracted little interest in the past, now became an important rock type to study. The structure of a typical oil field is such that it is necessary to drill through kilometers of shale layers to reach the reservoir, and most drilling problems occur in shale formations.

Shales are complex materials – even the definition of a shale is often questioned. From a rock mechanical point of view, it is natural to define a shale as a rock with more than 40% clay minerals, which implies that the clay minerals constitute a load-bearing framework. Due to their low permeability, typically in the nanoDarcy range, shales are excellent cap rocks – i.e. the top seal that has prevented the hydrocarbons from escaping from the reservoir and migrating up to the surface. Shales are highly water sensitive, hence the chemical content of the drilling mud plays an important role for avoiding drilling problems, although borehole stability as such is a rock mechanical problem. Understanding how fluid chemistry affects the mechanical stability of shales is therefore vital for avoiding drilling problems. For instance, osmotic effects induced by different water activities in the drilling mud and the pore fluid induces local variations in the pore pressure, and hence the effective stresses, in the rock around the hole. This may be utilized to ensure stability of a borehole if handled correctly – or cause severe problems otherwise. Thermal effects may also be utilized or make trouble as they also affect the effective stresses around the borehole, in particular because thermal expansion in water is much higher than in the rock framework. In both cases, the low permeability of shale enhances the magnitude and duration the consequences.

Borehole collapse (Figure 3) is usually a result of shear failure of the rock around the hole, induced by too low well pressure. The drill string may be stuck and in the worst case it has to be cut and abandoned, and the section has to be re-drilled. On the other hand, unintended hydraulic fracturing may occur if the well pressure is too high. This can be a hazardous situation, since mud will flow into the fracture and the mud column may no longer be able to uphold its hydraulic support and prevent flow into the wellbore. The result is uncontrolled leakage and in the worst case it could end up in a blowout, with serious consequences. Drilling into an existing, natural fracture may have similar consequences. The well pressure is mainly controlled by the density of the drilling mud – the mud weight – and the range between the collapse limit and the fracturing limit is called the mud weight window. While it is the job of the driller to keep the entire open section of the hole within the mud weight window, it is the job of the rock mechanics specialist to predict the location of the mud weight limits prior to drilling – a challenging job indeed, given the complexity of the problem and not the least the inherent lack of reliable data.

Today, the share of drilling costs that can be traced back to borehole instability problems is still about 5–10%, i.e. more or less the same fraction as it has been for many years. This does not mean that all the years of dedicated R & D on this subject have been in vain, however. It rather reflects that many of the wells drilled today are far more challenging than they were in the 1980'ies. For instance, high inclination wells are now common, while they were almost unthinkable some forty years ago. An often cited paper by Bradley published in 1979 analyzed the stability of non-vertical wells, and as an example it showed that under given conditions it might be impossible to maintain a stable borehole for well inclinations larger than 60 degrees, hence it was impossible to drill a horizontal well. This was just an example, but somehow the industry got stuck in the misconception that it was not possible to drill horizontal wells in general, and well planning was adjusted accordingly. Today, horizontal wells are drilled without major problems in formations with unconfined strength below 30 MPa.

The improved understanding of the rock mechanical aspects of drilling that has been achieved over the years does not imply that all problems have been solved once and for all, however. Each well is a unique case and needs individual analyses, but these are highly profitable investments. Stjern et al. (2003) estimated that rock mechanical analyses such as identification of troublesome zones, optimized mud weight and salt content etc., enabled savings in the order of 2.5 million USD for an average well on the Heidrun field offshore Norway. For a field with 50-100 wells, the potential for savings by including rock mechanical analyses is huge. Other operators report similar numbers.

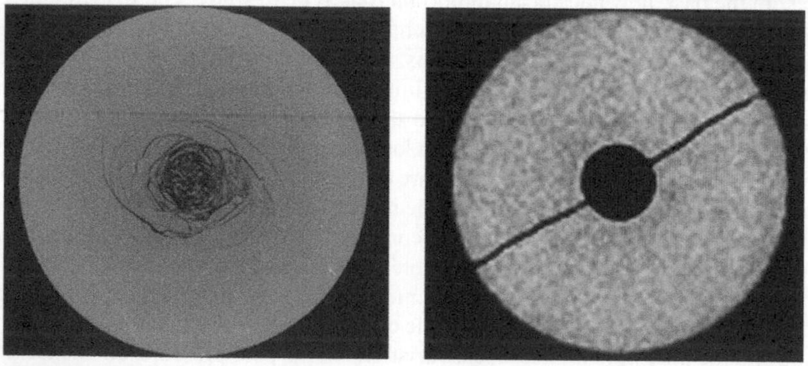

Figure 3. Borehole collapse (left) and fracturing (right).

5 PRODUCTION

When hydrocarbons are produced from a reservoir rock, pieces of the rock itself may be detached and follow the hydrocarbon flow to the surface. This "solids production" is an unintended byproduct of the hydrocarbon production and may generate substantial problems. Although it may happen in any kind of rock, it is the production of sand that has attracted most attention. This is partly because of abundance – sand production is, or is expected to become, a problem in most known oil fields. Initially, sand production was considered to be a minor problem. The sand was separated from the hydrocarbons at the surface and dumped in the vicinity of the rig without further concerns. However, due to their abrasiveness, sand grains are able to grind down steel, which may result in malfunctioning of valves or leakage of pressurized, high temperature hydrocarbons into the working areas of the rig crew. When it was realized that sand production was the real cause of some serious accidents, measures were taken to avoid sand production at any cost. And the costs are significant – safe prevention of sand production involves installment of screens and gravel packs which act as filters blocking the sand while allowing the hydrocarbon flow. These filters reduce the productivity of the well, and also need regular maintenance or replacement. Sand production do not always occur however, hence it was realized that money could be saved by installing such filters only when needed. As sand production originates from pieces of rock that have been broken off the rock around the well – in other words: rock failure – it was then a task for the rock mechanics community to develop methods for predicting the probability of sand production. Nowadays the policy regarding sand production is even more differentiated: Some sand production is acceptable as long as it does not exceed some specific limits. The challenge for the rock mechanics community is therefore not only to predict whether a given well may produce sand, but also how much sand it will produce under given conditions.

Sand production may be described as an erosion process from plastified rock around the hole. This implies that yield strength, in situ stress and pore pressure are the controlling parameters, in addition to the well pressure. Different failure modes may occur however, depending on the rock type, which may result in continuous, low intensity sand production, temporary large-scale sand production, or rapid, catastrophic sand production that fills the entire well.

Solids production may also occur in chalk reservoirs. In these cases, the rock around the hole is mobilized in a liquefaction-like process, making the chalk flow like toothpaste and clog the well. The economical consequences can be as severe as for sand production; however the safety aspect is less critical since the abrasiveness of chalk is much lower than for quartz.

6 DATA

Petroleum related applications of rock mechanics are typically dealing with relatively weak and porous sedimentary rocks. High stresses is a characteristic feature, but also high pore pressure which reduces the effective stresses. Access to data is a specific challenge for rock mechanical studies in oil fields. Direct access to rock samples that can be tested in the laboratory is only possible from sections where cores are taken during drilling. Taking cores slows down the drilling process, so this is usually done only in exploration wells, and only in the reservoir sections of the well, for petrophysical purposes. When the limited material is distributed among the labs, rock mechanical tests are not prioritized. The fact that rock mechanical tests are usually destructive do not make them more popular either. Some relevant information can be obtained from well logs however, although none of these are direct measurements of rock mechanical parameters. Short of other options, estimates of rock stiffness and strength are usually obtained from correlations with measurable parameters like sonic velocities, density and porosity. The precision can be rather questionable, however.

It is usually assumed that the vertical stress is given by the weight of the overlaying rock, which can be estimated with some confidence since the density is usually measured as a function of depth. In most cases, the vertical stress is also assumed to be a principal stress. The smallest principal

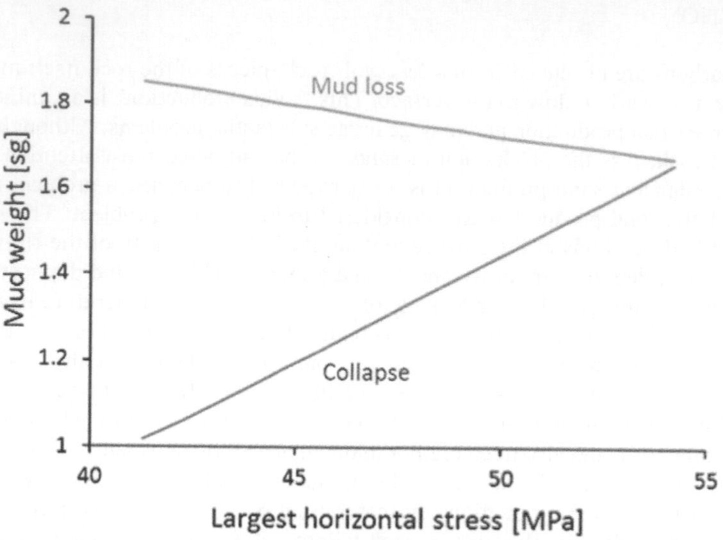

Figure 4. Example illustrating the significance of the largest horizontal stress on the mudweight window of a vertical well section.

stress – which is usually a horizontal stress – can be determined with relatively high precision at specific points along a well, by small-scale fracturing (XLOT) tests. Unfortunately, operators often settle for incomplete versions of these tests, which yields large uncertainties. The largest uncertainties are however related to the largest horizontal stress, which cannot be reliably determined by any existing procedure. This creates a significant problem for some quantitative predictions, as the largest horizontal stress can have a strong impact in some cases (Figure 4).

The considerations above are related to situations where data can be extracted from nearby wells. If no such wells exist, some relevant information can be obtained from structural maps and seismic velocity data, however quantitative estimates of rock mechanical parameters based only on such data are highly uncertain.

7 THE FUTURE

Today, the future of the petroleum industry is once again uncertain and unpredictable. The fight against global heating demands that the use of fossil fuel must be drastically reduced in the coming decades. This will also hit the petroleum industry. In the long run, it will naturally hit petroleum geomechanics as well. But we are not there yet. Over the years, millions of wells have penetrated the sealing rocks above the reservoirs in order to gain access to the resources. These wells are potential pathways for uncontrolled emission of methane into the atmosphere, and they have to be properly sealed off before they are abandoned. Also in this closing act, rock mechanics knowledge is in demand.

A typical oil well is a hole lined with a steel casing. Along some sections of the well, the gap between the casing and the surrounding rock is filled with cement. This is done in order to prevent leakage of oil or gas along the well. Along large sections of the well this gap remains open when the well is completed, however. The usual thing to do when a well is going to be abandoned is to fill critical sections of this gap with cement. This is time consuming, and expensive. It has been estimated that the total cost for abandonment of all wells in the Norwegian sector of the North Sea alone is in the order of 50 - 100 billion USD. The total cost for all wells worldwide is hardly imaginable. Only in the U.S. there are about 1.7 million active wells – no one really knows how many there are worldwide. Long term integrity is another problem with this traditional technique. Ordinary Portland cement is commonly used as a sealing material, as it has low permeability, easily fills all gaps, and remains stable for a long time. However, there are some serious questions asked about its sealing efficiency as well as possible deterioration over long time. The overall goal for plugging of a well is to restore the seal of the caprock. An ideal sealing material should therefore have the same sealing properties as the surrounding formation and be in chemical balance with the surroundings so that deterioration is eliminated. Interestingly, there are infinite amounts of such a material sitting right next to the well – it only needs to move a few inches to fulfil the purpose of sealing. This material is also exposed to strong forces acting in the right direction, the compressive in situ stresses, generously provided by nature itself. In some cases, the process happens all by itself. Statoil (now Equinor) reported already ten years ago (Williams et al., 2009) that long sections of uncemented, initially open annulus between rock and casing were found to be closed some weeks after the sections were completed – apparently as a result of shale creeping in and closing the gap. So far, Equinor has taken advantage of this generous assistance from Mother Nature in about 100 wells, bringing cost savings in the order of 1 billion USD (Carlsen, 2017). Rock mechanics is clearly an important element of the toolbox to optimize and take full advantage of this method for sealing of wells, which has a huge potential.

8 LAST WORDS

Hydrocarbons are created, stored and harvested from rock formations and rock mechanics has an impact on nearly all aspect of the production process. Therefore, rock mechanics has an important position in the petroleum industry, even though it does not belong to the core business of the operating companies. Founded on the fundamentals of rock mechanics, as described in classic textbooks like for instance Jaeger and Cook (1979), petroleum applications of rock mechanics has gradually developed into a discipline of its own, with its own set of textbooks, like Fjær et al. (1992), Charlez (1997) and Zoback (2006). This development continues. Poroelasticity, thermoelasticity and plasticity are key elements in this discipline. Thanks to the diversity of subject areas where rock mechanics has relevance for the petroleum industry – for enhanced production (hydraulic fracturing, solids production) as well as for cost cutting (drilling, P&A) – this discipline will continue to have important and interesting contributions to the industry also in the future.

REFERENCES

Bradley, W.B. 1979. Failure of inclined boreholes. *J. Energy Resources Tech.* 101: 232–239.

Carlsen, T. 2017 Experience with shale as a barrier. DrillWell Seminar, Stavanger.

Charlez, P.A. 1997. *Rock Mechanics. Volume 2. Petroleum Applications*. Éditions Technip, Paris.

Fjær, E., Holt, R.M., Horsrud, P., Raaen, A.M., Risnes, R. 1992. *Petroleum Related Rock Mechanics*. (2nd Ed. 2008). Elsevier, Amsterdam.

Jaeger, J.C., Cook, N.G.W. 1979. *Fundamentals of Rock Mechanics*, third ed. Chapman and Hall, London.

Stjern, G., Agle, A., Horsrud, P. 2003 Local rock mechanical knowledge improves drilling performance in fractured formations at the Heidrun field. *J. Pet. Sci & Eng*. 38: 83–96.

Williams, S., Carlsen, T., Constable, K., Guldahl, A. 2009. Identification and qualification of shale annular barriers using wireline logs during plug and abandonment operations. SPE/IADC 119321.

Zoback, M.D. 2007. *Reservoir Geomechanics*. Cambridge University Press, Cambridge.

Rock Mechanics for Natural Resources and Infrastructure Development – Fontoura, Rocca & Pavón Mendoza (Eds)

Innovative rock engineering solutions for deep tabular excavations

D.F. Malan
Department of Mining Engineering, University of Pretoria, South Africa

ABSTRACT: The shallow-dipping tabular excavations of the deep South African gold mines present unique rock engineering challenges. This paper discusses recent developments in terms of support designs and design criteria to improve safety in these mining excavations. Effective rockburst support designs are particularly problematic when the stoping width is small. A stiff support system, that is also yieldable to survive the convergence in the back areas, is required. Methods to improve the areal characteristics of the support system is discussed in the paper. In terms of design criteria, methods to simulate the stability of the fracture zone near the edges of tabular excavation layouts were recently developed. This work involved the extension of the classical ERR criterion to include dissipative mechanisms to allow for local on-reef failure. This enables engineers to simulate the effect of different mining rates and face advance increment lengths.

1 INTRODUCTION

The deep gold mines in the Witwatersrand Basin of South Africa have been in operation since 1886. More than 50 000 tons of gold have been extracted from this basin (Malan 2016). Surprisingly, even after this long period of production, the Witwatersrand Basin is still the world's largest gold resource (Minerals Council South Africa 2018). Unfortunately, much of this gold is contained in secondary reefs at a low grade and the depths of the existing mines are also increasing. Production can nevertheless continue for many years if the current poor profitability of the mines can be improved and a further improvement in safety can be achieved. The gold reefs are tabular in nature and a typical stoping width is approximately 1.6 m. Profitability of the mines is dependent on a small stoping width and in the early mines, it was attempted to keep the stoping width very small (Figure 1). In the modern mines, the minimum stoping width is maintained at a value of not less than 1 m – 1.2 m, regardless of the channel width of the reef, for safety reasons. The dip of the reef is small and typically vary from 25° to 35°.

As a result of the great depth, the mines are seismically active and rockbursts and falls of ground have been the cause of many fatalities. For the years from 1911 to 1943, the death rate per 1000 employees per annum caused by falls of ground varied from 0.72 to 1.2 (Jeppe 1946). Compared to modern standards, such as the guidelines provided by societal risk curves (Joughin 2011), these historic accident rates were unacceptable. The safety record of the mines has improved drastically in recent years and the additional improvement since 2007 is shown in Figure 2. In spite of this improvement, a number of rockburst accidents at the gold mines in 2017 and 2018 have resulted in an unacceptable loss of life. These accidents have emphasised the need for further research into methods to mitigate the risks of rockbursts. The production from the gold mines has also dropped significantly (Figure 3) and South Africa is now only ranked 8th in terms of gold production in the world. The gold production in 2017 was only 137 tonnes, whereas the peak production was a 1 000 tonnes in 1970. Mining depth, seismicity and an increase in costs were contributors to the recent decrease in production (Neingo & Tholana 2016).

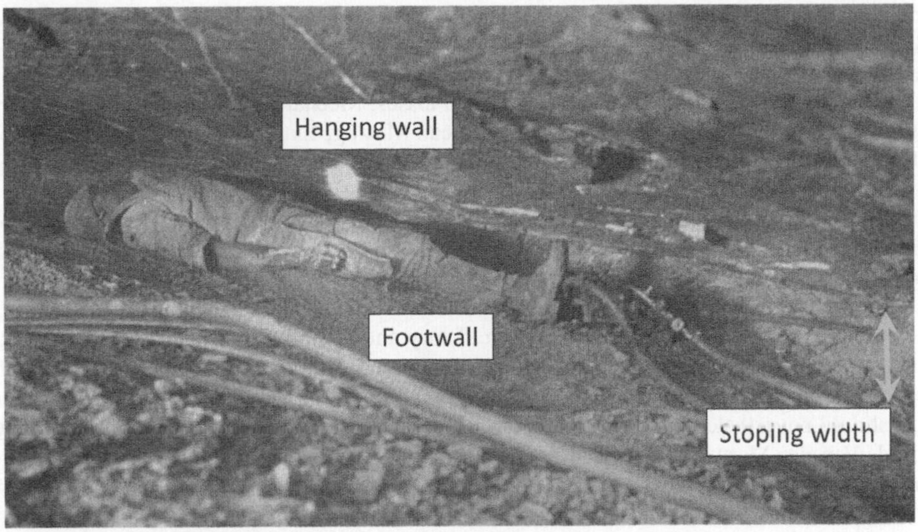

Figure 1. A photograph taken in the 1930s in a deep gold mine. This is a worker that squeezed into 30 cm of stoping width and pushed the jack-hammer with his feet. Note the absence of any support.

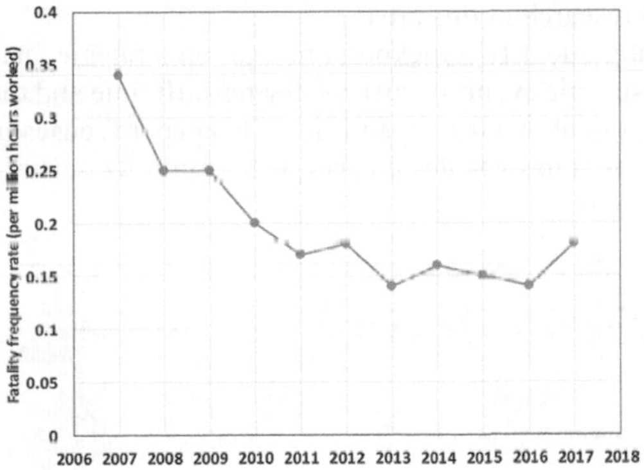

Figure 2. Fatality frequency rate in the South African gold mines from 2007 to 2017 (based on data from the Minerals Council South Africa 2018).

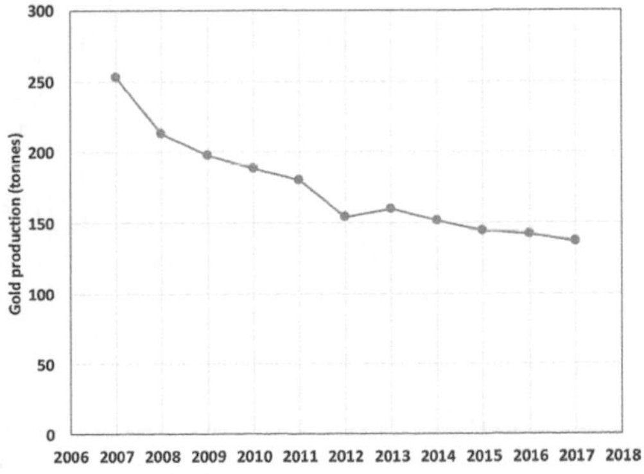

Figure 3. South African gold production from 2007 to 2017 (based on data from the Minerals Council South Africa 2018).

The gold industry still employs more than a 100 000 workers (Minerals Council South Africa 2018) and the social impact will be great if most of these workers lose their jobs owing to large-scale mine closures. Large scale mechanization is not only technically challenging, but is also not popular with the labour unions as they fear large scale job losses. Mechanisation is nevertheless currently being investigated by the SAMERDI (South African Mining, Extraction, Research, Development and Innovation) research programme as one possible solution to the current problems faced by the industry.

To achieve sustainability in this sector, an aspect requiring urgent attention is the development of updated design criteria for the very deep mines. Remnants and pillars are mined in many of the older operations (Figure 4) and most of these extractions are done safely. There is currently, however, a great need to improve the criteria used to select which remnants can be mined safely. Most of the current design criteria is based on elastic theory and the failure of the rock mass is not taken into account. This aspect require urgent research and the objective of this paper is to illustrate the recent developments that may lead to improvements in the layouts and better hazard identification. It should nevertheless be noted that a larger research programme focusing on this problem is required. Van der Merwe (2006) used the Coalbrook disaster as an example of lessons not learnt from disasters and deplored the dismal state of rock engineering research in South Africa. He stated: "*Is it conceivable that the most important lesson from Coalbrook, namely that in order to be effective at all, knowledge has to be generated before it is needed, was not learnt?*" Currently, rock engineers still grapple with what are the most appropriate design criteria to use when doing remnant extraction, but only very limited research on this topic was conducted during the last two decades. This paper highlights the shortcomings of the current design criteria and it will hopefully motivate the funding of additional research in this area.

The gold mines typically adopt two rockburst mitigation strategies namely measures to reduce the number of damaging seismic events occurring during shift time and the installation of rockburst resistant support with energy absorbing capabilities. This paper focuses on recent developments in both these two areas namely improved design criteria to minimize seismic events during shift time and improved support systems.

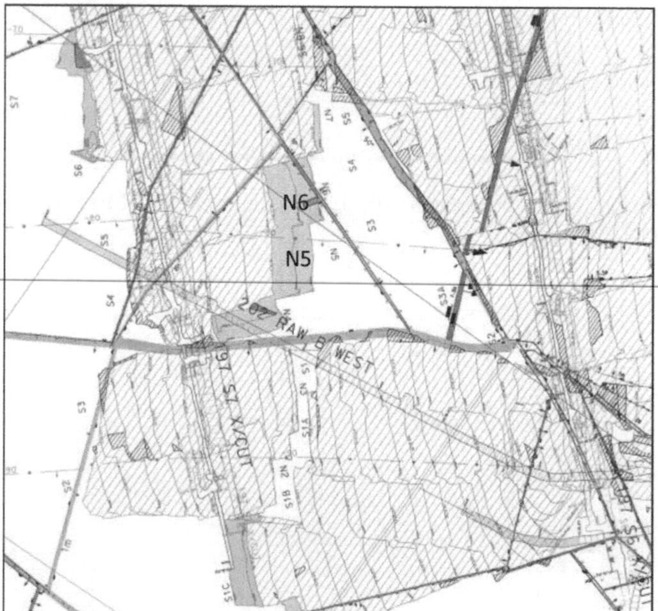

Figure 4. An example of remnant extraction (isolated white block in the centre) in the South African gold mining industry.

2 A REAPPRAISAL OF THE PRINCIPLES OF SUPPORT DESIGN

The implementation of effective rockburst support with appropriate areal coverage is particularly difficult in the South African gold mines. This is caused by the shallow dipping nature of the tabular orebodies (Figure 5). The stoping width can be as low as 1 m and the average dip of the reef is 25° to 35°. As the dip is less than the angle of repose of the broken ore, the rock needs to be cleaned using manual labour. Scraper and winch arrangements assist with this task, but the miners still need to enter the hazardous stope face areas and the systematic installation of on-reef support in the face area is required. In areas where the stoping width is very small, the support of the hanging wall is particularly problematic. In contrast, for steeply dipping vein deposits, mechanised mining methods such as sublevel open stoping or shrinkage stoping can be adopted and miners do not need to enter the stoped areas (Atlas Copco 1982). If rockbursts are encountered in these mines, the problem is simplified as only the access drifts and loading crosscuts need to be adequately supported. In the deep level South African gold mines, the angle of repose for broken ore is greater than the typical reef plane dip. The mined rock needs to be removed using miners and equipment that can be operated in the small stoping width.

Regarding the shallow dipping tabular stopes, a very large area needs to be supported as the extraction ratio is high and the miners need to enter the on-reef stopes. This support needs to be done in a cost effective manner. The occurrence of seismic events requires that the support system be rockburst resistant with the capacity to prevent the total collapse of the stope. The highly fractured nature of the hanging wall, caused by the elevated stress levels at the mining face, necessitates the use of areal support to prevent fallouts between support units. The stoping width, h_0, may be as small as 1 m to 1.6 m to make the mining operation economically feasible. This affect the ease of roofbolt installation and the length of bolts that can be used. The cleaning operations also affect areal support systems because the scraper and winch systems may pull out steel mesh installed on the hangingwall if the stoping width is small.

There are a number of reasons why the support units need to remain effective in the back areas of the stopes. The miners gain access to the working faces through a system of gullies that are excavated in the immediate footwall of the stopes. These gullies are the only access routes to the working faces and therefore need to remain operational in the back areas of the stopes. The support in the gullies and in the old stopes immediately adjacent to the gullies must remain effective to protect all personnel travelling in these excavations. In some stopes, significant convergence can occur in close proximity to the stope face. Jager & Ryder (1999) refer to a recorded example of 250

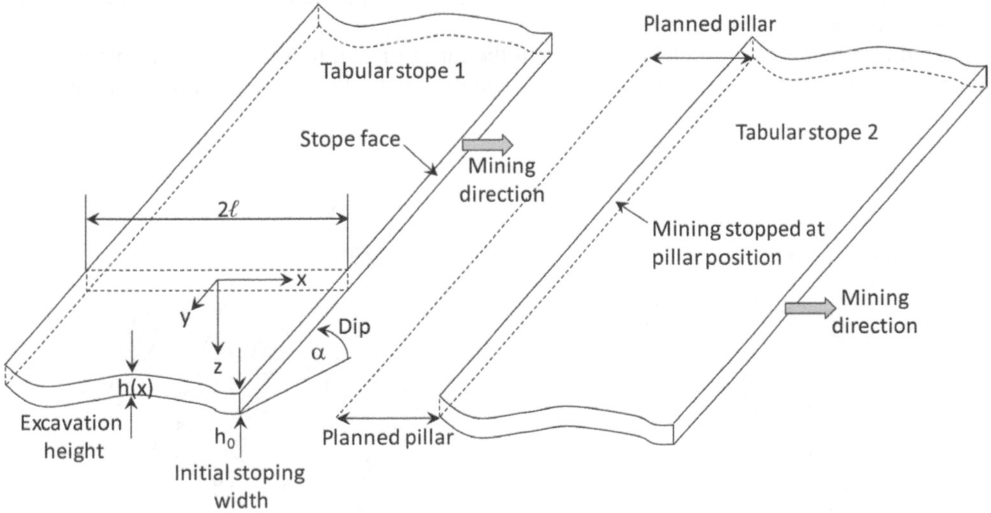

Figure 5. Mining of two shallow-dipping tabular stopes with a planned dip pillar between the two stopes. The planned maximum spans of the stopes along strike is typically 200 m and the width of the pillars are 30 m.

mm of convergence at a distance of 15 m from the stope face. The only support type that has the capacity to minimise stope convergence at these great depths is backfill and this has the beneficial effect of reducing the energy release rate and associated seismicity (Ryder & Jager 2002). Backfill is not used in all mines, however, owing to the difficulty of maintaining the piping infrastructure that is required to transport the fill over extensive distances to the stopes.

To provide additional insight into the difficulty of supporting the shallow-dipping tabular stopes, consider the simplified geometry shown in Figure 5. Assume a parallel-sided tabular excavation that dips at a small angle α. The span of the stope in the y-dimension is very large compared to the half span ℓ of the stope in the strike direction x. The original stoping width or mining height, h_0, is small compared to ℓ and it is assumed that $h_0 \ll \ell$. Typically, $h0 \approx 1.6$ m. In modern dip pillar layouts, a typical value of the span is $2\ell = 200$ m.

Assume that the stope is subjected to rockbursts and a rock block of a weight M_b is ejected from the hanging wall (Figure 6). The typical fallout height is $b \approx 1$ m. The stope is supported by yielding elongate support units, which allow a maximum deformation, d_m, before failure of these units occurs. Except for the blocks of rock in the immediate hanging wall, the rock mass is assumed to be elastic in nature. The blocks of rock remain in position and will be subjected to the overall elastic deformation of the rock mass until ejected during a rockburst. When considering the assumptions shown in Figures 5 and 6, the excavation height, *h(x)*, is dependent on the position in the stope and the effect of a possible rockburst and is given by:

$$h(x) = h_0 - [S(x) + d] \geq 0 \tag{1}$$

where h_0 is the original stoping width, S(x) is the elastic convergence in the stope and *d* is the distance the block of rock is ejected during a rockburst. The support unit can only undergo a maximum deformation, d_m, before failure and a further condition that must be applied to ensure the effective support of the excavation:

$$S(x) + d < d_m \tag{2}$$

Equation 2 applies an onerous restriction on the design of support units as the term *S(x)* increases away from the face of the stope into the back area and it reduces the deformation, *d*, that the support system can resist during rockbursts. It is known that *S(x)* is not only a function of the distance to the face, but it is also a time-dependent function, *S(x,t)*, owing to creep-like processes in the rock mass (Malan 1999). This aggravates the restrictive condition imposed by Equation 2 on support design as slow mining rates can result in failure of support units close to the stope face. The time-dependent nature of the rock is described in Malan & Napier (2018b) and is not explored further in this paper. The rock mass is assumed to be purely elastic to illustrate the problems associated with the support of shallow-dipping tabular excavations.

The mining height, h_0, which is typically in the range of 1 m to 1.6 m, imposes practical constraints on support selection. This restricts the length of bolt that can be installed and the drilling

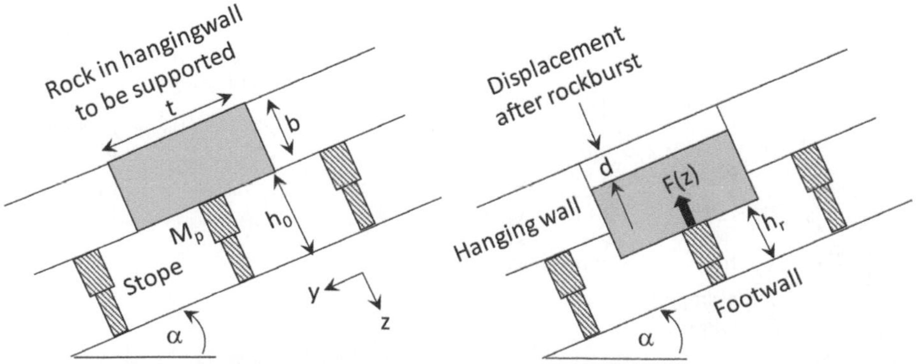

Figure 6. Support of a block of rock ejected during a rockburst (adapted from Malan & Napier 2018a).

of support holes normal to the plane of the reef is problematic. For the geometry shown in Figure 5, the maximum rigid bolt length, L_b, that can be used for installation normal to the plane of the reef is:

$$L_b < h_0 - S(x) \tag{3}$$

The typical bolt length currently used in the gold mining industry on the reef horizon is 0.9 m and the key function of these short bolts is to prevent small ground falls between pack and elongate support. The use of longer flexible cable anchors is a possibility, but drilling the deep holes by hand and grouting the anchors is not practical in the confined stoping height within the time constraints of the production cycle.

In the following discussion it is assumed that the *S(x)* component in Equations 1 and 2 corresponds to the elastic convergence $S_z(x)$ in the direction z normal to the reef plane. To simplify the problem, it is assumed the dip of the reef is zero. It is also assumed that "Tabular stope 1" in Figure 5 is an isolated stope and not in proximity to the second stope (Tabular stope 2 does not exist). If it is assumed that the excavation is at great depth, no contact occurs between the hanging wall and the footwall and the rock is isotropic and elastic, it can be shown that (Salamon 1968):

$$S_z(x) = \frac{4\left(1-v^2\right)\rho g H}{E}\sqrt{\ell^2 - x^2} \tag{4}$$

where 2ℓ is the span of the stope, ρ is the density of the rock, *g* is the gravitational acceleration, *H* is the depth below surface, ν is Poisson's ratio and *E* is Young's modulus. Substituting Equation 4 into Equation 2 yields the following inequality constraint that has to be met for support design:

$$\frac{4\left(1-v^2\right)\rho g H}{E}\sqrt{\ell^2 - x^2} + d < d_m \tag{5}$$

Equation 5 assumes that the elongate support is installed right on the face where the convergence is zero and the distance between the support units and the mining face increases gradually as the mining span is incrementally increased. It is also assumed that the force exerted by the support units is not large enough to affect the elastic convergence $S_z(x)$. The inequality constraint (5) will not be valid for support units installed at later stages in the back area of mature stopes or for very robust support, such as backfill, which can limit the elastic convergence occurring in the back areas. Malan & Napier (2018a) used Equation 5 to illustrate that in the case where a rockburst results in a dynamic deformation of d = 100 mm, at a distance of approximately 20 m from the face, the elastic convergence has already deformed the support unit to the extent that it will not be able to withstand the dynamic deformation during the rockburst. This illustrates the difficulty to support the tabular excavations as the cumulative convergence acting on the support units in the back areas render them ineffective as the yieldability of the units are diminished. Cai & Kaiser (2018) refers to this principle as "support capacity consumption" in relation to the support design of tunnels subjected to rockburst. The static and dynamic deformation of the rock mass consumes the energy absorption capacity of the tunnel support system. It is proposed that new rock bolts are installed in affected areas to provide additional displacement capacity.

During a rockburst in a tabular stope, the velocity, *v*, at which the rock is ejected is important to estimate the magnitude of dynamic deformation, *d*. The mass of the support M_p is ignored as it is much smaller than the mass of the ejected block M_b. The block of rock is brought to rest by the support resisting force *F(z)*. The work done by the resisting force is equal to the changes in kinetic energy and potential energy giving the following equation:

$$\frac{M_b v^2}{2} + M_b g d \cos(\alpha) = \int_0^d F(z)\,dz \tag{6}$$

This approach was also explored by Wagner (1982). To evaluate Equation 6, consider the special case of yieldable steel elongates that provides a near constant load when subjected to deformation. These props supplied a constant resisting force F_c during rockbursts. As the dip is assumed to be zero, Equation 6 can then be simplified to:

$$\frac{M_b v^2}{2} + M_b g d = F_c d \tag{7}$$

The deformation d is then given by:

$$d = \frac{v^2}{2\left(\frac{F_c}{M_b} - g\right)} \tag{8}$$

It follows from Equation 8 that the value of d becomes very large for a ratio F_c/M_b close to the value of g. It is therefore important to install supports units with a specific load bearing capacity, F_c, at the correct spacing to ensure a large value for the ratio F_c/M_b. Equation 8 is plotted in Figure 7 to illustrate the effect of support force and mass of the block on the deformation experienced in the stope. If larger blocks are ejected, a more robust support system is required to arrest the block movement within a specified distance. Based on the earlier information given in Equation 5, the ejection of large blocks will typically result in support failure in the back areas of the stope where significant elastic convergence has already taken place. By inserting Equation 8 into Equation 5, the general specification for a support unit with a constant deformation resistance force F_c in rockburst conditions is given as:

$$\frac{4\left(1-v^2\right)\rho g H}{E}\sqrt{\ell^2 - x^2} + \frac{v^2}{2\left(\frac{F_c}{M_b} - g\right)} < d_m \tag{9}$$

Based on this simple analysis, the support system will not be effective in rockburst conditions if the support units have either already failed as a result of excessive convergence in the stope or if the ejection velocity or mass of the ejected rock is too large to be contained by the force supplied by the support units. For effective support design, it is important to measure the ejection velocity, v, during rockbursts. After many years of research, it is not clear what ejection velocity should be used in support design (Wagner 1982). The value of 3 m/s was adopted in the South African gold mining industry in the 1980s and this value is rarely questioned for modern designs. As a further complication, the mass of the ejected rock, M_b, is typically based on prior experience of ejection thicknesses and may not be applicable in all cases.

In a stoping width as small as 1 m, the dynamic closure needs to be restricted to as small an amount as possible. Figure 7 illustrates that F_c/M_b must have a value of at least 3g to ensure that the

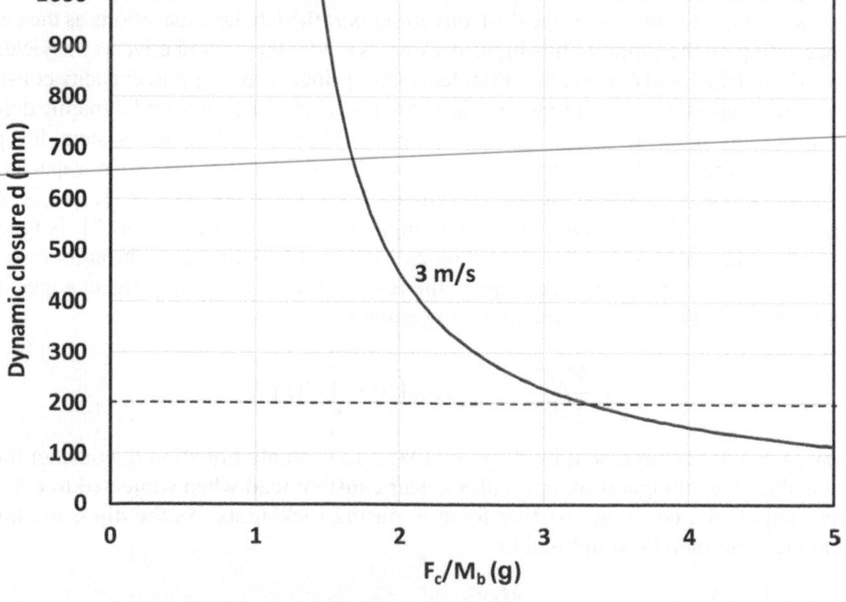

Figure 7. Amount of dynamic closure as a function of the ratio (in units of g) of the support force and the mass of the ejected rock.

dynamic closure is restricted to 200 mm for an ejection velocity of 3 m/s. This implies that support units with a large value of Fc must be installed if the fallout height is approximately 1 m.

The general constraint condition for support systems given in Equation 9 illustrates the need for a support system which combines units with different attributes. In the face area, a pre-stressed support type with enough yield capacity and sufficient force to minimise the damage caused by rockbursts is required. Practical support units that meet this requirement, typically fail some distance from the face where the cumulative stope convergence is large. A different back area support system is therefore also required. The support systems in the South African gold mining industry have developed over time to meet this requirement (Malan & Napier 2018a). Typical support systems consists of a combination of stiff pre-stressed elongates in the face area to minimize the rockburst damage. These are supplement by timber packs or backfill to provide support in the back area where the elongates have already failed due to excessive closure. The current designs and type of support units are unfortunately not well suited to provide adequate areal support in the face area and this is discussed in the next section.

2.1 *Developments in terms of robust areal support systems for tabular stopes*

A weakness of the current support design methodology is the difficulty to provide adequate areal support. Further research regarding this aspect is required. In 1924, the South African Government appointed a "Rock Burst Committee" to investigate the rockbursts in the early mines. This committee recognised "rock pressure" as the key problem facing mining engineers (Jeppe 1946). This committee appeared to have recommended areal support for rockburst conditions and they stated: "*...at 1200 to 1500 m, a ratio of 35 % support compared to the area excavated has been found adequate, as working depths increase the necessary ratio of support may also increase; so that at a depth of 2400 m, 70 % support may be necessary, and at 3000 m 80 % or more.*" Jager & Ryder (1999) defined a measure of areal coverage in the gold mines as the percentage of contact area of the support units compared to the total area of the hanging wall. For elongate systems this can be as small as 1% and for timber pack systems it only increases to approximately 10%. Ideally it should be 100% and the low percentages are problematic where the hanging wall is intensely fractured. They also state: "*The predominant cause of falls of ground in stopes is inadequate areal coverage or interaction between support units.*" This is illustrated in Figure 8 illustrating a fall of ground after a seismic event between the rows of support.

Figure 8. Falls of ground occurring between support units as a result of a seismic event.

Modern studies place great emphasis on the need for areal support support in rockburst conditions. Kaiser & Cai (2012) discussed the three key functions of rockburst support namely reinforce (strengthen and control bulking), retain (prevent fractured blocks falling between reinforcing elements) and hold (anchor reinforcing elements in stable ground). Of significance to the South African gold mines is their statement: *"Under high stress conditions, fractured rocks between reinforcing or holding elements may unravel if they are not properly retained."* In the gold mines, historically the emphasis was placed mainly on "reinforce" and "hold" aspects and less on the "retain" function. In Figure 6, the assumption of each yielding elongate supporting a block of width t dominated the support design methodology for many years (Roberts 1999). Ryder & Jager (2002) also describes this methodology and a graphical method is included to determine whether the support units meet the design criteria at a specified distance from the face. The requirement of areal support is not explicitly included in the methodology. A partial solution is to reduce the spacing between support units, but unfortunately the spacing in the strike direction is typically dictated by the scraper-winch equipment used for the cleaning operations after blasting. The scraper needs to fit between adjacent rows of elongates.

In recent years, measures, such as headboards, roof bolting between elongates and packs and temporary nets, were introduced to ameliorate this problem. Bolting nevertheless remains problematic as the installation of roof bolts is difficult in the small stoping widths. Special rigs to drill the holes have been developed in recent years and one model is shown in Figure 9. Based on the restriction given in Equation 3, the bolts are limited to short lengths and these are typically 0.9 m. These lengths are typically less than the fallout height b. The support design methodology is therefore based on the premise that the elongate and pack support cater for the typical fallout height and the bolting is installed to prevent smaller falls between the elongates or packs. Although bolting has been a major improvement in terms of the support design methodology, fallouts between the bolts during rockbursts still occur. An example is shown in Figure 10. Louchnikov & Sandy (2017) emphasised that "*...the weakest link in a ground support system is often the surface support, including the connections with the rock bolts.*"

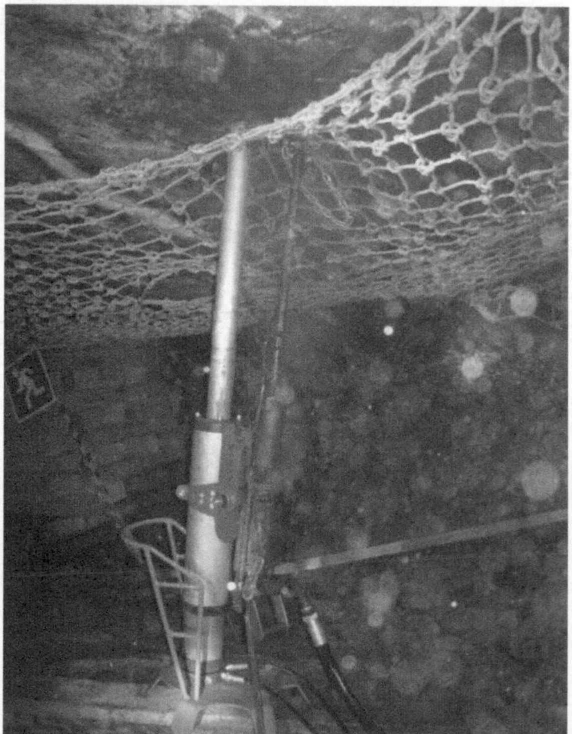

Figure 9. A special drill rig to install roof bolting in the small stoping widths of the deep gold mines.

Figure 10. Falls of ground occurring between the roof bolts as a result of a seismic event. It is evident that improved areal support is required.

Various types of temporary nets have been used in recent years to provide for improved areal support. A rope net is shown in Figure 9. Although these nets are not robust, they have been useful during some collapses as shown in Figure 11. Observations have indicated that steel rope nets can be left permanently in place, but these are susceptible to damage by the scrapers in low stoping width environments. Although some success has been achieved with these measures, a more robust permanent areal support system is required for rockburst conditions. A promising new development is the use of high-tensile steel wire mesh used as permanent areal support installations. This application is currently being trialed and this mesh installed in a stope is shown in Figure 12.

Figure 11. Small falls of ground caught by a rope net attached to the roof bolts.

Figure 12. The installation of high-tensile steel wire mesh manufactured by Geobrugg (courtesy of M. du Plessis).

3 NEW DEVELOPMENTS IN TERMS OF DESIGN CRITERIA

3.1 *Numerical modelling to include the effect of on-reef failure*

A key problem in deep level tabular mine design is to optimize the layouts to minimize high stress concentrations. The displacement discontinuity boundary element method (DDM) is the preferred method used for the computational analysis of stress distributions in the vicinity of gold mine tabular excavations. In the early 1960s, this approach was pioneered by Salamon in South Africa (Salamon 1963). This approach was made possible by the experimental work that indicated that the intact rock mass behaviour can be simulated by using elastic theory (Ryder & Officer 1964). The simulation package MINSIM, based on the DDM method, was developed in South Africa and became one of the popular design tools used in the gold mining industry and details are given in Deist et al. (1972), Ryder & Napier (1985) and Napier & Stephansen (1987). Although the DDM method was used with success for more than forty years, a major drawback of this approach was that it could not simulate the ubiquitous fracture zone found in the immediate vicinity of the deep level stopes.

Napier & Malan (2018) proposed a time-dependent limit equilibrium model to simulate on-reef failure. It gives a representation of the on-reef horizontal and vertical stress distribution adjacent to a pillar edge as shown in Figure 13. The reader is referred to Napier & Malan (2018) for additional details regarding the model. In summary, the model assumes that if the reef material fails, a relationship exists between the reef normal and parallel stress in the seam ahead of the face. The reef-parallel stress $\sigma_s(x,t)$ at position x and time t is balanced by a frictional shear traction $\mu_I\sigma_n(x,t)$ at the interfaces between the fractured material and the intact reef. The presence of these interfaces in the model makes it such an attractive concept to study the failure of the rock ahead of stope faces. The parameter μ_I is the interface friction coefficient and $\sigma_n(x,t)$ is the traction component normal to the reef. The limit strength model specifies that a relationship exists between $\sigma_n(x,t)$ and $\sigma_s(x,t)$. This is specified as:

$$\sigma_n(x,t) = \sigma_c(x,t) + m(x,t)\sigma_s(x,t) \tag{10}$$

The strength envelope parameters $\sigma_c(x,t)$ and $m(x,t)$ are functions of position and time. This allows the time-dependent failure of the reef material to be studied. From Equation 10 and the model shown in Figure 13, the average reef-parallel confining stress $\sigma_s(x,t)$ at position x and time t can be given by the differential equation of the form:

$$H\frac{\partial\sigma_s(x,t))}{\partial x} = 2\mu_I\left[\sigma_c(x,t) + m(x,t)\sigma_s(x,t)\right] \tag{11}$$

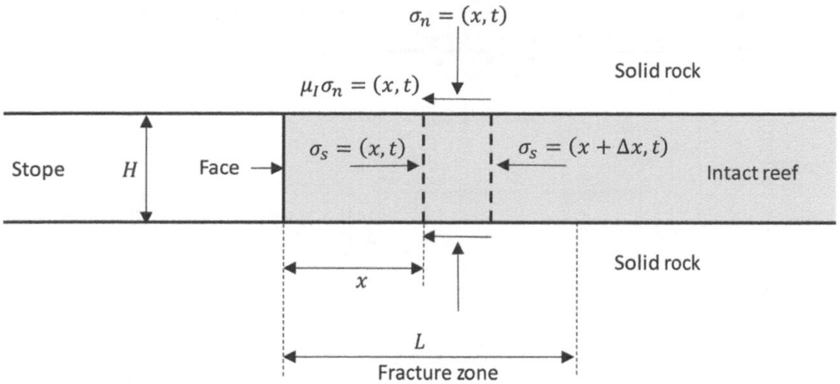

Figure 13. Force balance for the average reef-parallel confining stress and reef-normal stress at position x and time t in a plane section at right angles to the mining face. The stoping width is H and the distance of the fracture zone edge from the pillar face is L.

The implication of this model is that the stress increases exponentially towards the edge of the fracture zone. A drawback of the model is that the failure is restricted to the plane of the reef. It is nevertheless an elegant model to simulate failure on the reef horizon in a large scale mine-wide model.

Malan & Napier (2018b) used this model to simulate the time-dependent closure behavior in a tabular stope after a blast. Typical experimental results from an underground stope is shown in Figure 14. For many years, it was difficult to simulate this behaviour. The elastic DDM simulations could not simulate the time-dependent behaviour and simple creep models such as, viscoelastic theory, erroneously predicted that the rate of deformation will increase into the back area. In contrast, when using the time-dependent limit equilibrium model described above, the rock mass behaviour shown in Figure 14 could be replicated by the model. This is shown in Figure 15. A remaining challenge is to calibrate the parameters of the model.

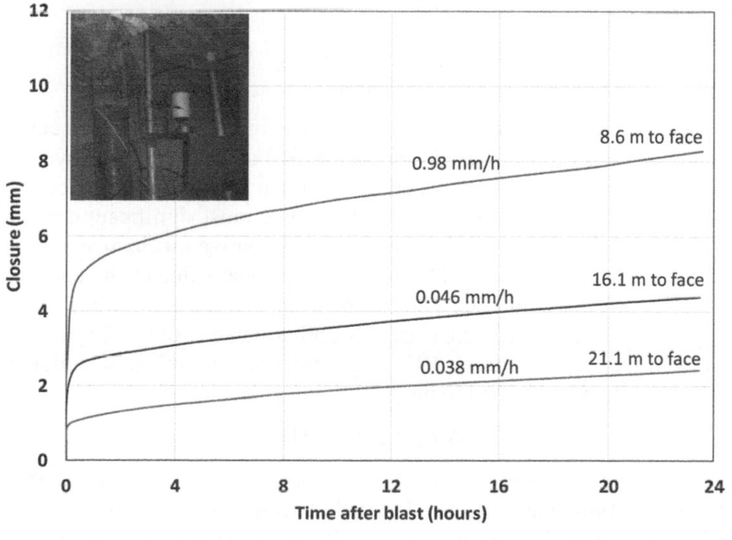

Figure 14. An example of time-dependent closure data recorded in a deep gold mine after a blast. This was recorded by three instruments (an example shown in the photograph) installed at different distances to face. The traditional elastic DDM modelling approach could not simulate this behavior (adapted from Malan & Napier 2018b).

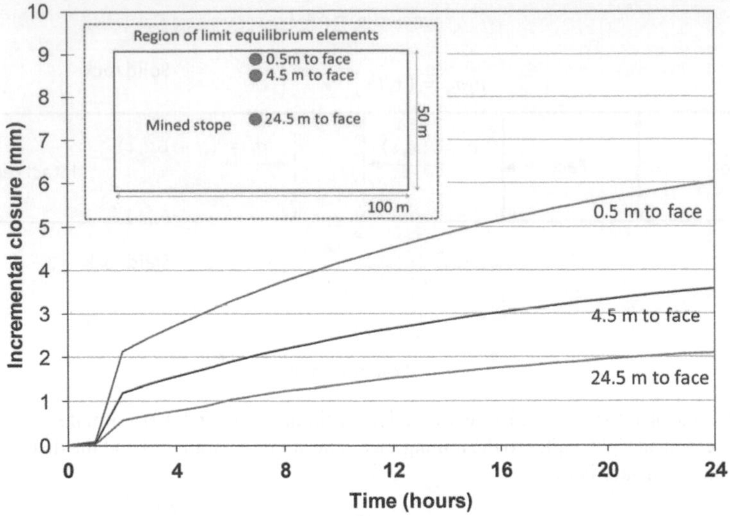

Figure 15. An example of modelling conducted on a simplified geometry with a time-dependent limit equilibrium model. The model qualitatively agrees with the actual stope behaviour shown in Figure 14 (adapted from Malan & Napier 2018b).

3.2 *Extended energy release rate criterion*

A common criterion that is used in the design of the deep level gold mine layouts is the energy release rate (ERR) (Cook 1963). In the definition of this criterion, the energy release increment, ΔWA, represents the difference between the incremental work done by gravity forces acting on the rock mass, ΔW, and the incremental change in the strain energy, ΔU, that is stored in or released from the rock mass when mining an increment (Napier & Malan, 2014). This is given by:

$$\Delta W_A = \Delta W - \Delta U \tag{12}$$

For tabular excavations, the mining increment is typically expressed in terms of the area mined with respect to the plan view of the stope. If the incremental area mined is designated by ΔA, then the energy release rate is given as

$$ERR = \lim_{\Delta A \to 0} \frac{\Delta W_A}{\Delta A} \tag{13}$$

For practical ERR calculations, stress analysis programs based on the displacement discontinuity boundary element method is typically used. The use of ERR as a criterion for layout design has been extensively discussed. It has a number of practical shortcomings as a measure of the rockburst hazard as described by Salamon (1984) and Napier (1991). The most significant drawback of the ERR criterion when used with the elastic models is that no dissipative mechanisms are incorporated to allow for local on-reef failure. Equation 13 represents the local value of the energy release at each point of the tabular excavation boundary and can be used as a measure of the local stress concentration at the stope face. An improved measure of stability can be obtained by modifying Equation 12 to include an energy dissipation term ΔW_D. A general measure of incremental mining stability, designated as ΔW_R, can then be defined to be:

$$\Delta W_R = \Delta W_A - \Delta W_D \tag{14}$$

The incremental stability measure, defined by Equation 14, is associated with each incremental change to the excavation shape. It can also include released energy from explicitly modelled faults or other discontinuities. Napier & Malan (2018) proposed that the fracture zone adjacent to the edges of tabular excavations can be represented by a simple time-dependent limit equilibrium model. In this case, the energy dissipation term ΔW_D can be computed explicitly in a series of discrete time steps with imposed face advance increments corresponding to a given mining schedule.

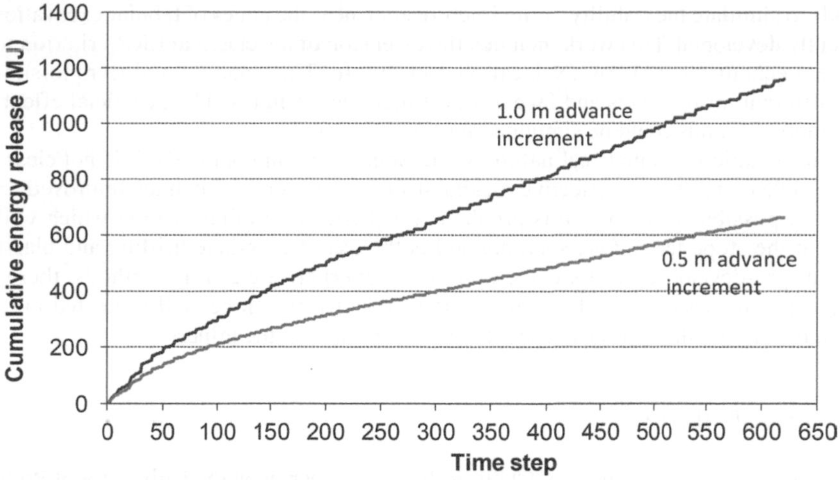

Figure 16. Simulating the effect of different face advance increments using the time-dependent limit equilibrium model and using the revised measure of incremental mining stability (adapted from Napier & Malan 2018).

As an illustration of the value of this new approach, Figure 15 illustrates the effect of different mining increment sizes when mining two adjacent panels using the limit equilibrium approach. An actual example of two such adjacent panels are panels N5 and N6 in Figure 4. For the 1 m increment, only one panel per day was blasted whereas for the 0.5 m increment, both panels were blasted every day. The overall advance rate was therefore similar. The geometry for this preliminary simulation was simplified and arbitrary parameters for the limit equilibrium model was selected (Napier & Malan 2018). For the simulation with the smaller mining increment, the cumulative energy release (ΔW_R in Equation 14) for the 0.5 m mining increments is significantly less. This qualitatively agrees with what has been known for many years in the deep gold mines. In areas prone to seismicity, it appears to be beneficial to use short production hole lengths (0.5 m) compared to the typical normal hole length of 0.9 m. This modelling is encouraging as it was not possible to simulate this behaviour with the classical DDM approach. Although encouraging, calibration of the limit equilibrium model is required for this approach to be used as a practical tool.

4 CONCLUSIONS

The deep South African gold mines present unique rock engineering challenges. These challenges are mainly caused by the shallow-dipping tabular nature of the reefs and a small stoping width. Mechanisation is difficult and workers need to enter the reef horizon for support installation, blasting operations and the removal of broken ore. The entire hanging wall needs to be supported to protect the miners. This paper discusses recent developments in terms of support design and design criteria to improve safety in these mining excavations. Effective rockburst support designs are particularly problematic when the stoping width is small. An analysis presented in the paper illustrated the need for a support system which combines units with different attributes. In the face area, pre-stressed support types with a large yield capacity and sufficient force to minimise the damage caused by rockbursts, is required. Practical support units that meet this requirement typically fail some distance from the face if the stope convergence is large. A different back area support system is therefore also required. The support systems in the South African gold mining industry have developed over time to meet this requirement. Typical support systems consists of a combination of stiff pre-stressed elongates in the face area to minimize the rockburst damage and this is supplemented by timber packs or backfill to provide support in the back area. The current designs and type of support units are unfortunately not well suited to provide adequate areal support in the face area. A combination of roof bolting and steel mesh is a possible solution to provide improved areal support.

Methods to simulate the stability of the fracture zone near the edges of tabular excavation layouts were recently developed. This work included the extension of the classical ERR criterion to include dissipative mechanisms to allow for local on-reef failure. This enables engineers to simulate the effect of different mining rates and face advance increment lengths. The beneficial effect of using short production rounds could be simulated for the first time.

Based on the unique geometrical nature of the South African gold reefs, it is not clear if it will ever be possible to design cost-effective rockburst support systems with much improved areal characteristics. A possible solution to this problem is a change in mining method which will remove miners from the stope faces. One such method is the use of long-hole drilling and blasting techniques from pre-developed access drives. By using alternative cleaning methods, there may not be the need for miners to enter the stopes. The access drives can be well protected by rockburst resistant support with the appropriate areal coverage to protect the miners.

ACKNOWLEDGEMENTS

This work forms part of the Harmony Chair of Rock Engineering at the University of Pretoria. The author would also like to thank Harmony Gold for funding this research. Dr Michael du Plessis and Geobrugg South Africa is thanked for supplying photographs of the steel mesh currently being tested in the mining industry.

REFERENCES

Atlas Copco Manual. 1982. Ljungforetagen AB, Orebro: Atlas Copco.

Cai, M. & Kaiser, P.K. 2018. *Rockburst Support Reference Book – Volume 1: Rockburst phenomenon and support characteristics*. Unpublished limited distribution manuscript.

Cook, N.G.W. 1963. The basic mechanics of rockbursts. *Journal of the Southern African Institute of Mining and Metallurgy* 64:71–81.

Deist, F.H., Georgiadis, E. & Moris, J.P.E. 1972. Computer applications in rock mechanics. *Journal of the Southern African Institute of Mining and Metallurgy* 72:265–272.

Jager, J.A. & Ryder, A.J. 1999. *A handbook on rock engineering practice for tabular hard rock mines*, SIMRAC, Johannesburg.

Jeppe, C.B. 1946. *Gold Mining on the Witwatersrand Vol. 1*. Johannesburg: The Transvaal Chamber of Mines.

Joughin, N.C. 2011. Engineering considerations in the tolerability of risk. *Journal of the Southern African Institute of Mining and Metallurgy* 111: 535–540.

Kaiser, P.K. & Cai, M. 2012. Design of rock support system under rockburst condition. *Journal of Rock Mechanics and Geotechnical Engineering* 4(3): 215–27.

Louchnikov, V. & Sandy, M.P. 2017. Selecting an optimal ground support system for rockbursting conditions, In: Deep Mining 2017: *Eight International Conference on Deep and High Stress Mining*. Perth 613–23.

Malan, D.F. 1999. Time-dependent behaviour of deep level tabular excavations in hard rock. *Rock Mechanics and Rock Engineering* 32(2):123–55.

Malan, D.F. 2016. *Krugerrand – Golden Jubilee*. Johannesburg: Prestige Bullion.

Malan, D.F. & Napier, J.A.L. 2018a. Rockburst support in shallow-dipping tabular stopes at great depth. *International Journal of Rock Mechanics and Mining Science* 112: 302–312.

Malan, D.F. & Napier, J.A.L. 2018b Reassessing continuous stope closure data using a limit equilibrium displacement discontinuity model. *Journal of the Southern African Institute of Mining and Metallurgy* 118:227–233.

Minerals Council South Africa. 2017. Facts and Figures.

Napier, J.A.L. 1991. Energy changes in a rockmass containing multiple discontinuities. *Journal of the Southern African Institute of Mining and Metallurgy* 91:145–157.

Napier, J.A.L. & Malan, D.F. 2014. A simplified model of local fracture processes to investigate the structural stability and design of large-scale tabular mine layouts. *48th US Rock Mechanics/Geomechanics Symposium*, Minneapolis, USA.

Napier, J.A.L. & Malan, D.F. 2018. Simulation of tabular mine face advance rates using a simplified fracture zone model. *International Journal of Rock Mechanics and Mining Sciences* 109:105–114.

Napier, J.A.L. & Stephansen, S.J. 1987. Analysis of Deep-level Mine Design Problems Using the MINSIM-D Boundary Element Program, *APCOM 87. Proceedings of the Twentieth International Symposium on the Applications of Computers and Mathematics in the Mineral* Industries, *vol. 1: Mining*. Johannesburg, SAIMM 3–19.

Neingo, P.N. & Tholana, T. 2016. Trends in productivity in the South African gold mining industry. *Journal of the Southern African Institute of Mining and Metallurgy* 116: 283–290.

Roberts, M.K.C. 1999. The design of stope support systems in South African gold and platinum mines, *PhD thesis*. University of the Witwatersrand. Johannesburg.

Ryder, J.A. & Jaeger, A.J. 2002. *A textbook on rock mechanics for tabular hard rock mines*. SIMRAC, Johannesburg.

Ryder, J.A. & Napier, J.A.L. 1985. Error analysis and design of a large-scale tabular mining stress analyser. *5th International Conference on Numerical Methods in Geomechanics*, Nagoya, Japan 1549–1555.

Ryder, J.A. & Officer, N.C. 1964. An elastic analysis of strata movement observed in the vicinity of inclined excavations. *Journal of the Southern African Institute of Mining and Metallurgy* 64(6): 219–244.

Salamon, M.D.G. 1963. Elastic analysis of displacements and stresses induced by the mining of seam or reef deposits—Part I: Fundamental principles and basic solutions as derived from idealised models. *Journal of the Southern African Institute of Mining and Metallurgy* 63: 128–149.

Salamon, M.D.G. 1968. Two-dimensional treatment of problems arising from mining tabular deposits in isotropic or transversely isotropic ground. *International Journal of Rock Mechanics and Mining Sciences* 5:159–185.

Salamon, M.D.G. 1984. Energy considerations in rock mechanics: fundamental results. *Journal of the Southern African Institute of Mining and Metallurgy* 84:233–246.

Van der Merwe, J.N. 2006. Beyond Coalbrook: what did we really learn? *Journal of the Southern African Institute of Mining and Metallurgy* 106: 857–868.

Wagner, H. 1982. Support requirements for rockburst conditions. *Proceedings of the First International Congress on Rockbursts and Seismicity in Mines* 209–18.

Rock Mechanics for Natural Resources and Infrastructure Development –
Fontoura, Rocca & Pavón Mendoza (Eds)
© 2020 ISRM, ISBN 978-0-367-42284-4

Translating the micro-scale to the macro-scale: Signatures of fracture evolution

L.J. Pyrak-Nolte, L. Jiang, A. Modiriasari
Purdue University, West Lafayette, Indiana

H. Yoon
Sandia National Laboratories

A. Bobet
Purdue University, West Lafayette, Indiana

ABSTRACT: In this paper, the role of micro-scale properties and behavior on the detection of crack initiation, propagation and geochemical alteration is examined through three topics: (1) identification of a geophysical precursor for a system transitioning from meta-stability to unstable behavior with specific focus on crack nucleation, propagation and coalescence; (2) demonstration of acoustic emissions from geochemically-induced fractures; and (3) understanding the role of depositional layers and mineral fabric on tensile crack formation. The results from these studies advance current understanding of which microscopic properties of evolving fracture systems are most useful for predicting macroscopic behavior and the best imaging modalities to use to identify the seismic signatures of time evolving fracture properties.

1 INTRODUCTION

The recent growth in shale gas extraction, geothermal energy development and storage of anthropogenic gases and fluids has led to increased human interaction with the Earth's subsurface. A management challenge for these subsurface sites is to optimize extraction/storage approaches to yield maximum potential while minimizing risks. Fractures are one of the dominant factors that influence the success or failure of these management tasks because all subsurface activities perturb fluid pressures and stresses in rock, causing mechanical discontinuities to open, close, initiate, coalesce and/or propagate, while natural and engineered fluids can result in geochemical alterations that lead to crack growth. With the goal of sustaining production/isolation throughout the life-cycle of a subsurface site, it is necessary to detect and image fracture systems to monitor alterations as well as to link geophysical measurements to mechanical and hydraulic integrity of the subsurface rock.

Failure in rock is a progression of energy transfers from the smallest scales (lattice or microstructure) to potentially the full scale of a system under consideration. At the smallest scale, the mineral composition, distribution, orientation and bonding among minerals are known to affect the engineering properties of a rock in addition to the presence of structural features such as micro-cracks, layers and other sources of porosity. For example, Brace (1965) demonstrated experimentally that the anisotropy in the intrinsic elastic property of linear compressibility is effected by mineral crystal orientation based on measurements on rock with oriented mica, calcite and quartz at pressures of 0.2 to 0.9 GPa to remove the effects of micro-cracks. Agliardi et al. (2014) found that failure modes and uniaxial compressive stress depended on the orientation of loading relative to foliation in gneissic rock but with significant scattered in the values of UCS. While Chandler et al. demonstrated for shale the effect of the orientation of fine scale layering relative to loading direction on fracture toughness. They found that layer orientation alone was not sufficient to explain the observed anisotropy in fracture toughness.

One difficulty in performing experiments to link macro-scale behavior to micro-scale features and the composition of rock is the inherent presence of heterogeneity in mineral composition and structural features that can occur among samples taken from the same formation or block of rock, and even within a single sample. To overcome this difficulty, synthetic or analog rock has been used in the past to study the effect of specific structural features on rock behavior. Gypsum has been used as a model soft rock (Nelson, 1968) to study crack initiation, propagation and coalescence with designed flaws to generate specific crack types (Bobet & Einstein, 1998; Wong & Einstein, 2006; Park & Bobet, 2009), slip along mechanical discontinuities (Mutlu and Bobet, 2006; Hedayat et al., 2014a&b); and the effects of surface roughness on normal and shear stiffness of a fracture (Choi et al.2014). In many studies on the effect of fractures or mechanical discontinuities on compressional, P, and shear, S, wave propagation, planar or machined fractures in aluminum and other synthetic materials have been used to investigate the transition from displacement discontinuity behavior to resonant scattering (Nolte et al., 2000); cross-coupling stiffness (Nakagwa et al., 2000), and wave propagation in a medium with two orthogonal fracture sets (Shao & Pyrak-Nolte, 2016).

With the advent of 3D printing, also known additive manufacturing (Burns, 1993), the ability to control structural features and compositional homogeneity has improved such that samples can be "geo-architected". The concept of a geo-architected rock is a synthetic analog that is fabricated and structured in the laboratory using natural or synthetic constituents to develop controlled features and/or geochemistry/mineralogy in specimens that promote repeatable experimental behavior rock (Pyrak-Nolte & DePaolo, 2015; Mitchell & Pyrak-Nolte, 2018). Much work has been performed on controlling the structure in analog rock such as 3D printing in plastics or with polymers to reconstruct porosity or fracture surface roughness from X-ray tomographic scans and laser surface profiling (e.g. Ishutov et al., 2015; Jiang et al., 2016; Head & Vanorio, 2016; Suzuki et al., 2017). Polymer/resin-based and gypsum-based printers have been used to study dynamic crack coalescence (e.g. Jiang et al., 2016; Gell et al., 2018); mechanical behavior of fractured rock (Zhu et al., 2018); and elastic properties of material with penny-shaped inclusions (Huang et al., 2016). Some studies have combined 3D printing and analog materials to generate molds of fracture surfaces from which casts of fracture surfaces are made with cement (Woodman et al., 2017) or transparent acrylic to aid visualization (Boomsma et al., 2015). Casting processes enable control of the matrix properties in terms of strength, porosity and brittleness/ductility. While other research has focused on using powder forms of the constituent minerals in shale to form synthetic layered shale by subjecting the structured powder mixture to high pressures and temperatures (e.g. Luan et al., 2016; Gong et al., 2018).

In this paper, the role of micro-scale properties and behavior on the detection of crack initiation, propagation and geochemical alteration is examined through three topics where specific structural and/or compositional features are controlled to: (1) identify of a geophysical precursor for a system transitioning from meta-stability to unstable behavior with specific focus on crack nucleation, propagation and coalescence; (2) demonstrate the existence of acoustic emissions from geochemically-induced fractures; and (3) understand the role of depositional layers and mineral fabric on tensile crack formation. The results from these studies advance current understanding of which microscopic properties of evolving fracture systems are most useful for predicting macroscopic behavior and the best imaging modalities to use to identify the seismic signatures of time evolving fracture properties.

2 SIGNATURES OF ORIENTED MICRO-CRACKS DURING CRACK FORMATION

2.1 *Background*

The formation and growth of discontinuities in rock with natural and engineered processes cause instabilities such as infrastructures failure, rock slope instability, or slip along faults and earthquakes. Monitoring the time and location of crack formation and coalescence, and determination of the crack type (tensile or shear) are the challenges in the rock mass evaluation. Seismic wave imaging has been used in previous studies as a precursory method to crack formation and failures (Pyrak-Nolte & Roy, 2000; Modiriasari et al., 2017) and to slip along fractures (Chen et al., 1993; Nakagawa et al., 2000; Nagata et al., 2008; Hedayat et al., 2014; Rouet-Leduc et al., 2017). Modiriasari et al. (2017) have identified precursors associated with the initiation and propagation of tensile cracks with significant changes in the amplitude of the compressional (P) and shear (S) waves. However, the shear crack initiation could not be detected from the P- or S-wave amplitudes. Nakagawa et al. (2000) showed that during shearing along a synthetic fracture, which can be idealized with an array of oriented microcracks, an incident S- or P-wave is converted to P- or S-wave, respectively, even for normal incident waves. Here, we show that converted waves emerge at the onset of damage in the form of shear cracks and provide precursory signatures to shear crack formation. The precursory signatures arise from the formation of oriented micro-cracks prior to coalescence.

2.2 *Experimental set-up*

In the laboratory experiments, uniaxial compression loading was applied on several prismatic Indiana limestone specimens to investigate seismic precursors to the formation of shear cracks. The specimens contained designed flaws with specific geometries to induced shear cracks. The dimensions of the specimens were 203.2 × 101.6 × 38.1 mm (Figure 1a). The intact specimens had an average density of 2,326 kg/m^3, unconfined compressive strength of 47 MPa, Young's modulus of 7.4 GPa, and porosity of 15-25%. The average P- and S-wave velocities measured in the samples were 4,380 and 2,570 m/s, respectively. Two parallel designed flaws (cracks) were cut into the thickness of the specimen (Z direction in Figure 1a) using a scroll saw. Different geometries of the flaws were tested in the experiments (Modiriasari, 2017). The flaw geometry that is discussed in this paper is shown in Figure 1a. This geometry was selected because it is associated with the formation of shear cracks between the two flaws.

The formation of cracks on the surface of the specimen was detected using two-dimensional Digital Image Correlation (DIC) (Chu et al., 1985; Pan et al., 2009; Sutton et al., 2009) to measure surface displacements. DIC images were taken from the Region of Interest (ROI), gray region on the specimen surface in Figure 1b. A Grasshopper (Point Gray) CCD camera (with 2,248 × 2,048 pixels) and a Fujinon lense (Model HF50SA-1, with a focal length of 50 mm) were used to take images in the experiments. The details of the DIC analysis are explained in Hedayat et al. (2014), Modiriasari et al. (2017), and Modiriasari et al. (2018). An array of source and receiver transducers were placed on the sample's right and left side, respectively, to generate and propagate P- and S-waves into the specimen. The transducers were held in place using two steel plates and four springs with a tension of 70 kPa (Figure 1b). The lateral load was small enough to ensure that no lateral loading was applied on the specimen. The transducers had a central frequency of 1 MHz. Two pairs of transducers were used at the external tips of the flaws to monitor the formation of tensile cracks. In addition, two pairs of transducers were used in the intact material for reference signals. The signals from transducer 3S will be analyzed in the next section because they probed between the two flaws, where the shear crack formed and the crack coalescence occurred. The 3S transducer pair was polarized vertically, in the Y direction.

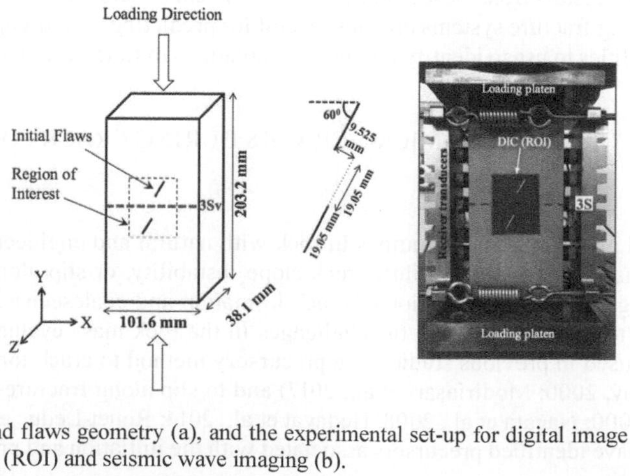

Figure 1. Sample and flaws geometry (a), and the experimental set-up for digital image correlation (DIC) on the Region of Interest (ROI) and seismic wave imaging (b).

The transducers were coupled on the specimen surface using a thin layer of oven-baked honey (at 90°C for 90 minutes). The specimen surface was covered by plastic tape to prevent the penetration of honey into the sample pores. Initially, the sample was loaded in the uniaxial direction (using Instron loading machine and two steel loading plates as shown in Figure 1b) at a stress of 2 MPa for four hours to stabilize the seismic signals in a constant amplitude. Then the uniaxial load increased at a rate of 0.04 mm/min until the specimen failed. A thin Teflon film and petroleum jelly was used between the loading plates and the specimen surfaces to reduce the friction between the plates and the rock and the concentration of stresses on a sample. During loading, the applied load and displacements in the Y direction, two DIC images, and one full array of transmitted and reflected signals were recorded concurrently at a sampling rate of 1 Hz.

2.3 *Experimental results*

2.3.1 *Digital Image Correlation (DIC)*

The path of crack formation interpreted from DIC results at different loads is shown in Figure 2. In the analysis, the crack initiation is identified when there was a minimum discontinuity of 5 μm in the horizontal displacement of two adjacent points. This threshold is greater than the noise in the measured displacements from the DIC and is large enough to determine the location of crack tips. The colorbar in the figure shows the aperture of the cracks (mm) inferred from the displacement discontinuity values. As shown in Figure 2b, the initiation of shear cracks was detected at the internal tips of the flaws at 88 kN. With uniaxial compression, the shear crack propagated until the coalescence occurred at 92 kN (Figure 2d). The aperture of the crack reached 60 μm at 105 kN (Figure 2f) and the specimen failed at 110 kN. Figure 2c shows that the signals from transducer 3S were roughly perpendicular to the path of shear crack formation.

2.3.2 *Seismic wave imaging*

2.3.2.1 *Transmitted waves*

Normalized amplitude of transmitted waves from traducer 3S with uniaxial load is shown in Figure 3. The amplitudes were extracted from wavelet analysis (Combes et al., 1989; Nolte et al., 2000; Polikar, 1999; Sheng, 1995) at a frequency of 440 kHz and were normalized with respect to the initial amplitude of the transmitted signal before increasing the load. The inset shows the waveforms of the transmitted signals arriving at ~42-44 μs at different loads. The normalized amplitude of the transmitted signals gradually decreased with load until ~90 kN. This is associated with the opening of microcracks inside the intact rock. The first significant change in the normalized wave amplitude occurred at ~92 kN, at the time of crack coalescence. This supports the previous finding that the shear crack initiation cannot be detected using the seismic wave transmissions (Modiriasari et al., 2017).

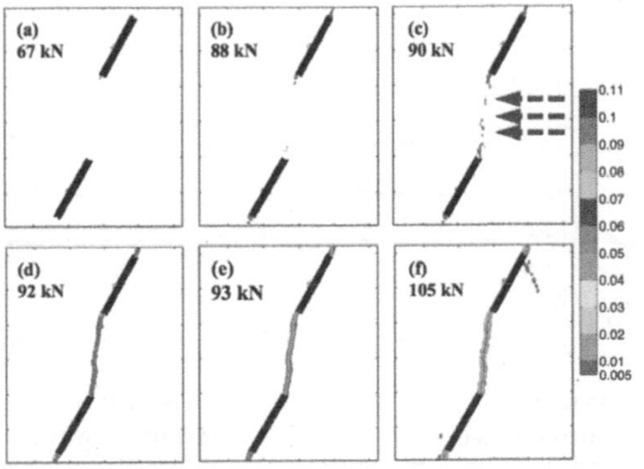

Figure 2. Initiation and propagation of cracks at the flaw tips, obtained from DIC, with load. The blue arrows denote that the incident wave between the two flaws were roughly perpendicular to the path of shear crack evolution.

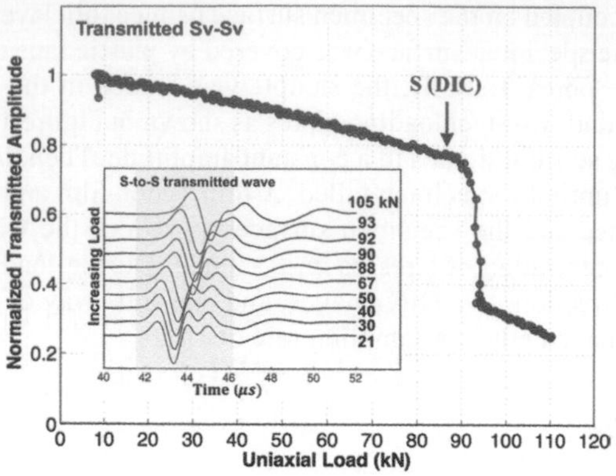

Figure 3. Waveforms and normalized amplitude of the transmitted signals from transducer 3S with uniaxial load. The blue arrow shows the load of the shear crack initiation detected using the DIC results.

However, the initiation of shear cracks was detected with the DIC at 88 kN. A minimum transmission of signals across the shear crack occurred after the crack coalescence and increase in the crack aperture.

2.3.2.2 *Converted waves*

Full waveforms of the signals were recorded during the experiment. The waveforms of the transmitted signals arriving between 22-40 μs at different loads are shown in Figure 4. S-wave transducers also generated small amplitude P-wave signals that had an arrival time of ~25 μs. At a load of ~67 kN, a new transmitted signal emerged in the waveforms and increased in amplitude with load up to 92 kN. The arrival time of this signal (34.7 μs) corresponded to the arrival time of a converted S- to P-wave, knowing the location of shear crack from the DIC results.

The normalized amplitudes of the transmitted (blue) and reflected (green) waves (in the left axis), and converted (purple) waves (in the right axis) from transducer 3S, with uniaxial loads are shown in Figure 5. The reflected and converted wave amplitudes are normalized with respect to the initial

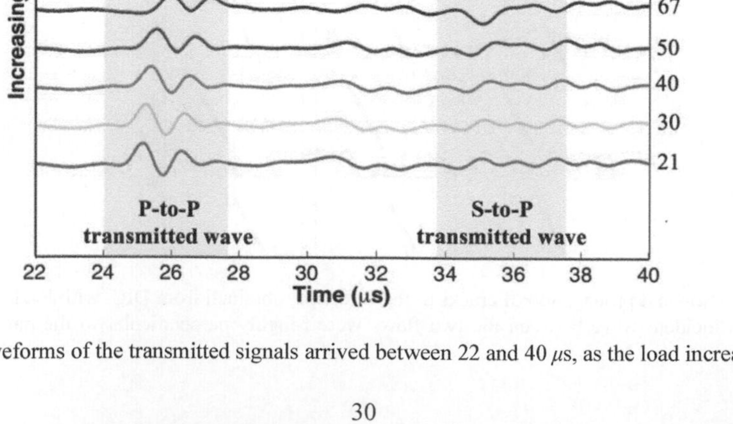

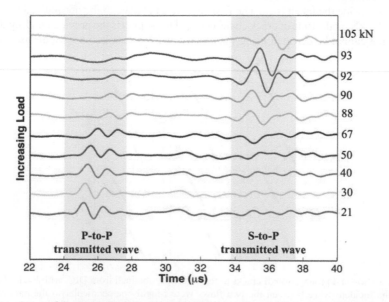

Figure 4. Waveforms of the transmitted signals arrived between 22 and 40 μs, as the load increased.

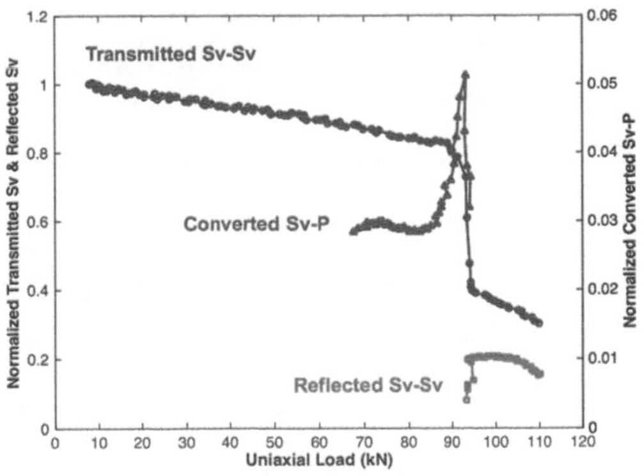

Figure 5. Normalized amplitude of the transmitted (blue), reflected (green), and converted (purple) waves from transducer 3S, as a function of load.

value of the transmitted wave amplitude. The emergence of converted waves occurred at 67 kN, 76% of the crack detection load with DIC and 60% of the failure load. The converted wave amplitudes increased with shear crack propagation. With the crack coalescence, a significant drop in the transmitted wave amplitudes was observed. At this load, most of the energy of the wave was reflected from the surface of shear cracks and the reflected signals emerged and increased in amplitude.

2.4 *Discussion*

The emergence of the converted waves is associated with the formation of an array of similarly oriented microcracks during shear crack growth. This is supported by the work of Nakagawa et al. (2000) that showed experimentally, theoretically, and numerically that information on the fracture void geometry is contained in P-S or S-P converted modes (P-compressional waves, S-shear waves). The conversions arise from a cross-coupling stiffness that can occur along rough surfaces. In their study, the microcracks in the array were oriented at 45° and -45°, and the cross-coupling stiffness was controlled through the application of shear and normal stress. They explained the experimental observations by extending the displacement discontinuity theory (Pyrak-Nolte et al., 1990; Schoenberg, 1980, 1983), and defining the cross-coupling fracture specific compliance (i.e. when a shear crack dilates, normal and shear stresses on a crack plane induce tangential and normal displacements, respectively). As a fracture is subjected to shearing, an array of similarly oriented microcracks forms that causes the cross-coupling compliance greater than zero and the emergence of converted S-P or P-S waves even at normal incidence. This affects the partitioning of energy to the transmitted, converted, and reflected wave phases. The converted phase is a function of the local-induced microcrack orientation prior to coalescence, the stiffness of the evolving damage zone, and the signal frequency. Jiang & Pyrak-Nolte (2018) used 3D printed arrays of micro-cracks with orientations (relative to the horizontal axis) from -90° to 90° to examine energy partitioning among the transmitted, reflected and converted modes using ultrasonic P and S wave transducers (Figure 6). In their study, the maximum amplitude in the converted mode occur for micro-crack orientations of +/- 45°. As shown by Nakagawa et al (2000) and observed in Figure 6, rotating the micro-crack 180° results in a signal with the same amplitude but 180° out of phase.

Here we observed the emergence of the S-P converted mode during shear crack formation but prior to coalescence for nearly normal incident waves. DIC showed that when the S-P converted mode emerges, micro-crack widths were on the order of 10 micrometers, a small fraction of the shear wavelength of 2.3 mm at 1 MHz. This suggest that a distributed array of oriented micro-cracks may form prior to coalescence and the extent of the array is sufficient to partition energy from the transmitted wavefront into a converted-mode but is not sufficiently open to partition energy into a reflected mode. Modiriasari et al. (2018) showed that relative amplitudes among the transmitted, reflected and converted modes observed in the data presented here is consistent with theoretical predictions for wave incident on a fracture at 10° and a cross-coupling factor of 0.75. The converted

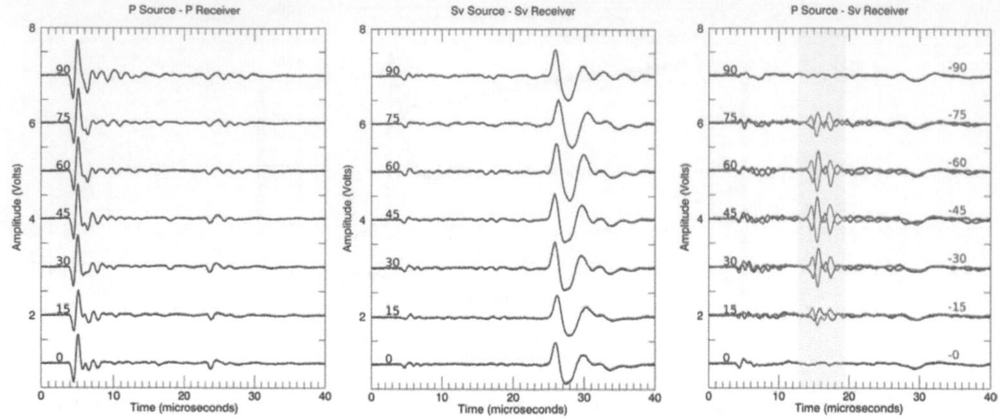

Figure 6. Example of received (left) compressional wave, (center) shear wave, and (right) S-P converted mode on 3D printed samples with oriented micro-cracks. The numbers represent the orientation of the micro-cracks relative to the horizontal. (right) Signals are shown from the sample positive (blue) and negative (red) orientations.

mode was also observed in the experiments when waves are obliquely incident on a fracture (details are discussed in Modiriasari (2017)). The orientation of the inclined microcracks and stiffness of shear cracks were estimated using the theory and the relative changes in the wave amplitudes. This works illuminates how microstructural evolution during the crack formation affects macroscopic geophysical measurements.

3 GEOCHEMICAL-GEOMECHANICAL INTERACTIONS

3.1 *Background*

Volume expansion and contraction occur as minerals hydrate and dehydrate in response to changes in temperature, chemical absorption/desorption or other physical processes. This is of key importance for subsurface sites that contain shale with clay embedded in the matrix. Depending on the clay structure, certain clay minerals and clay bearing rock are capable of swelling as a function of the relative humidity, salt concentration or temperature (Tambach, et. al., 2004; Sone and Zoback, 2014). The charged layered structures and large reactive surfaces of clays introduce a complex relationship because of their reactivity with fluids, and play an important role in the mechanical behavior of the rock. Of the different types of clays, illite and montmorillonite-smectite are the most water-sensitive. Here, we use geo-architected rock to investigate geochemical-geomechanical coupling during dehydration.

3.2 *Geo-architected synthetic rocks*

3.2.1 *Sample preparation*

Geoarchitected rocks were fabricated to determine the influence of clays and their respective geochemically-geomechanically driven volumetric alterations on geophysical properties, fracturing, and damage evolution in the microstructure of clay-rich structures during drying. The baseline for all samples is Ordinary Portland Cement (OPC Type I-II) and Ottawa Sand (SCS 250). All synthetic rocks are cured in a hot-water bath heated to +25° C. Samples were designed with localized clay volumes and with clay distributed throughout the matrix to examine the effect of clay distribution on fracture formation. For these two types of 20% clay-rich samples, the same proportions (by weight) of material used to create the baseline sample (referred to as the cement mortar sample) was used with the addition of clay. The Montmorillonite clay was hydrated before adding it to the baseline mixture (Mitchell and Pyrak-Nolte, 2018). The clay-rich mixture was casted using a cylindrical mold with an average diameter of ~38 mm and an average length ~76 mm. Samples were de-molded after 24 hours and subjected to curing at 25° C for a number of days.

3.3 *Experimental methods*

Samples were monitored during drying to determine if the dehydration of clay induced any physical changes in the geo-architected samples. The methods of characterization are described below.

3.3.1 *Sample characterization*

X-ray micro-tomography was performed to determine if any changes in the structure of the samples occurred during dehydration. A Zeiss Xradia 510 Versa – 3D X-Ray Microscope (3D XRM) was used to monitor the curing and drying of 1" diameter clay-rich samples over a 7-day period. The Versa system maintains an in-situ temperature of ~28°C. Object Research Systems (ORS) Dragonfly Pro software was used to visualize and quantify the porosity and fracture network generated in the samples. FIJI (ImageJ) open source image processing package was also used for image analysis. The voxel edge length from the imaging was 40 micrometers.

3.3.2 *Acoustic emission monitoring*

Clays are known to shrink or swell upon interaction with fluids (Wagner, 2013). Clay swelling or shrinkage can occur in the presence of pore-fluids and/changes in moisture content. Shrinkage can also occur upon drying. The extent of swelling depends on the amount and type of clay minerals, their fabric and the stress conditions, as well as the permeability (Prinz and Strauß, 2006; Wagner, 2013). Local volumetric changes caused by clay minerals in rock can cause crack initiation and growth, which in turn generates acoustic emissions. In addition to the behavior of clays in a matrix, drainage of fluids from porous media has been known to produce acoustic emissions (Michlmayr, 2012).

Acoustic-emission measurements were monitored to detect time-dependent crack formation in a sample during drying. The samples were monitored over a 7-day period using an array of six (6) broadband transducers (with flat frequency response between 20-400kHz (Physical Acoustics Corporation – F15-α). The sensors were connected via preamplifiers to a Mistra/Physical Acoustics AE measurement system with a 5MHz sampling frequency. The threshold amplitude was set at 25dB. During monitoring, an AE signal was recorded when the signal amplitude exceeded an ascribed threshold.

3.4 *Results*

From the acoustic emission (AE) measurements, acoustic events were recorded on the cement-mortar sample and two samples with clay. In Figure 7a, acoustic emission events are shown for all of the samples. From X-ray tomographic samples (Figure 7d), no fractures or cracks were observed in the cement mortar samples. Unconnected porosity and the lack of any fractures for the cement mortar samples are observed in the 3D reconstruction shown Figure 8c and the AE events are attributed to movement of the air-water interface during drying. Moebius et al. (2012) demonstrated on glass bead packs that the movement of fluid-fluid interfaces resulted in measurable acoustic emissions. They attributed the AE events to Haines jumps, reconfiguration of fluids behind a drainage front, and also grain re-arrangement. They observed larger amplitude AE during drainage than during imbibition.

When a sample contained clay that was localized as discrete balls, cracks were observed within the clay balls and around the clay balls as shown in the 2D cross-sections from the X-ray tomography data (Figure 7b). As the clay dehydrated, the shrinkage of the clay resulted in debonding from the cement-mortar matrix and cracking within the clay. However, from the 3D reconstruction of the X-ray tomograms (Figure 8 red regions), fractures connecting the clay balls are observed. The increase in acoustic emission around 600 minutes (Figure 7a) for the localized clay sample relative to the cement mortar sample is attributed to the initiation and propagation of the fractures that link the clay balls as water is released from the clay structure during dehydration.

Images of a geo-architected sample with distributed clay are shown in Figures 7c and 8b. The fracture network generated during the drying of the clay-rich geo-architected sample is extremely

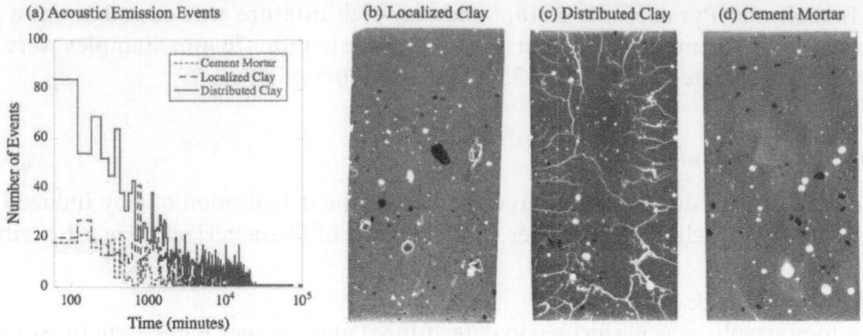

Figure 7. (a) Number of acoustic emission events as a function of time for the cement mortar, localized clay and distributed clay samples. 2D images from reconstructions of the X-ray tomograms for (b) localized clay, (c) distributed clay and (d) cement mortar samples. Air represents air.

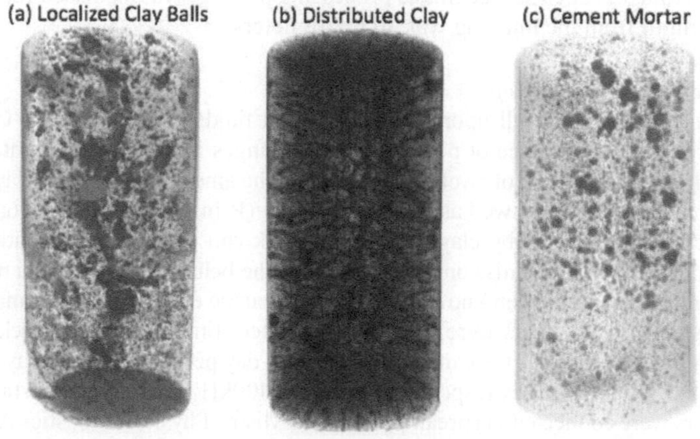

Figure 8. 3D reconstruction for X-ray tomograms from 25 mm diameter samples showing porosity (blue) and fractures induced from the shrinkage of clay (red) for (a) localized clay and (b) distributed clay. (c) The cement mortar samples contained no induced fractures, only unconnected porosity.

extensive extending from the exterior all the way through to the interior of the sample. This is consistent with the observed number events for the distributed clay sample which in the first 300 minutes had a factor of 2 to 4 times as many events as observed in the localized clay or the cement sample during that same period. These fractures were generated as a result of the chemo-mechanical behavior of clays, inducing fractures as the clay compacts and shrinks in volume.

3.5 *Discussion*

Examining the microstructure of geo-architected rocks after a period of drying with micro-CT has shown that the presence of clay can induce damage in a rock which would thereby affect the geophysical properties such as elastic moduli. The crack network formation is sufficient to generate measurable acoustic emission events. Previous work by Mitchell & Pyrak-Nolte (2018) has shown that these events have frequencies in 60 kHz – 250 kHz. Though these frequencies are much higher than traditional borehole sensors (20kHz), fiber optic-based distributed acoustic sensors with sensitivities on the order of nano-strains may have the potential to detect acoustic signatures from cracking induced by volumetric changes in minerals. The potential for geochemical-geomechanical coupling is high given the amount and different types of fluids that are pumped into the ground during fracturing activities. Future work will examine the effect of matrix strength on cracking and clay type using geo-architected samples.

4 TENSILE FRACUTRING: THE ROLE OF MINERAL TEXTURE AND LAYERING

A recognized challenge in quantifying the strength of rock is effect of layering which is well known to result in anisotropic rock properties and behavior. Anisotropy in rock arises from natural geological processes such as sedimentary processes where sediments are sequentially deposited in layers, and also from metamorphic processes that align minerals. For example, the complex response of shale is, in part, caused by anisotropy from layering within the rock and also from the mineral and organic components that compose the layers. The orientation of layers/bedding relative to in-situ and induced stresses can significantly affect rock strength and deformation because layers or interfaces between layers often behave as natural planes of weakness that can debond when stresses attain a failure condition.

From previous studies, rock strength has been shown experimentally to depend on the direction of fracture propagation relative to layering (Gao et al., 2017, Chandler et al., 2016, Na et al., 2017). In describing the relative orientations, the nomenclature often used is divider, short traverse and arrester (Figure 9). From experiments, fracture toughness is sometimes greatest for divider samples (e.g. Gao et al., 2017), while others have observed that the short traverse samples can exhibit both the largest and smallest values of fracture toughness (Chandler et al., 2016, Na et al., 2017). Variation in behavior relative to layering is often attributed to heterogeneity within a layer or among samples, the presence of clays, fluids, or organic material. A challenge in determining the effect of layer orientation on fracture toughness is rock variability. Here, geo-architected laboratory samples with repeatable mineral fabric and structural features were used to improve current understanding of the role of layering on peak failure load or fracture toughness in layered material. In this study, we found that predictions of fracture resistance based on layer orientation alone is insufficient. The presence of an oriented mineral texture within the layers also influences fracture toughness.

4.1 *Experimental set-up*

4.1.1 *Sample preparation*

A 3D printing process (ProJet CJP 360 printer) was used to create layered geoarchitected rock. A proprietary water-based binder (ProJet X60 VisiJet PXL) bonded layers of calcium sulfate hemi-hydrate (0.1 *mm* thick bassanite powders), resulting in a gypsum (calcium sulfate hemi-hydrate) reaction product. The direction of gypsum mineral growth is strongly affected by the direction of the binder application. The gypsum crystals form a bond between bassanite layers as one layer of bassanite is deposited on the previous layer. An oriented mineral texture forms because stronger bonds are formed amongst gypsum crystals than between gypsum crystals and bassanite powder.

With the aid of 3D printing, orientation between the bassanite layers and gypsum texture were controlled to examine the effect of texture direction relative to layer direction on mode I crack growth. Specimens were printed with dimensions of 25.4 x 76.2 x 12.7 *mm* with a 5.08 *mm* long 1.27 *mm* wide central notch. Tensile fractures were induced through three-point bending (3PB) experiments to induce a mode I crack from the central notch. In this paper, the 3D printed samples used

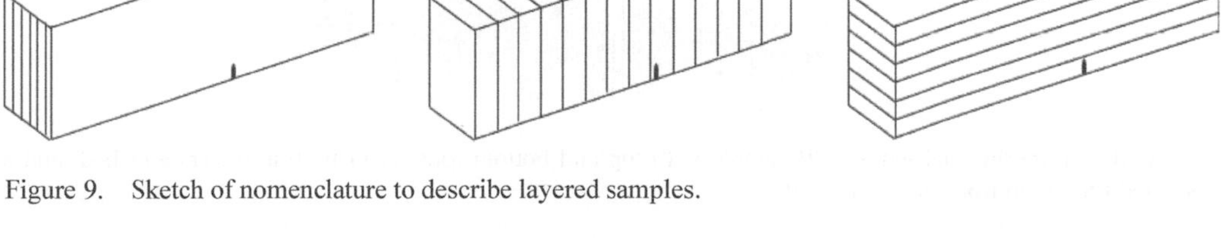

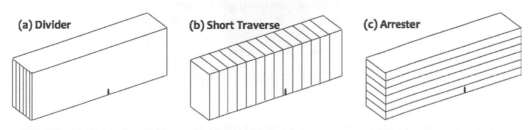

Figure 9. Sketch of nomenclature to describe layered samples.

in the study included: two short traverse samples with mineral textures perpendicular (sample V) and parallel (Valt) to the direction of fracture propagation, and two arrester samples with mineral textures parallel to layering but perpendicular to the fracture plane (H) and parallel to the fracture plane (Halt). For additional information on these samples, please see Jiang et al (2019).

For comparison, a cast gypsum sample was created to act as a standard. First, a resin-based 3D printer (FormLabs 2) produced a solid resin sample with the same dimensions as the 3D printed gypsum samples. The solid resin sample was used to fabricate a silicon rubber mold that was filled a mixture of gypsum and water and then vibrated to minimize the amount of trapped air. The cast gypsum samples were then cured in an oven at 40 °C for 4 days. The resulting cast gypsum sample had no preferred direction of mineral orientation.

4.1.2 *Tensile fracturing*

Figure 10 shows a digital image of the three-point bending (3PB) method for inducing a tensile fracture (Mode I) in the 3D printed and cast samples. A rod is centered at the top of the sample and two additional rods are place along the bottom of the samples.

Load was applied to a sample using an ELE International Soil Testing load frame with a 2000 *lbs* (8896 *N*) applied a load to a sample that was recorded with a capacity S-shaped load cell. A loading rate of 0.03 *mm/min* was applied to the sample while load and displacement were recorded at a 5 *Hz* MSamples/sec.

4.2 *Results*

4.2.1 *3 Point bending results*

Figure 11 provides the load-displacement data for four 3D printed geometries and for a cast gypsum sample. The strongest sample was the arrester sample H that had a mineral texture perpendicular to the direction of fracture propagation. As a tensile fracture to propagate in sample H, gypsum-to-gypsum bonds were broken. However, for the second arrester sample, Halt with mineral texture parallel to the induce fracture plane, the peak strength was 20% less than that observed for sample H. The two short traverse samples (V and Valt) exhibited similar peak loads that were the lower in magnitude than those for samples H and Halt. Cast gypsum samples that did not have a preferred mineral texture orientation were stiffer than the 3D printed samples and exhibited post-peak brittle behavior. The 3D printed sample, on the other hand, had relatively ductile post-peak behavior (Figure 11).

Three cohorts of cast gypsum and 3D printed samples H, Halt, V and Valt were tested (Figure 12). Peak load, which is linearly related to fracture toughness (Whittaker et al., 1992), is observed to depend on both the orientation of the bedding relative to the load and on the orientation of the mineral texture relative to the layering (Figure 11). The short traverse samples (V and Valt) exhibited

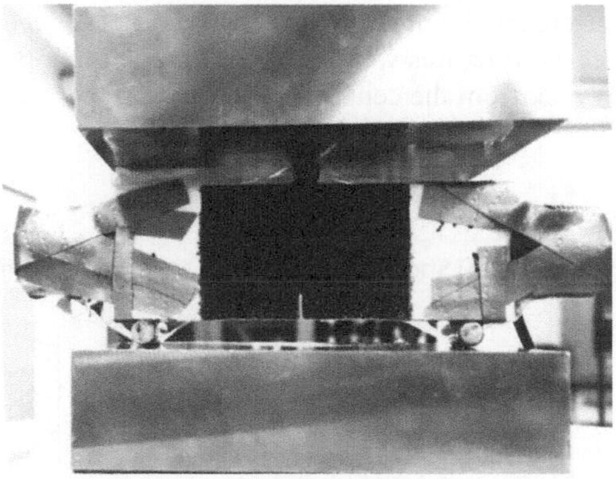

Figure 10. Experimental setup: 3PB sample with top and bottom rods, acoustic transducers attached, and a speckled region on front surface for DIC.

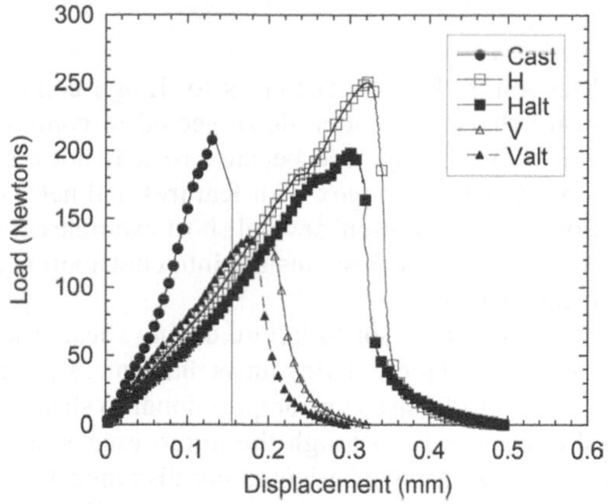

Figure 11. Load vs. displacement for cast gypsum and 3D printed gypsum samples.

consistently lower peak strengths than the arrester samples (H and Halt). Sample H exhibited the largest values of peak load and is used to normalize the peak load for all samples. In sample H both the mineral texture orientation and layering were perpendicular to the failure plane. Failure in the short traverse samples is dominated by debonding of the bassanite layers with mineral orientation slightly affecting the peak load. Peak loads at failure were on average found to be greater for the Valt than sample V which is attributed to the difference in the direction of mineral texture between the two samples. Halt samples, with mineral texture parallel to the induced tensile fracture plane, exhibited peak failure loads that were on average 15% lower than those for Halt.

4.3 *Discussion*

Geo-architected and analog gypsum samples enabled exploration of the effect of layer and mineral texture orientations on peak load achieved as a tensile fracture was induced. Additional testing has shown that the observed trends in peak failure load are consistent in terms the ranking of peak strength and the relative orientation between the layers and mineral texture. The greatest failure loads where obtained when the mineral texture direction was perpendicular to the fracture plane, and the smallest when the layering is parallel to the fracture plane. The 3D printed and cast gypsum samples exhibit some variability (Figure 12) but less variation than that observed in natural rock samples. Future directions will include testing on divider samples (Figure 9a) and examination of the effect of mineral texture on tensile fracture geometry.

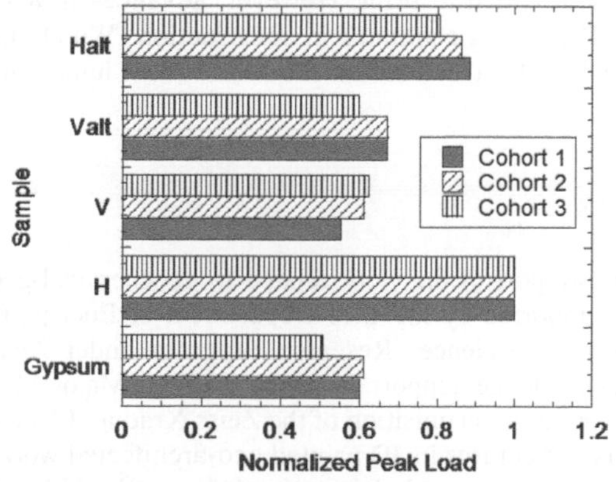

Figure 12. Normalized peak load by sample for 3 testing cohorts.

5 CONCLUSIONS

The goal of experimental laboratory-based research is to design controlled experiments to illuminate fundamental physical behavior of materials subjected to controlled conditions. The field of laboratory rock physics is a fascinating field because rock is not a simple, uniform material, but is structurally and compositionally complex with features and heterogeneity occurring over a range of length scales. In this paper, we present several short examples of the use of crafted or geo-architected samples to explore and provide basic insight into contribution of microscale texture and structure on macro-scale measurements.

The study on the emergence of a precursor to failure during shear crack initiation and propagation was possible through the design of pre-existing flaws that generate shear cracks under uni-axial loading conditions. The precursor to failure is a compressional to shear wave converted mode that is associated with oriented micro-cracks. Though the micro-cracks are 1/100 of the scale of the wavelength of the probing signal, an array of oriented but disconnected micro-cracks is sufficient to generate a signature in the macro-scale measurements by elastic waves. This occurs because a surface of discontinuity in material properties, even if the discontinuity arises from small-scale heterogeneity much smaller than a wavelength, generates energy partitioning among macroscopic scattered fields.

The work on geo-architected rock with clay provides a method for examining the effect of volumetric shrinkage of minerals on the formation of fractures in a sample. The distribution of clay affects both the geometry of the fracture network that is formed and the acoustic emissions that are released during the fracture network formation. As part of an ongoing study, these geo-architected samples are enabling us to study how clay type (i.e. degree of swelling/shrinkage, chemical composition of the clay), clay distribution, fluid chemistry, and matrix strength effect the formation of fracture networks during dehydration processes. The goal is to provide insight into role of fluid chemistry during fluid injections/withdrawals into subsurface rock on the formation and accumulation of damage in samples with minerals that swell and shrink.

Finally, the 3D printed gypsum samples enabled experiments to determine why the strength of a layered sample differs from the expected behavior. Often, rocks are viewed as transversely isotropic because of layering in the sample. This study showed that in addition to layering, mineral texture orientation can have a significant role in the peak failure load. A determining factor is whether bonding between layers or bonding among oriented minerals within a layer is stronger. This suggests that mineralogy studies of specimens to determine if the minerals with layers are randomly oriented or align could provide insight into potential failure behavior of a rock specimen prior to testing.

3D printing has many advantages for studying specific structural and compositional distributions found in rock such as repeatable dimensions, composition and access to numerous samples. However, there are several drawbacks that include: the smallest feature size is often limited by the printer resolution, material behavior might not be purely brittle as some rock types, and currently limitations in materials with which to print. However, advances in additive manufacturing are ongoing as the need for 3D printing continues in construction (Wu et al., 2016) and especially in the area of 3D printing habitable structures on the Moon from lunar soils (Cesaretti et al., 2014; Meurisse et al., 2018).

ACKNOWLEDGEMENTS

The authors acknowledge support of the work related to geochemical-geomechanical coupling in geo-architected samples supported by the U.S. Department of Energy, Office of Science, Office of Basic Energy Sciences, Geosciences Research Program under Award Number (DE-FG02-09ER16022). We also acknowledge support from the EVPRP Major Multi-User Equipment Program 2017 at Purdue University for acquisition of the Zeiss Xradia 510 Versa 3D X-ray Microscope. From work related t o tensile fractures in 3D printed geo-architected work we acknowledge Sandia National Laboratories. Sandia National Laboratories is a multi-mission laboratory managed and operated by National Technology & Engineering Solutions of Sandia, LLC, a wholly owned subsidiary of Honeywell International, Inc., for the U.S. Department of Energy's National Nuclear Security Administration under contract DE-NA0003525. This work is supported by the Laboratory Directed Research and Development program at Sandia National Laboratories. This paper describes objective technical results and analysis. Any subjective views or opinions that might be expressed in the paper do not necessarily represent the views of the U.S. Department of Energy or the United States Government. The research presented in this paper related to the emergence of converted elastic wave modes during shear crack initiation and formation has been supported by the National Science Foundation, Geomechanics and Geotechnical Systems Program, with award number 1162082-CMMI. All of this support is gratefully appreciated.

REFERENCES

Agliardi, F., Zanchetta, S., and G.B. Crosta. 2014. Fabric controls on the brittle failure of folded gneiss and schist. *Tectonphysics*. v.637. 150–162.

Bobet, A. and Einstein, H.H. 1998. Fracture Coalescence, In Rock-Type Materials Under Uniaxial and Biaxial Compression. *International Journal of Rock Mechanics, Min. Sci. and Geomechanics Abstract*, Vol. 35, No. 7, pp. 863–889.

Boomsma, E. and L.J. Pyrak-Nolte. 2015. Chapter 5: Particle swarms in smooth-walled fractures. in *American Geophysical Union Monograph: Fluid Dynamics in Complex Fractured-Porous Systems*. Eds. B. Faybishenko, S.M. Benson and J.E. Gale. John Wiley & Sons, Inc. Hoboken, NJ. 65-84.

Brace, W.F. 1965, Relation of elastic properties of rocks to fabric. *Journal of Geophysical Research*. v. 70. no. 22. 5657–5667.

Burns, M. 1993. *Automated fabrication: improving productivity in manufacturing*. Prentice Hall. Englewood Cliffs. 369 p.

Cesarettie, G., Dini, E., De Kestelier, X., Colla, V. and L. Pambaguian. 2014. *Acta Astronautic*. 93. 430–450.

Chandler, M.R., Meredith, P.G., Brantut, N., Crawford, B.R. 2016. Fracture toughness anisotropy in shale. *Journal of Geophysical Research: Solid Earth*, 121 (3), pp. 1706–1729.

Chen, Q.-Y., Lovell, C.W., Haley, G.M., & Pyrak-Nolte, L.J. 1993. Variation of shear-wave amplitude during frictional sliding. *International Journal of Rock Mechanics and Mining Sciences*. 30(7). 779–784. https://doi.org/10.1016/0148-9062(93)90022-6.

Choi, M.-K., Pyrak-Nolte, L.J. and A. Bobet. 2014, The effect of surface roughness and mixed-mode loading on the stiffness ratio Kx/Kz for fractures. *Geophysic*. 79:5: D319–D331.

Chu, T.C., Ranson, W.F., & Sutton, M.A. 1985. Applications of digital-image-correlation techniques to experimental mechanics. *Experimental Mechanics*. 25(3). 232–244. https://doi.org/10.1007/BF02325092.

Combes, J., Grossmann, A., & Tchamitchian, P. 1989. Wavelets: Time-frequency methods and phase space. *The Journal of the Acoustical Society of America*. 89(5). 2477–2478. https://doi.org/10.1121/1.400986

Espinoza, D.N. and J.C. Santamarina. 2012. Clay interaction with liquid and supercritical CO2: The relevance of electrical and capillary forces. *International Journal of Greenhouse Gas Control*. 10: 351–362.

Galan, E. and R.E. Ferrell. 2013. Genesis of Clay Minerals. *Developments in Clay Science*. ch. 3: 83–126.

Gao, Y., Liu, Z., Zeng, Q., Wang, T., Zhuang, Z., Hwang, K.-C. 2017. Theoretical and numerical prediction of crack path in the material with anisotropic fracture toughness. *Engineering Fracture Mechanics*. 180. pp. 330–347.

Gell, E.M., Walley, S.M., and C.H. Braithwaite. 2018. Review of the validity of the use of artificial specimens for characterizing the mechanical properties of rock. *Rock Mechanics and Rock Engineering*. https://doi.org/10.1007/s00603-019-01787-8.

Gong, F., Di, B., Wei, J., Ding, P., Li, H. and Dingyuan Li. 2018. Experimental investigation of the effects of clay content and compaction stress on the elastic properties and anisotropy of dry and saturated synthetic shale. *Geophysics*. https://doi.org/10.1190/geo2017-0555.1, **83**. 5. (C195-C208).

Head, D., and T. Vanorio. 2016. Effects of changes in rock microstructures on permeability: 3-D printing investigation. *Geophys. Res. Lett*. 43. 7494–7502. doi:https://doi.org/10.1002/2016GL069334.

Hedayat, A., Pyrak-Nolte, L.J., & Bobet, A. 2014. Detection and quantification of slip along non-uniform frictional discontinuities using digital image correlation. *Geotechnical Testing Journal*. 37 (5).20130141. https://doi.org/10.1520/GTJ20130141.

Hedayat, A., Pyrak-Nolte, L.J. and A. Bobet. 2014a. Multi-modal monitoring of slip along frictional discontinuities. *Rock Mechanics Rock Engineering*. DOI https://doi.org/10.1007/s00603-014-0588-7.

Hedayat, A., Pyrak-Nolte, L. and Bobet, A. 2014b. Detection and Quantification of Slip along Non-uniform Frictional Discontinuities using Digital Image Correlation. *ASTM Geotechnical Testing Journal*. DOI: https://doi.org/10.1520/GTJ2013.

Huang, L., Stewart, R.R., Dyaur, N. and J. Baez-Franceshi. 2016. 3D-printed rock models: Elastic properties and the effects of penny-shaped inclusions with fluid substitution. Geophysics. v. 81, no. 6. (NOVEMBER-DECEMBER 2016); P. D669–D677.

Huggett, J.M. and P.J.R. Uwins. 1994. Observations of waterclay reactions in water-sensitive sandstone and mudrocks using an environmental scanning electron microscope. *Journal of Petroleum Science and Engineering*. 10: 211–222.

Ishutov, S., Hasiuk, F.J., Harding, C. and J.N. Gray. 2015. 3D printing sandstone porosity models. *Interpretation*. 3(3),SX49-SX61.

Jiang, C., Zhao, G-F., Zhu, J., Zhao, Y-X., and L. Shen. 2016. Investigation of dynamic crack coalescence using a gypsum-like 3D printing material. *Rock Mechanics and Rock Engineering*. v. 49. 3983–3998.

Jiang, L. and L.J. Pyrak-Nolte. 2018. Elastic wave conversions from fractures with oriented voids. *52nd US Rock Mechanics/Geomechanics Symposium*. Seattle, Washington, USA. 17–20 June 2018. ARMA 18-1204.

Jiang, L., Yoon, H., Bobet, A., and L.J. Pyrak-Nolte. 2018, Effect of Mineral Orientation on Roughness and Toughness of Mode I Fractures. *53rd US Rock Mechanics/Geomechanics Symposium*. New York, USA. 17–20 June 2019. ARMA 19-0483.

Jiang, Q., Feng, X., Gong, Y., Song, L., Ran, S. and J. Cui. 2016. Reverse modelling of natural rock joints using 3D scanning and 3D printing. *Computers and Geotechnics*. v. 73. p210–220.

Luan, X., Di, B., Wei, J., Zhao, J. and X. Li. 2016. Creation of synthetic samples for physical modelling of natural shale. *Geophysical Propsecting*. v64. 898–914.

Meurisse, A., Makaya, A., Willsch, C. and M. Sperl. 2018, Solar 3D printing of lunar regolith. *Acta Astronautica*. v.152, 800–810.

Mitchell, C.A. and L.J. Pyrak-Nolte. 2017. Microstructural controls on the macroscopic behavior of geo-architected rocks. *Proceedings of the 52th US Rock Mechanics/Geomechanics Symposium*. Seattle, WA, USA. June 17–20, 2017. ARMA 18-1235.

Mitchell, J.K. and K. Soga. 2005. *Fundamentals of Soil Behavior*. 3rd ed. Wiley.

Modiriasari, A. 2017 *Geophysical Signatures of Fracture Mechanisms*, Ph.D. Thesis. Purdue University. West Lafayette, IN. 135.

Modiriasari, A., Bobet, A., & Pyrak-Nolte, L.J. 2017. Active seismic monitoring of crack initiation, propagation, and coalescence in rock. *Rock Mechanics and Rock Engineering*. 50(9). 2311–2325. https://doi.org/10.1007/s00603-017-1235-x.

Modiriasari, A., Pyrak-Nolte, L.J and A. Bobet. 2018. Emergent wave conversion as a precursor to shear crack initiation. *Geophysical Research Letters*. 45. 18, 9516–9522.

Moebius, F., Canone, D. and D. Or. 2012. Characteristics of acoustic emissions induced by fluid front displacement in porous media. *Water Resources Research*. v.48, W11507, doi:https://doi.org/10.1029/2012WR012525.

Mutlu, O. and Bobet, A. 2006. Slip Propagation along Frictional Discontinuities. *International Journal of Rock Mechanics and Mining Sciences*. Vol. 43. pp. 860–876.

Na, S., Sun, W., Ingraham, M.D., & Yoon, H. 2017. Effects of spatial heterogeneity and material anisotropy on the fracture pattern and macroscopic effective toughness of Mancos Shale in Brazilian tests. *Journal of Geophysical Research: Solid Earth*. 122(8). 6202–6230.

Nagata, K., Nakatani, M., & Yoshida, S. 2008. Monitoring frictional strength with acoustic wave transmission. *Geophysical Research Letters*. 35. L06310. https://doi.org/10.1029/2007GL033146.

Nakagawa, S., Nihei, K.T.T., & Myer, L.R. 2000. Shear-induced conversion of seismic waves across single fractures. *International Journal of Rock Mechanics and Mining Sciences*. 37(1-2). 203–218. https://doi.org/10.1016/S1365-1609(99)00101-X.

Nelson, R. 1968. *Modeling a jointed rock mass*. M.S. Thesis. Massachusetts Institute of Technology, Cambridge.

Nolte, D.D., Pyrak-Nolte, L.J., Beachy, J., & Ziegler, C. 2000. Transition from the displacement discontinuity limit to the resonant scattering regime for fracture interface waves. *International Journal of Rock Mechanics and Mining Sciences*. **37**. 219–230.

Pan, B., Qian, K., Xie, H., & Asundi, A. 2009. Two-dimensional digital image correlation for in-plane displacement and strain measurement: A review. *Measurement Science and Technology*. 20(6). 062001. https://doi.org/10.1088/0957-0233/20/6/062001.

Papaliangas, T.T. 1996. *Shear behavior of rock discontinuities and soil-rock interfaces*. PhD. Thesis. University of Leeds.

Park, C.H., and Bobet, A. 2009. Crack coalescence in specimens with open and closed flaws: a comparison. *International Journal of Rock Mechanics and Mining Sciences*, Vol. 46, pp. 819–829.

Polikar, R. 1999. The wavelet tutorial. I*nternet Resources:* http://engineering.rowan.edu/polikar/WAVELETS/WTtutorial.html.

Prinz, H., Strauß, R. 2006. Abriss der Ingenieurgeologie, fourth ed. Spektrum Akademischer Verlag, Oxford, 671
Sposito, G. 1989. *The Chemistry of Soils*, Oxford University Press, New York.

Pyrak-Nolte, L.J., Myer, L.R., & Cook, N.G.W. 1990. Transmission of seismic waves across single natural fractures. *Journal of Geophysical Research*. 95(B6). 8617. https://doi.org/10.1029/JB095iB06p08617.

Pyrak-Nolte, L.J., & Roy, S. 2000. Monitoring fracture evolution with compressional-mode interface waves. *Geophysical Research Letters*. 27(20). 3397–3400. https://doi.org/10.1029/1999GL011125.

Pyrak-Nolte, L.J., and D.J. DePaolo. 2015. *Controlling subsurface fractures and fluid flow: a basic research agenda*. Department of Energy Roundtable Report. DOI: https://doi.org/10.2172/1283189.

Rouet-Leduc, B., Hulbert, C., Lubbers, N., Barros, K., Humphreys, C.J., & Johnson, P.A. 2017. Machine learning predicts laboratory earthquakes. *Geophysical Research Letters*. 44. 9276–9282. https://doi.org/10.1002/2017GL074677.

Schoenberg, M. 1980. Elastic wave behavior across linear slip interfaces. *The Journal of the Acoustical Society of America*. 68(5). 1516–1521. https://doi.org/10.1121/1.385077.

Scrivens, W.A., Luo, Y., Sutton, M.A., Collette, S.A., Myrick, M.L., Miney, P., & Li, X. 2007. Development of patterns for digital image correlation measurements at reduced length scales. *Experimental Mechanics*. 47(1). 63–77.

Shao, S. and L.J. Pyrak-Nolte. 2016. Elastic wave propagation in isotropic media with orthogonal fracture sets. *Rock Mechanics and Rock Engineering*, DOI https://doi.org/10.1007/s00603-016-1084-z.

Sone H. and M.D. Zoback. 2014. Time-dependent deformation of shale gas reservoir rocks and its long term effect on the in situ state of stress. *International Journal of Rock Mechanics & Mining Sciences*. 69: 120–132.

Srodo'n, J., V.A. Drits, D.K. McCarty, J.C.C. Hsieh, and D.D. Eberl. 2001. Quantitative XRD analysis of clay-rich rocks from random preparations. *Clays, Clay Miner*. 49: 514–528.

Srodo'n, J. 2010. Evolution of mixed-layer clay minerals in prograde alteration systems. In: Fiore, S., Cuadros, J., Huertas, F.J. (Eds.), Interstratified Clay Minerals. Origin, Characteristics and Geochemical Significance. *AIPEA Educational Series*, vol. 1. Digilabs, Bari, Italy, pp. 114–175.

Suzuki, A., N. Watanabe, K. Li, and R.N. Horne. 2017. Fracture network created by 3-D printer and its validation using CT images. *Water Resour. Res*. 53. 6330–6339. doi:https://doi.org/10.1002/2017WR021032.

Tambach, T., P.G. Bolhuis, and B. Smit. 2004. Angew. *Chem. Int*. ed. 2650–2652.

Wagner, J.F. 2013. Mechanical Properties of Clays and Clay Minerals. Ch. 9: pp. 347–381. *Developments in Clay Science*. vol. 5A.

Woodman, J., Murphy, W., Thomas, M.E., Ougier-Simonin, A., H. Reeves and T.W. Berry. 2017. A novel approach to the laboratory testing of replica discontinuities: 3D printing representative morphologies. *American Rock Mechanics Association - ARMA 17–049*.

Whittaker, B.N., Singh, R.N., Sun, G., 1992. *Rock Fracture Mechanics—Principles, Design and Applications*. Elsevier, Amsterdam.

Wong, L., and Einstein H.H. 2006. Fracturing Behavior of Prismatic Specimens Containing Single Flaws. *Proceedings of the 41st U.S. Symposium on Rock Mechanics*. Golden, Colorado. Paper 06–899.

Wu, P., Wang, J. and X. Wang. 2016. A critical review of the use of 3-D printing in the construction industry. *Automation in Construction*. v. 68, 21–31.

Zhu, J.B, Zhou, T., Liao, Z.Y., Sun, L., Li, X.B., and R. Chen. 2018. Replication of internal defects and investigation of mechanical and fracture behaviour of rock using 3D printing and 3D numerical methods in combination with X-ray computerized tomography. *International Journal of Rock Mechanics and Mining Sciences*. v. 106, 198–212.

Rock Mechanics for Natural Resources and Infrastructure Development – Fontoura, Rocca & Pavón Mendoza (Eds)

Conventional and advanced monitoring of tunnels with selected case histories

G. Barla
Politecnico di Torino, Italy

ABSTRACT: "Conventional" monitoring instruments installed in and around tunnels and on the ground surface are introduced together with "advanced" equipment, e.g. robotic total station, interferometric synthetic aperture radar, and 3D laser scanning. It is shown that the monitoring data can be analyzed with numerical methods and back analysis to assess the ground behavior and system response. The cases of three large size TBM tunnels are presented: (1) a twin-tunnel (each tube is 15.1 m diameter), in clay and clay marl. (2) A twin tunnel (each tube is 15.6 m diameter) in claystone and clayish sandstone under a deep-seated landslide. (3) A single tunnel (13.6 m diameter), in paragneiss micaschist under two old highway tunnels excavated by conventional methods. These cases are discussed with reference to the monitoring systems adopted. Attention is given to the technical problems met during excavation.

1 INTRODUCTION

Performance monitoring is a fundamental component of the Observational Method when applied to tunnels. The objective is to find if the design predictions on the ground behavior and system response, including the stabilization and reinforcement systems adopted during face advance and final lining installation are within the expected limits. The need is to continuously update the geological and geotechnical models in order to carry out the required back analysis and understand, in near real time, if the overall tunnel performance is as predicted at the design stage. It is implied, where needed, to take the appropriate actions on the excavation and construction methods adopted.

With the objectives of this lecture well in mind, the "conventional" monitoring equipment used for displacement measurements in and around the tunnel and on the ground surface (where needed) are outlined. Also considered is the instrumentation that is generally installed for determining the stresses in the support and reinforcement components. In view of the specific applications in three case studies, which form the main core of the lecture, "advanced" monitoring equipment and methods are presented such as in particular: Robotic Total Station (RTS), Interferometric Synthetic Aperture Radar (InSAR), and 3D Laser Scanning (TLS).

Attention is dedicated in the lecture to the evaluation of the monitoring data, including the use of numerical methods for purpose of back analysis. Emphasis is placed on the need to use such methods so that the calculations performed lead to realistic results, which can be carefully checked and reflect the in situ behavior and tunnel response realistically. It is pointed out that the ability to do so has been enhanced through the years given that the development of new monitoring tools has been accompanied by an increased ability to obtain and evaluate the ground characteristics.

Based on the experience gained, three case studies are presented as follows: (1) a twin-tunnel (each tube is 15.1 m diameter), excavated with an EPB TBM in difficult ground (clay and clay marl). (2) A twin tunnel (each tube is 15.6 m diameter) excavated with an EPB TBM in claystone and highly clayish sandstone under a deep-seated landslide. (3) A single tube tunnel (13.6 m diameter), excavated through a paragneiss micaschist with a Single Shield TBM under two old tunnels constructed more than 60 years ago by conventional methods. The three cases are discussed with attention to the monitoring systems adopted of "conventional" or "advanced" type, and in view of the technical problems met during excavation and in order to put the tunnels in service, allowing for long-term performance monitoring, where needed.

2 MONITORING EQUIPMENT

2.1 *"Conventional" equipment (OGG 2014)*

Displacement monitoring during tunnel excavation has become a common practice by using a Total Station (TS) and targets. If reference is made (e.g.) to conventional tunneling, precise prism targets as well as bireflex-targets (reflectors) are used and their spatial position determined within the global or project coordinate system of reference. In this way, absolute displacement monitoring has replaced the previously used tape convergence measurement, with significant advantages in the appropriate interpretation of the monitored data and understanding of the tunnel response during face advance.

With tunnels driven at great depth, as in the case of the Alpine Base Tunnels in Europe, in particular in squeezing ground conditions, monitoring of the extrusion, i.e. the longitudinal displacement ahead of the face, is as well important for understanding the ground behavior. This is possible by using displacement monitoring of targets installed directly at the face. In addition, the Reverse Head Extensometer (RHE) was developed, which allows for a continuous monitoring of the ground extrusion (Steiner 2007).

If the attention moves to the surrounding ground, extensometers with single point or multiple points of different types and characteristics are available for measuring elongation or shortening between the head and a fixed point or between different points within the ground, with either manual or remote readout with different transducers. The interest is to monitor the amount of displacement occurring in the ground behind the extensometer head, a very important component for understanding the ground behavior away from the tunnel contour.

At the same time, in particular for tunnels near to the ground surface or interacting with landslides (Barla. 2018), single probe inclinometers and/or in-place inclinometers (chain inclinometers) are installed for an appropriate understanding of the problem. In such cases, one should underline the importance of providing the head with a displacement-monitoring target and positioning it immediately after the installation and parallel to the zero measurement of the inclinometer.

Keeping with the ground surrounding the tunnel, ground water pressure monitoring by water level gauges and piezometers in boreholes is as well to be recalled. With attention paid to tunnel performance monitoring of the support system, finally the use of strain gages and sensors of different types, attached to reinforcement and fully embedded in the shotcrete lining and in the final concrete lining is a well recognised practice.

2.2 *"Advanced" equipment*

The Robotic Total Station (RTS) is an electronic transit theodolite integrated with electronic distance measurement of both vertical and horizontal angles and the slope distance from the instrument to a particular point. An on-board computer collects data and performs triangulation calculations. In essence, a remote operation is possible and the equipment is used in automated setups, i.e. without the need of operators.

The Interferometric Synthetic Aperture Radar (InSAR) is used for monitoring surface ground movements. A high density of measurement points over large areas is possible and, based on advanced techniques, such as PSInSAR™ (Ferretti et al. 2001) and SqueeSAR™ (Ferretti et al. 2011), high precision time series of movement are obtained and typical displacement patterns highlighted such as changes in ground movement over time. In recent years, an increase in the use of the InSAR for monitoring tunnels has taken place (ITAtech 2015; Barla et al. 2016), with application in all phases of tunneling projects.

Radar sensors mounted on specific satellites transmit radar signals toward the earth, some of which reflect off objects on the ground, bouncing back to the satellite. Basic InSAR techniques

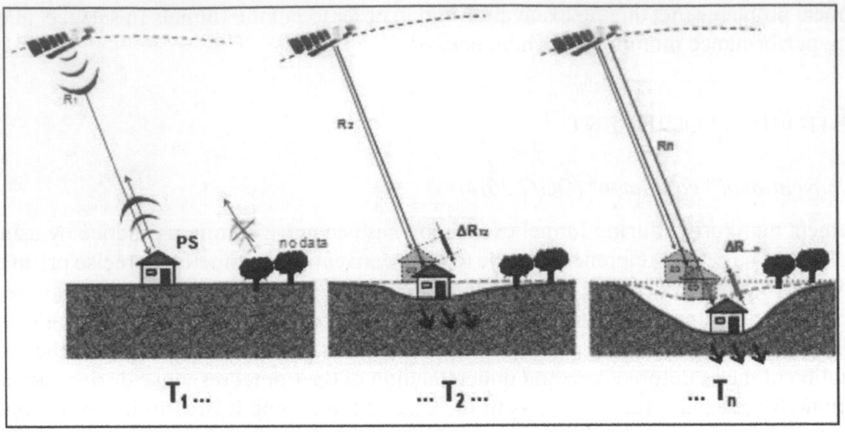

Figure 1. Relationship between ground displacement and signal phase shift. Basic principle of InSAR for measuring ground movement (courtesy of TRE ALTAMiRA).

consist of comparing the phase values of two SAR images, acquired at different times with similar looking angles. The phase difference is proportional to the target motion occurring along the sensor-target line-of-sight (LOS) direction during the time interval of interest (Figure 1).

3D Laser Scanning, also known as TLS (Terrestrial Laser Scanning) or LIDAR (Light Detection and Ranging), with growing range and resolution, is today available. In particular, the pulse-based scanner, also called time-of-flight (TOF) scanner, which uses laser light to probe the object of interest is available. The scanner object distance is measured by calculating the round-trip time a laser pulse takes to reach the object surface from the point of emission and return.

The field of interest is scanned by changing the view directions of the laser rangefinder through a system of rotating mirrors. The related horizontal and vertical angles are measured at a very high data acquisition rate (thousands points per second). The Cartesian coordinates of each point of the scanned object surface are computed given the measured distance and scan angle.

The data obtained from laser scanning is the so-called ‘point cloud’, i.e. a set of vertices in a 3D coordinate system x-y-z. The point cloud is a digital rapresentation of the field being investigated. It is noted that a reference survey is carried out by using a Total Station to set up a local coordinate system around the scanning area which is linked to a global coordinate system used for reference.

The point cloud representing the scanned data can be combined with color photographs, which are captured by a digital camera coupled with the laser scanner equipment. In this manner, a true 3D color model of the rock mass investigated can be obtained. For example, Figure 2 shows the case of a tunnel face with the point cloud superimposed on the corresponding optical image. By using TLS, the primary characteristics of the rock mass can be well extracted from the 3-D point clouds (Barla 2019).

With the above in mind, the use of TLS data for displacement monitoring is of great interest in tunnelling given that significant advantages can be obtained with respect to conventional techniques (Feng and Röshoff 2014). This is the case, in particular, if one could move from the monitoring of a limited number of points on the tunnel surface/perimeter (e.g. targets and reflectors, when using a TS equipment) to the high density scanning points at the milllimiter level which are available with the TLS.

A number of studies have been carried out with this specific objective in mind (e.g. Xie and Lu 2017). 2D and 3D modelling algorithms and fitting methods have been developed successfully and the results obtained show that the topographic measurements compare satisfactorily with TLS data. It is clear however that this is only possible for finite time intervals and continuous near real time monitoring is not yet achievable.

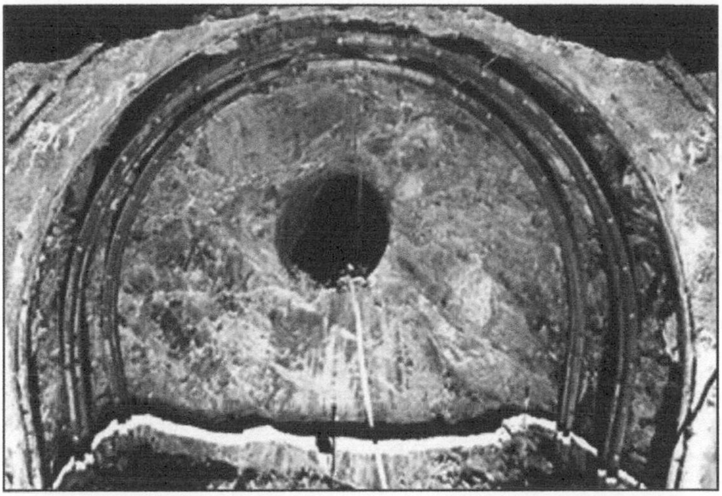

Figure 2. Large size tunnel face with a small size exploratory pilot tunnel. TLS colored point cloud (*Barla 2019*).

Keeping with the "advanced" equipment and monitoring tools in tunneling, it is also of interest here to point out selected and novel equipment, which have been introduced in the last few years, when dealing with the ground behavior and tunnel response during excavation and in the long term, when assessing performance versus time. An example regards sensors and transducers, i.e. the components that monitor a physical quantity and convert it into a signal to be sent to an acquisition and processing system. This is the case of the MEMS (Micro Electro Mechanical Systems) technology.

The MEMS sensors are today used in many geotechnical applications with the advantages due the miniature size and high sensitivity, also associated with wireless technology. Inclinometer-accelerometer instrumentation systems, used to evaluate inclinations can be mentioned, as is the case of inclinometers and tiltmeters. Keeping with tunneling, convergence and joint openings monitoring in segmental lining can be recalled with significant advantages gained with respect to conventional monitoring (Huang et al. 2017).

The magnetostrictive measurement principle is used in borehole extensometers (MagX) for simultaneous measurement of high precision displacements, with the instrument configured as a manual or automatic device. This is also the case, as previously pointed out, of the reverse head extensometer (RHE) for continuous monitoring of deformation ahead of the tunnel face during excavation, to control the process. In this case, the data logging head is located at the deepest point of the borehole instead of at the surface.

Another "advanced" equipment that is of interest to mention here the multi packer column (DMS) which is intended for monitoring in one single borehole the x-y-z displacement components, in addition to temperature, acceleration, groundwater level, etc.. This equipment implies the use of packers, which can work inside uncased or cased boreholes.

3 CASE STUDY 1

3.1 *Preliminary remarks*

The "Caltanissetta" twin tunnel, in Caltanissetta (Italy), has a length of 4 km approximately. The inter-axis of the two tubes ranges between 35 m minimum and 80 m maximum. Prefabricated, steel reinforced concrete segments 60 cm thick and 2 m long form the lining. Excavation took place by using an EPB-TBM with excavation diameter of 15.1 in weak Complex Formations (AGI 1977) under a maximum cover of 130 m.

During the excavation of the two tubes, settlements of significant magnitude did occur and concerns were raised because the tunnel was excavated under an inhabited area (Caltanissetta city) with a number of important buildings for size and use under passed. Settlement monitoring was performed during excavation by conventional topographic measurements (TS) and target points installed on the buildings. In addition, a systematic InSAR analysis was carried out before the excavation of the second tube for gaining insights into the settlements.

The induced horizontal displacements (extrusions) during face advance were studied in a selected tunnel section, where surface displacement monitoring was also undertaken. The interest here stems from the use of "conventional" and "advanced" monitoring methods, together with three-dimensional (3D) numerical modelling and back analysis of the results, with the final objective to find out a likely motivation for the observed response in the tunnel and at the ground surface.

3.2 *Geological conditions*

As shown in Figure 3, the tunnel is located in the Terravecchia Formation (TRVb) consisting of clay breccias and grey-blue brecciated clays, with inclusions of 'varicoloured' clay-shales. Weathered zones of clay and sandy-clay silt are also present. From the ground surface down to the tunnel alignment, fine sands intercalated with cemented layers (SLN), well evidenced due to selective erosion processes, are found.

The following GER complex is formed of prevailing marl clays, silty in cases, with thin intercalated silty layers. Then, a transition takes place to marl limestone and calcareous clay, in general a weak rock complex formation (TRB). The water table is 30 m below the ground surface. The geotechnical parameters were known based on the laboratory tests performed during the tunnel design stages.

3.3 *Monitoring methods and results*

The length of the tunnel shown in Figure 4 was selected to be a monitoring test site in order to test the TBM performance during excavation of the second tube (the right one) and relate the ground response on the surface (settlements) and in the tunnel ahead of the face (extrusions). The interest stems from the presence of a number of buildings on the ground surface (e.g. numbered 28, 29, etc. in the inset of Figure 4), where displacement monitoring was to take place during excavation of the right tube, with the left tube excavation being already completed.

Prior to excavation, a horizontal inclinometer (also shown in Figure 4) was installed from the left tube to cross the right tube, with the purpose to monitor the horizontal displacements (extrusions) ahead of the face. Monitoring of the surface displacements was performed systematically during tunnel excavation in conjunction with real time continuous TBM parameters trend monitoring. As anticipated, during the excavation of the second tube, also InSAR monitoring data of the surface ground movements over time were obtained.

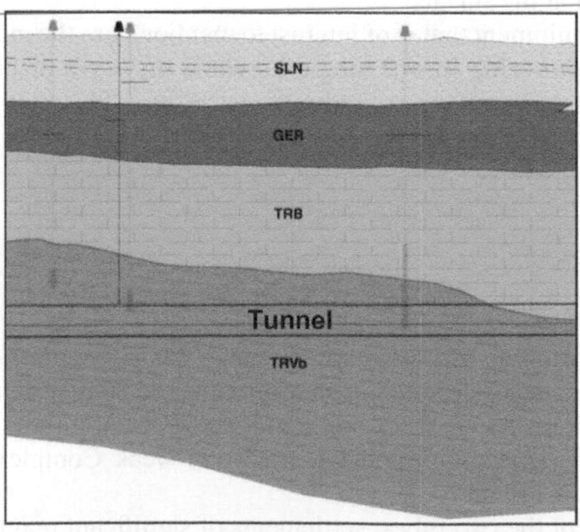

Figure 3. Typical geological profile along the "Caltanissetta" twin tunnel.

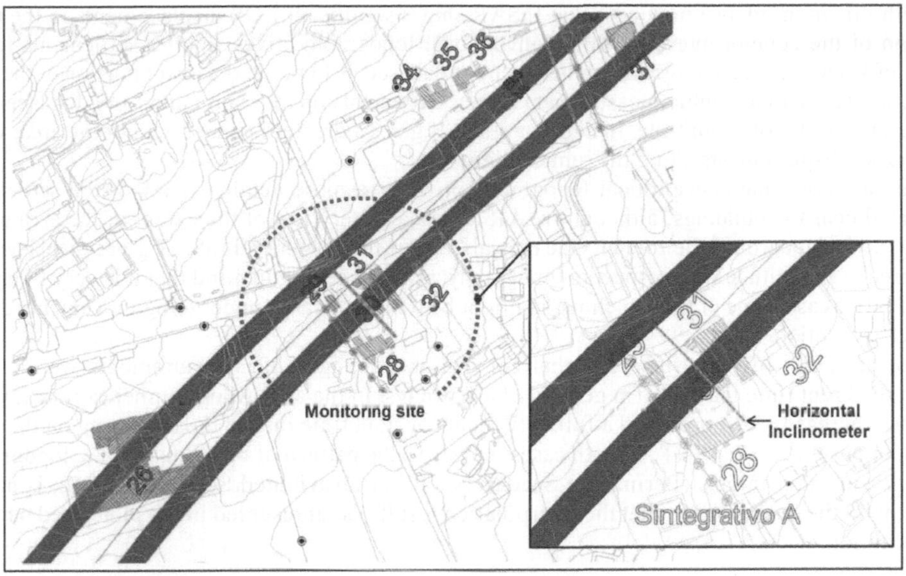

Figure 4. Monitoring test site.

The InSAR data of interest here are from the COSMO-SkyMed (CSK) dataset, in descending geometry, and cover a time interval of about 3 years. Given that preliminary analyses showed that the CSK dataset did not allow the complete description of the displacements occurred in the area of interest, in particular during the excavation of the right tube, an additional dataset of images obtained with the SENTINEL (SNT) satellite was used. Note that while CSK has in Italy a visitation time of a given area of 16 days, with SNT following September 2016 this was reduced down to 6 days.

The time span covered is between 2013 (i.e. prior to excavation) and September 2016. It is of interest to compare in Figure 5, for a selected representative area, where data are available for both the methods, the results of the data analysis performed with InSAR with those obtained with conventional topography. It is clear that in doing so the need is to account for the substantial difference between the sources of the data. With topography, one measures a displacement vector of a target installed on a building and derives the settlement, while with InSAR the LOS displacement is obtained.

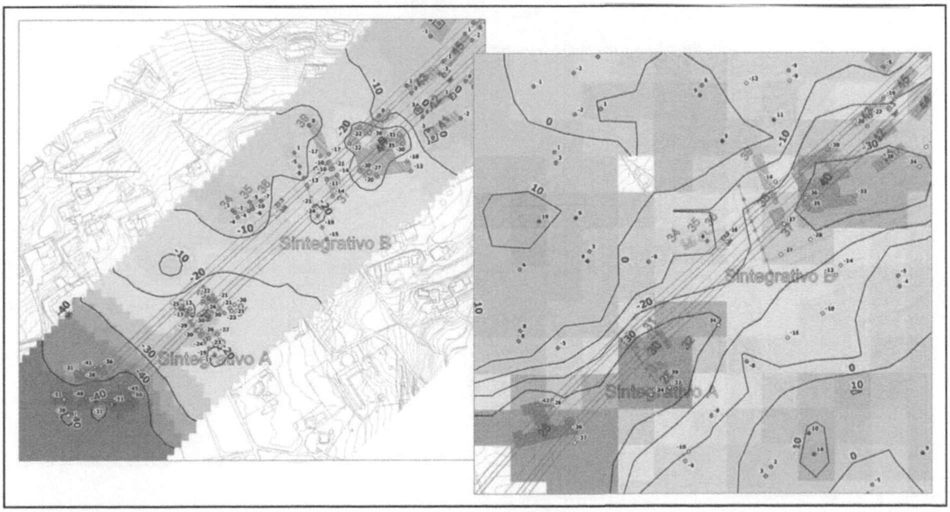

Figure 5. Settlement (right) and LOS displacement (left) distributions obtained with conventional topography and InSAR respectively during excavation of the right tube.

With this in mind, one may note that InSAR shows significant LOS displacements in the upper portion of the zone of interest, while conventional topography gives smaller settlements (of the order of 1-3 mm). A better agreement between the data sets is found in the lower portion of the same zone, where the topographic measurements show settlements up to 50 mm and LOS displacements of the same order of magnitude. In addition, InSAR gives a greater lateral extent of the area, which undergoes displacements along the tunnel alignment.

It is observed that conventional topography provides a more detailed view of the settlements occurred near the buildings, although InSAR allows a better view of the extent of the area undergoing movement around the tunnel. One may also argue that in general InSAR gives displacements of a greater magnitude with respect to topography, always keeping in mind that it is very likely that in the latter case some of the benchmarks might be within the area undergoing movements during tunnel excavation.

Another point of interest here is that conventional topographic measurements were carried out at different time intervals (in general of a few days, in the area where tunnel excavation took place), whereas InSAR provided a full data set at constant time intervals, depending on the satellite being used, CSK or SNT. In both cases, however, the pattern of benchmarks, i.e. topographic targets and radar targets (Permanent Scatterers - PS or Distributed Scatterers - DS), is not the same, with the consequence that the interpolation results as represented in the plots need be taken with caution.

3.4 *Monitoring in the test area during face advance*

As anticipated, the length of the tunnel shown in Figure 4 was selected as a test area for analyzing the TBM performance during excavation and monitoring of the ground response ahead of the tunnel face and on the ground surface. It is of interest to describe in the following the results obtained with conventional topography and InSAR monitoring respectively. Then attention is also paid to the displacements along the tunnel axis provided by a horizontal inclinometer installed prior to excavation of the right tube.

Figures 6 and 7 show the results of ground surface monitoring for two different time intervals. The first time interval is between July 31, 2014 and November 7, 2015, i.e. up to the completion of the left tube excavation. The second time interval is between April 6, 2016 and November 11, 2016, i.e. following the passage of the TBM along the right tube under the test site. Note that on the latter date the TBM had reached a distance from the site of more than 200 m approximately.

In the test site, the data provided in the first time interval (Figure 6) with both the topographic measurements and InSAR monitoring are in significant number. In comparison, very few data are

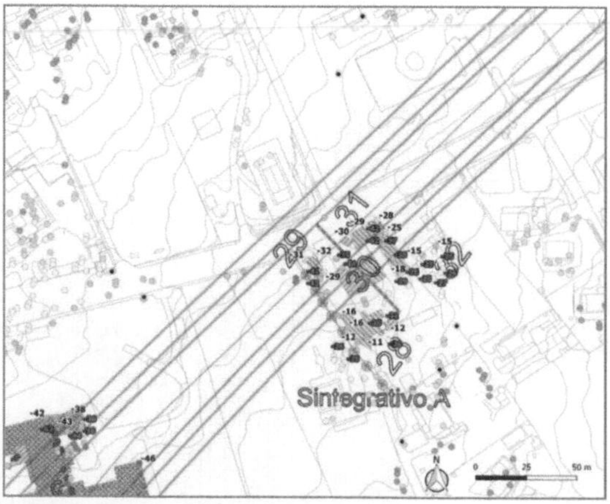

Figure 6. Test site. Settlement and LOS displacements superposed following excavation of the left tube (between July 31, 2014 and November 7, 2015). Note that the shaded number gives the LOS displacement from InSAR monitoring. The not shaded number gives the settlement from conventional topographic measurements.

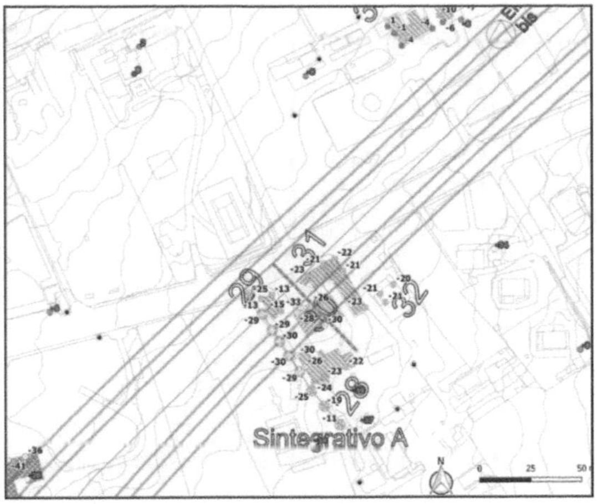

Figure 7. Test site. Settlement and LOS displacement values superposed following the passage of the TBM along the right tube (between April 6, 2016 and November 11, 2016). Note that the shaded number gives the LOS displacement from InSAR monitoring. The not shaded number gives the settlement from conventional topographic measurements.

available with InSAR in the second time interval (Figure 7). The maximum settlement following excavation of the left tube results to be 33 mm approximately, which is in the same range of the LOS displacement. In the second time interval, i.e. during excavation of the right tube, the settlement is approximately of the same order of magnitude. With reference to InSAR monitoring it is to be kept in mind that in the first time interval the satellite data were obtained with CSK, while in the second time interval the SNT satellite was used.

To complete the understanding of the data obtained at the test site, the results provided by the horizontal inclinometer installed along the right tube are of interest. Figure 8 gives a plot of the monitored horizontal displacement distribution during face advance as the TBM approaches the inclinometer. No horizontal displacement occurs when the TBM head is 47 m approximately from the same inclinometer. Then, the displacements increase to 275 mm approximately when the TBM head is near to overcome the inclinometer position.

Significant settlements of the ground surface were experienced for the tunnel during the left and right tube excavation. This raised concern on the operating conditions during face advance in relation to the head pressure being applied and the measures adopted for filling the gap between the installed lining and the excavation profile. The most attention was therefore placed in improving the operating conditions in the tunnel and in particular on a close control of the TBM head pressure being applied, which during excavation of the right tube along the test site zone ranged between 3.4 bar (head-top) and 4.9 bar (head-bottom).

It is also important to underline that systematic and continuous monitoring of the extracted volume of the ground with respect to the theoretical volume being excavated during face advance was done. For example, taking again the test site as an example, an excess in the extracted volume of 13.4 per cent was experienced systematically, this being equivalent to 47 m^3 extracted volume for a face advance of 2 m. This was related to the unexpectedly high horizontal displacements (extrusions) measured along the right tube shown in Figure 8.

3.5 *Numerical modelling and interpretation*

In order to analyse the ground response observed at the test site, a 3D model was implemented with the Finite Difference Method (FDM) by using the FLAC3D code (Itasca 2019a). Following preliminary simplified analyses in axisymmetric conditions, 3D studies were carried out as briefly illustrated in the following. The model in Figures 9 and 10 extends 260 m along the tunnel axis and 240 m orthogonally to it, in order to comprise the two tubes. The height is 217.5 m and the model

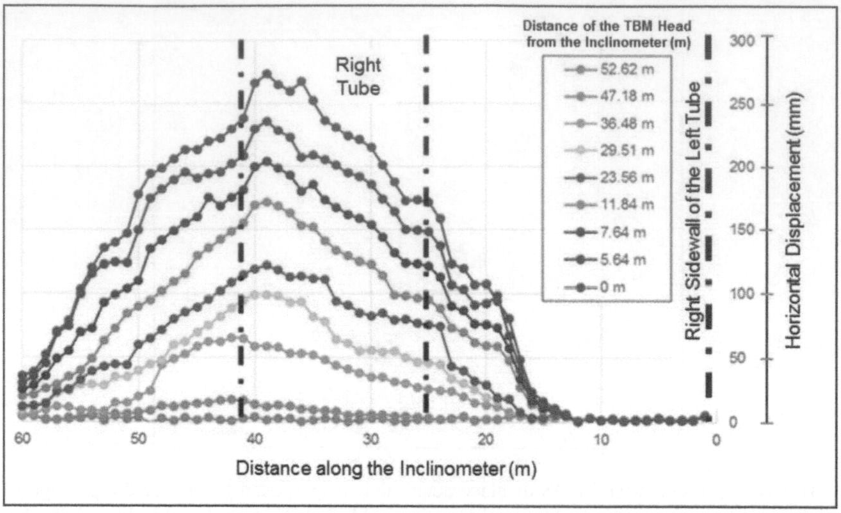

Figure 8. Test site. Horizontal displacements (extrusions) as the TBM head.

extends to the ground surface. The geological-stratigraphic conditions represented in the model are those shown in Figure 2.

With the intent to keep the model simple, the excavation of each tube was simulated with a step-by-step approach with 4 m excavation length each step. The TBM was represented by a given pressure distribution around each tube and decreasing it in a controlled manner. This pressure was progressively reduced up to obtaining a closure of the gap between the shield and the ground. The shield length was assumed 12 m. The gap between the segmental lining and the ground was taken to be 4 cm in excess with respect to the shield conicity. With the desired radial displacement achieved, the lining was activated, this being represented as an equivalent cylinder of assigned stiffness.

The geotechnical parameters used for modelling are given in Table 1. The analyses were performed in undrained conditions with the elastic plastic Mohr-Coulomb model for the ground. The computation steps were as follows: 1. initialization of the model to obtain the initial state of stress in the ground. 2. Excavation of the left tube with a 4 m length excavation step for 65 steps. 3. Excavation of the right tube with 4 m steps. Figure 11 shows the excavation process simulation, with the left tube completed and the right tube under excavation.

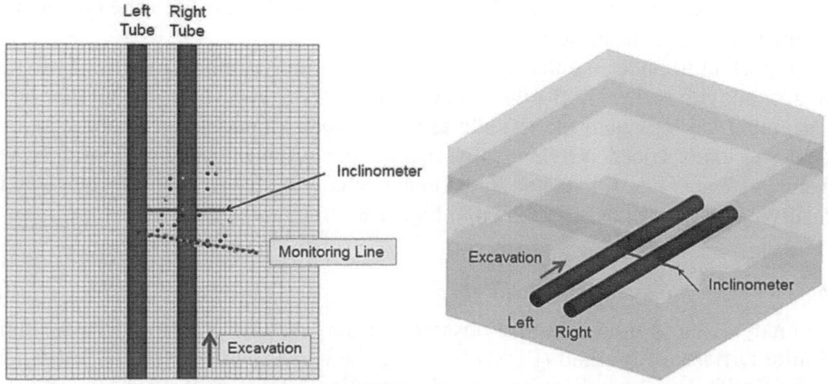

Figure 9. View of the 3D FDM model showing the ground surface (left) and the zone of interest with the tunnel and the inclinometer, including a surface monitoring line (right).

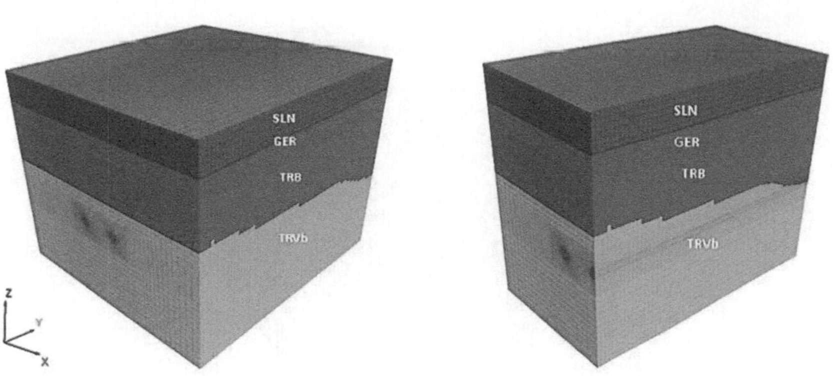

Figure 10. 3D FDM model showing the complete model (left) and a portion of it across the right tube (right).

Table 1. Geotechnical parameters.

Complex Formation	Drained Parameters					Undrained Parameters			
	γ (kN/m^3)	c' (kPa)	φ' (°)	E' (MPa)	ν (-)	γ_{sat} (kN/m^3)	c_u (kPa)	E_u (MPa)	ν_u (-)
SLN	19.5	10	23	100	0.3	–	–	–	0.48
GER	19.5	30	24	175	0.3	23.5	400	200	0.48
TRB	19.5	30	24	310	0.3	23.5	400	350	0.48
TRVb	19.5	23	24	175	0.3	23.5	400	200	0.48

γ and γ' unit weight, c' and c_u cohesion, φ' friction angle, E' and E_u deformation moduli ν Poisson's ratio

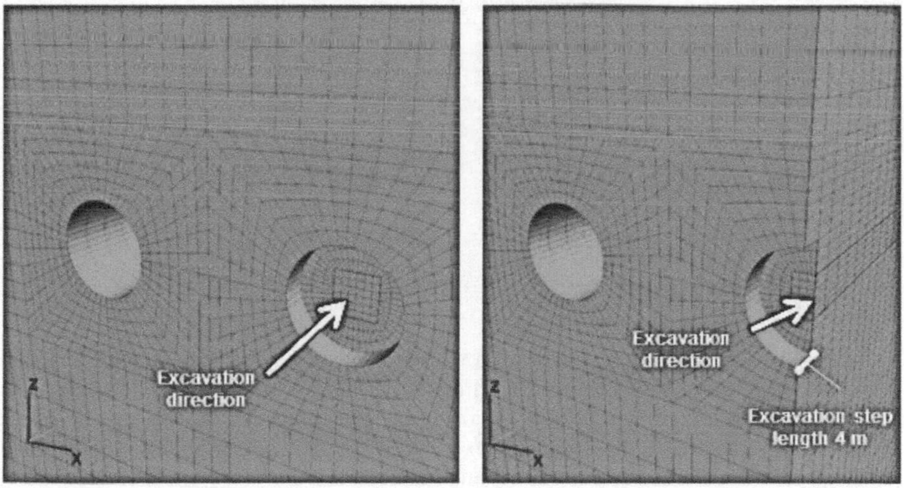

Figure 11. 3D FDM model showing the simulation of the excavation process.

Two different assumptions were made for the angle of dilation ψ' of the TRVb complex formation, in order to represent with a dilatant plastic volumetric strain activated the observed excess in the extracted volume and the related horizontal displacements (extrusions) measured along the horizontal inclinometer.

The tunnel was assumed to be at 130 m depth as for the test site and the corresponding vertical stress σ_v was computed based on gravity in total stress conditions. The horizontal σ_h stress in the tunnel cross section was taken equal to σ_v and the horizontal σ_H stress along the tunnel axis was set equal to 1.25 σ_v. This is to say that an anisotropic in situ state of stress was assumed.

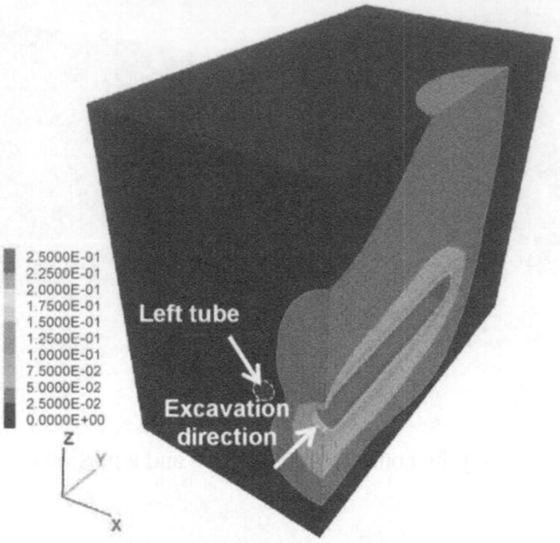

Figure 12. Displacement plot during the right tunnel excavation. Section along the right tube axis.

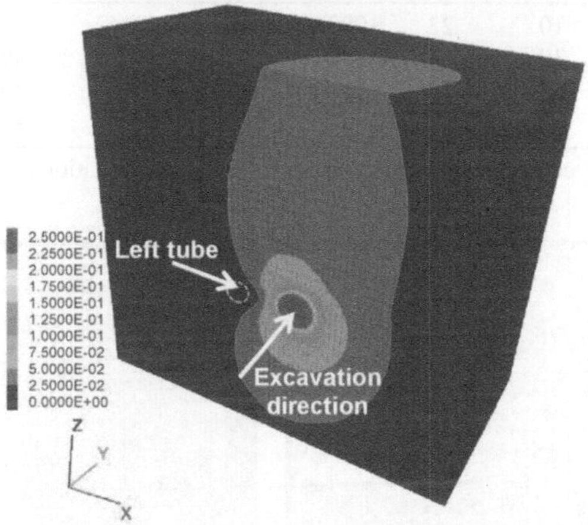

Figure 13. Displacement magnitude plot during the right tunnel excavation. Cross section near the face.

The results obtained are described in the following by considering the excavation of the right tube with the left one completed. For example, Figures 12 and 13 illustrate the displacement magnitude along a longitudinal section and in a cross section near the opening face respectively. It is noted that only the displacements due to the right tube excavation are shown in the plots. It is clear that the induced displacement do extend up to the ground surface.

A comparison of the maximum computed and monitored horizontal displacements (extrusions) ahead of the tunnel face as the right tube is excavated and the TBM head approaches the horizontal inclinometer installed from the left tube is shown in Figure 14. The horizontal displacements are for the angle of dilation ψ' of the TRVb Formation equal to 0° and 10° respectively. The computed maximum displacement is in general smaller than the monitored displacement, with significantly greater values attained as the TBM approaches the inclinometer.

If the attention moves from the tunnel up to the ground surface, always with reference to the test site, the computed and the monitored settlements can be compared at the targets along the monitoring line as shown in Figure 15. In the time interval comprised between July 19, 2016 and January 9, 2017, as the right tube excavation takes place, the settlement of the different points along the monitoring line increases, with the model giving in general greater values.

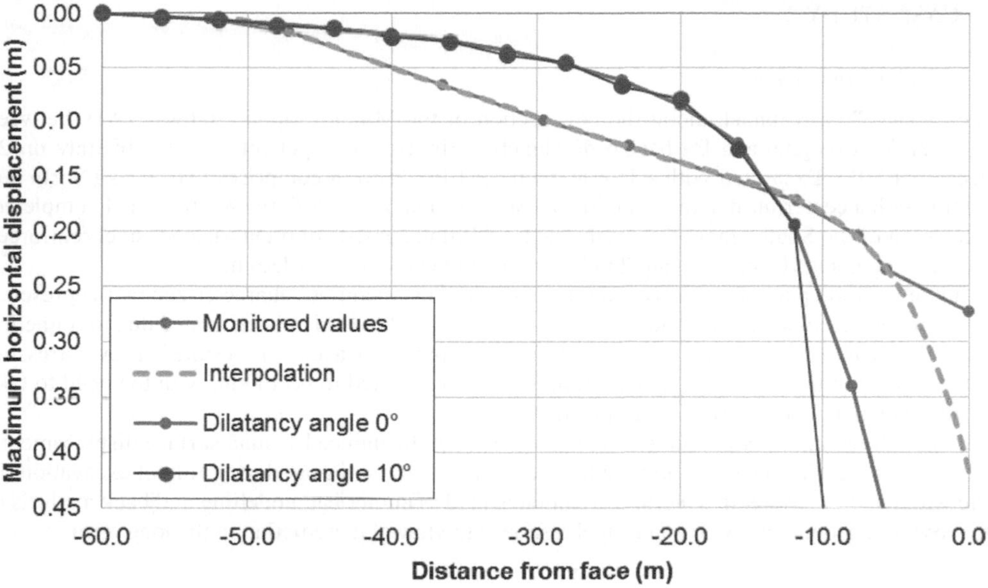

Figure 14. Comparison of monitored and computed maximum horizontal displacement ahead of the TBM head during excavation of the right tube.

It is important to point out that the simulation of the excavation of the right tube under the assumption of an elastic plastic behavior, with the angle of dilation ψ' of the TRVb Formation equal to 10°, results in the 13.4 per cent increase in extracted volume, equivalent to 47 m^3 for a face advance of 2 m. As well illustrated in Figure 14, the computed horizontal displacements (extrusions) are greater than the values measured by the inclinometer in the first 12 m near the TBM head.

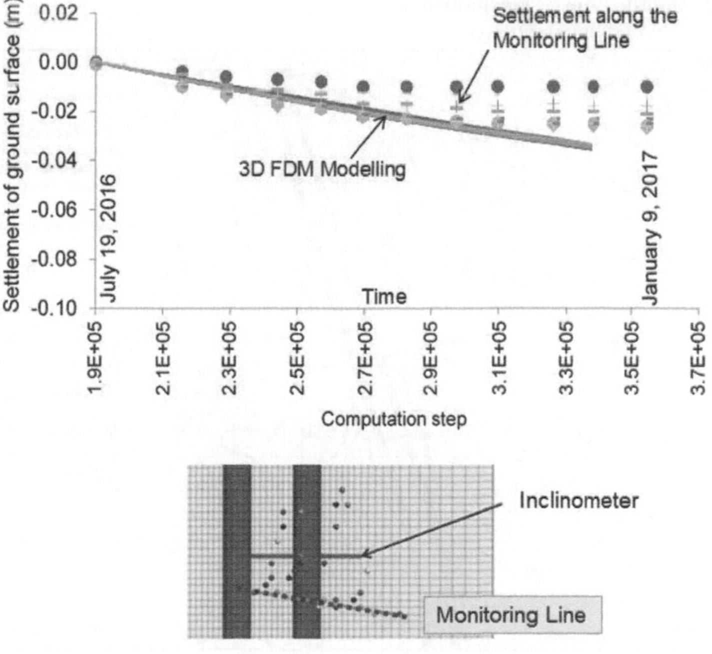

Figure 15. Comparison of monitored and computed settlement for benchmarks along the monitoring line. at the test site. The plot is for the right tube excavation. The angle of dilation ψ' of the TRVb formation is equal to 10°.

4 CASE STUDY 2

4.1 *Preliminary remarks*

The “Sparvo” twin tunnel, along the new section of the Milano-Napoli Highway (A1), between La Quercia (Bologna) and Barberino di Mugello (Firenze), was put into service in Italy on 23 December 2015. Excavated with a 15.6 m diameter EPB TBM, it comprises two tubes (North and South), with a centerline distance ranging between 30 and 60 m. With the North tunnel completed, excavation of the South tunnel followed. The final lining consists of prefabricated, steel reinforced concrete segments 70 cm thick and 2 m long. The tunnel diameter is 13.6 m.

Tunneling took place along a typical area of the Italian Apennines characterized by the presence of “dormant” deep-seated landslides. During excavation, a landslide was activated due to tunneling with significant displacements taking place at the ground surface. This resulted in overstressing of the linings in the two tubes, along a tunnel length estimated to be 250 m, with the need to take appropriate actions for putting the tunnel into service.

In the following, the attention is on the monitoring of the induced ground surface displacements, the activation of the landslide, and the induced stresses in the lining, during tunnel excavation. In line with the objective of this lecture, attention is paid to numerical modelling and back analysis of the observed ground behavior and tunnel response in view of a prediction in the long term.

4.2 *Geological and geotechnical conditions*

The slope along the Apennines under passed by the “Sparvo tunnel” is characterized by the presence of a number of “dormant” and “locally active” landslides. The Complex Formations crossed by the tunnel are the Palombini clay-shales and the Scabiazza snadstone, which are well known for the difficulties met when tunneling.

In order to give a view of the surface area, also for providing a reference framework for the following, Figure 16 shows a geological-geomorphological map where different landslides are underlined based on a classification provided in SGSS (2018). Of particular relevance for the tunnel are the landslides defined as “active/dormant” and “rotational/planar” to identify in simple terms the state of activity and the type of mechanisms.

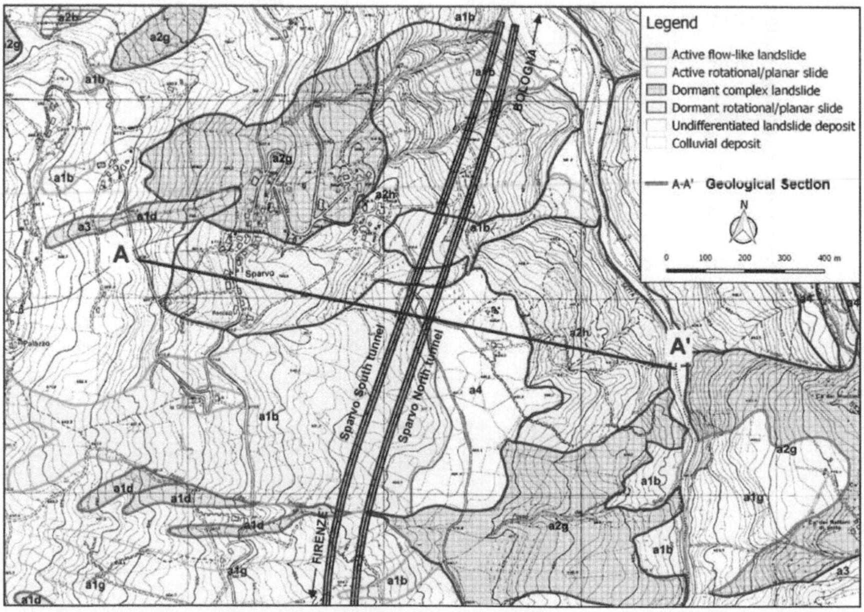

Figure 16. Geological-geomorphological map of the area with the “Sparvo” tunnel alignment shown (prepared on the basis of the data available in SGSS 2018).

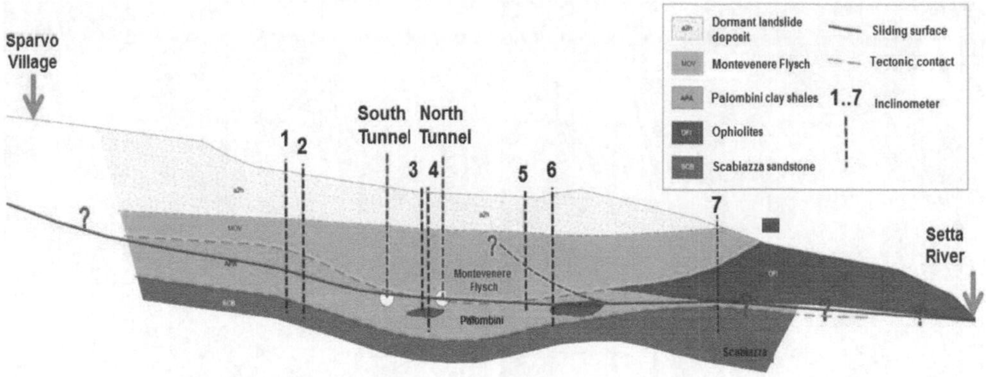

Figure 17. Selected geological section (AA' in Figure 16) crossing the "Sparvo" tunnel.

The geological cross section AA' (see Figure 16 for the location) is illustrated in Figure 17. It is pointed out that this is the result of extensive geological and geotechnical investigations performed through the years (at the design stage, during and following tunnel excavation, up to the present time), including surface geological mapping, borehole drilling with detailed geological stratigraphy derived, geophysical investigations, and inclinometer monitoring.

Paying attention to the geological cross section AA', the Complex Formations shown are, from the ground surface down to the tunnel, the Montevenere Flysch, the Palombini clay shales. Following this, the Scabiazza sandstone is encountered. Well outlined in this cross section is the sliding surface of the deep-seated landslide, inferred on the basis of inclinometer monitoring through the years of observation, in particular following tunnel excavation.

The Montevenere Flysch (MOV) consists of dominant sandstone layers alternating with shales, siltstones, clay marls and rare calcilutites. The Palombini clay shales (APA) are alternating shales and fine-grained calcareous or siliciclastic turbidites. Thin, fine-grained sandstones and siltstone characterize the Scabiazza sandstones (SCB), with shale interlayers, with prevalent embedded arenaceous blocks.

Table 2 gives representative geotechnical parameters of the different Complex Formations of interest shown in Figure 17. The values reported are based on in situ and laboratory tests performed at the tunnel design stage, not disregarding the experience gained in a number of tunnels excavated along the Italian Apennines.

4.3 *Monitoring methods and results*

Surface and subsurface displacement monitoring was carried out as TBM excavation of the two tunnels was taking place. In addition to the RTS measurements of a number of targets on the ground surface and on selected buildings, in conjunction with the inclinometer data obtained from a series of boreholes in the area, as already illustrated above in Figure 17, also the results of a SqueeSAR analysis of the RADARSAT S3 dataset were used.

Figure 18 illustrates the points (PSs) monitored between May 2003 and September 2015 for four different areas (Zones 1–4) on the ground surface, while Figure 19 gives the corresponding displacement history. It is clear that the excavation of the North tube induced a sudden increase of the displacements along the LOS in Zones 1 and 2, which apparently continued with the same displacement rate when the South tube reached the same zones.

Table 2. Geotechnical parameters.

Complex Formation	Drained Parameters				
	γ (kN/m^3)	c' (kPa)	φ' (°)	E' (MPa)	ν (–)
MOV	25	150–120	24–20	800	0.25
APA	22	50–35	16–11	160	0.35
SCB	25	250–200	30–24	310	0.25

γ unit weight, c' cohesion, φ' friction angle, E' deformation modulus, ν Poisson's ratio

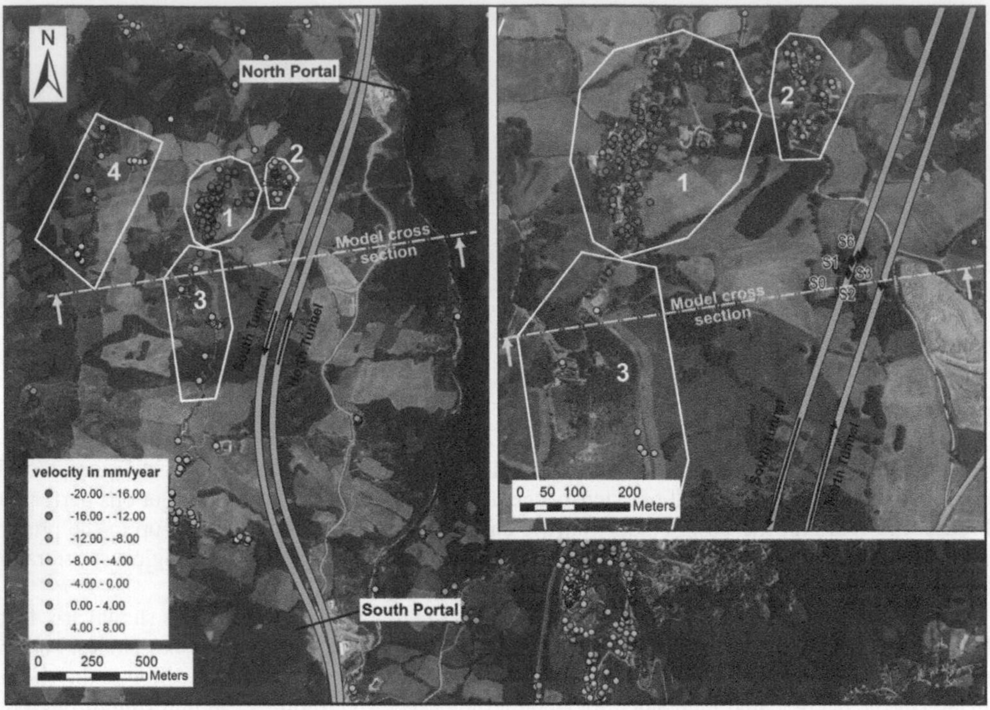

Figure 18. RADARSAT S3 dataset. May 2003 to September 2015 (Barla et al. 2015, Barla 2018).

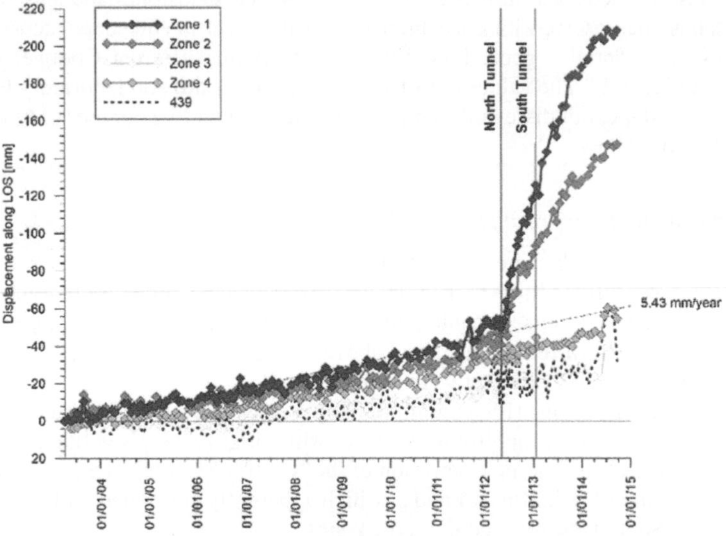

Figure 19. Displacement history based on InSAR data (Barla 2018).

It is of interest to point out that with the InSAR data shown in Figure 19 one is in position to infer the trend of behavior of the deep-seated landslide prior to tunnel excavation, characterised by a rate of displacement of 5.4–4.1 mm per year for the same Zones 1 and 2. With the excavation of the South tube, the rate of displacement measured following the transition of the tunnel face is up to 50–100 mm per year.

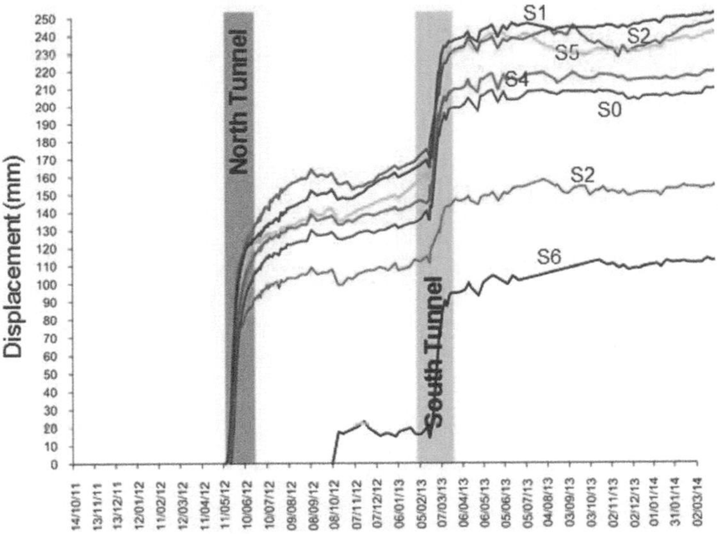

Figure 20. Displacement history at selected targets (Figure 21).

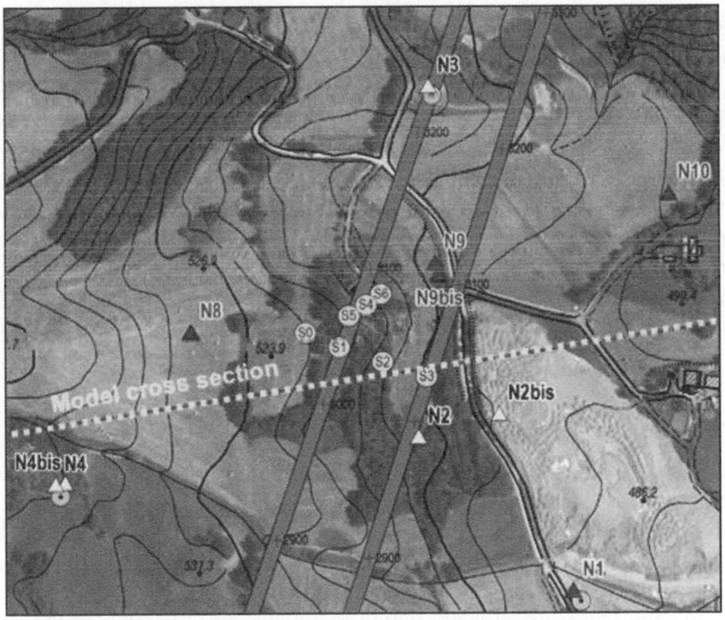

Figure 21. Selected targets S0, S1, S2, S3, S4, S5, and S6 with surface displacement monitoring.

Keeping in mind the observed trend of behavior, based on the displacement history from the RADARSAT S3 dataset, which demonstrates the reactivation of the deep-seated landslide, the plot of Figure 20 brings in full evidence the phenomenon that is taking place. The observed surface displacement history measured at the topographic targets S0, S1, S2, S3, S4, S5, and S6 shown in Figure 21 underlines the deep-seated landslide response to tunnel excavation.

In both cases, surface displacements occur downslope with a maximum of 130 mm following the excavation of the North tube and 70 mm after the transition in the same zone of the South tube. In addition, one is now able to see that as the North tube moves ahead with respect to the target points

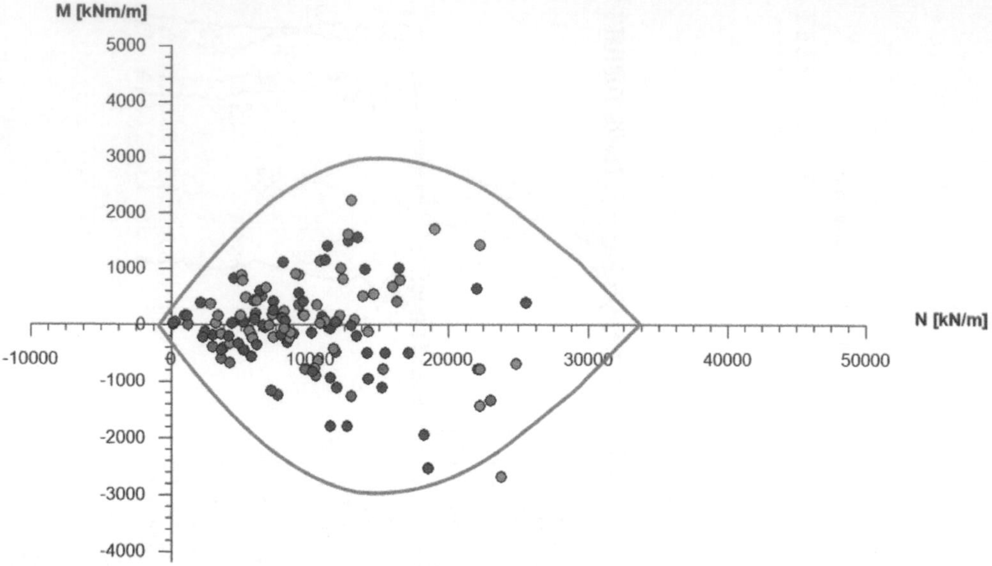

Figure 22. North tube lining capacity plot – bending moment M vs. axial thrust N (Barla 2018).

and prior to the arrival of the South tube, the displacement rate is 7-8 mm per month, significantly greater than that observed prior to tunnel excavation.

As anticipated, the reactivation of the deep-seated landslide resulted in a stress increase in the segmental linings, with evidence of overstressing. The hoop stresses in the lining of both the tunnels were therefore determined by the over coring method with doorstopper measurements and four strain gage rosettes, positioned at 45° one from the other one. Measurements were performed in the lining segments at 15 cm and 45 cm depths from the intrados. In general, measurements were done at the tunnel sidewalls, hinges and crown.

With the hoop stresses known, under the assumption of a linear internal stress distribution in the lining, the internal stress characteristics, bending moment M and axial thrust N could be computed. Typically, the M-N plot of Figure 22 points out a few points near the strength domain (for a safety factor of 1).

4.4 *Numerical modelling and interpretation*

Numerical analyses in plane strain conditions were carried out by using the FDM and the FLAC code (Itasca, 2019b). Although numerical analyses were also performed in view of the tunnel design, the attention in the following is dedicated to the back analysis of the observed interaction with the deep-seated landslide with the North and South tubes. This was done by using the time-dependent elasto-viscoplastic model SHELVIP described in Debernardi and Barla (2009).

Figure 23 shows the FDM model used, which is a clear simplification of the geological cross section plotted in Figure 17. This model comprises the deep-seated landslide resting on a 10 m thick shear zone, inferred on the basis of the inclinometer monitoring data available. As shown in the inset, the grid size in this zone and in the near vicinity of the tunnels was carefully defined. Attention was in particular paid to the model extent, in order to avoid the influence of boundary constraints. In addition, care was taken not to impair the mobilization of the landslide along the shear zone.

In the cross section considered, the twin tunnel is at 100 m depth approximately. With the in situ gravity stress imposed, the stress ratio ($K_0 = \sigma_h/\sigma_v$) in the model was set equal to 1.5, i.e. the vertical stress σ_v is gravity-dependent, while the horizontal stress σ_h is 1.5 times the vertical stress. The groundwater table is 10 m below the ground surface and the analyses were run without considering any water flow toward the tunnel.

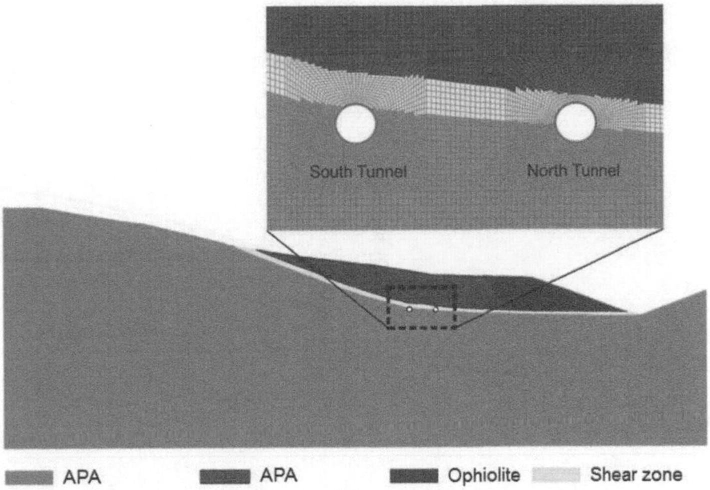

Figure 23. Details of the plane strain FDM model (Barla et al. 2015, Barla 2018).

The details of the modelling studies, including the simulation stages, are described in Barla et al. (2015) and Barla (2018). It is underlined that the time-dependent behavior was activated in the shear zone only and the ground above (the landslide body) and below it (Figure 23) was assumed to follow an elastic perfectly plastic model. The analyses were run first under gravity only, up to obtaining a rate of displacement for the landslide of 5 mm per year based on the RADARSAT S3 dataset, i.e. the rate of displacement prior to landslide reactivation due to tunnelling.

Subsequently, the excavation of the North tunnel was simulated allowing for 35% stress relief on the tunnel contour prior to the segmental lining installation, when a complete stress relief was permitted. Then, in line with the work schedule, also the excavation of the South tunnel was simulated by using the same stress relief sequence as for the North tunnel. The time-dependent elastic viscoplastic analysis was then activated with the intent to represent the longtime system behavior.

The results obtained are illustrated in Figure 24. Back analyses were performed first by obtaining the time dependent parameters of the shear zone material, which allowed a rate of displacement for the entire deep-seated landslide of 5 mm per year as observed prior to tunnel excavation. Then, with the same parameters adopted during tunnel excavation (the North tunnel followed by the South tunnel), a sudden increase of the rate of displacement is shown to take place as observed in practice.

However, as illustrated in Figure 24, following excavation these same parameters assumed for the shear zone would result in no movement of the landslide, which is unlikely. It is clear that the assumption introduced with the presence of a single, constant thickness, shear zone represents an over simplification of the problem. One would reasonably expect, as is the case, that following tunnel excavation the rate of displacement of the deep-seated landslide would be the same as prior to it or more likely would increase.

It is clear that this aspect of the modelling work need be further investigated. At present, with the tunnel completed and in service, this is being done with care based on the results of monitoring data, which are being obtained. As shown in Figure 24, with the purpose to follow up in the time dependent analysis and obtain a first estimate of the long term stresses in the tunnel lining, the parameters of the time dependent model which would allow to simulate a rate of displacement of 5 mm per year were obtained.

The computed and monitored displacement histories are compared in Figure25 . It is noted that the monitored displacements are the surface displacements of a number of target points (S0, S1, S2, S3, S4, S5, and S6) resulting from the topographic measurements. Also shown in Figure 25 is the head displacement history at the inclinometer N4.

Figure 26 shows the computed plastic strains in the shear zone. Plastic shearing occurs around the North tunnel at the crown, at about the hinges and the invert, whereas for the South tunnel this takes place prevalently at the crown. It is the intersection of the openings with the shear zone which determines the most significant structural response of the segmental lining.

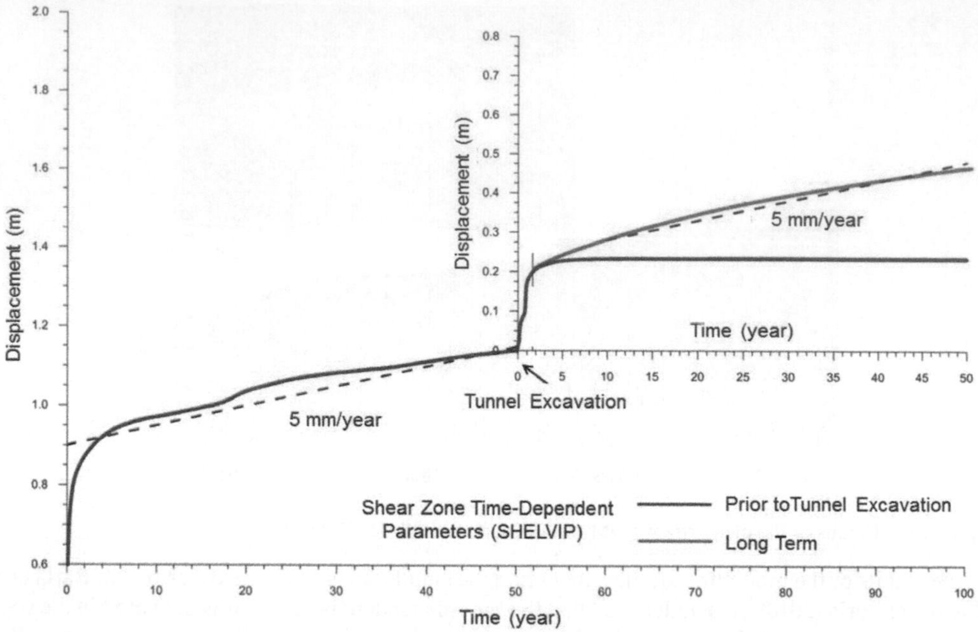

Figure 24. Computed displacement history of the deep-seated landslide interacting with the tunnel. Back analysis and long term results.

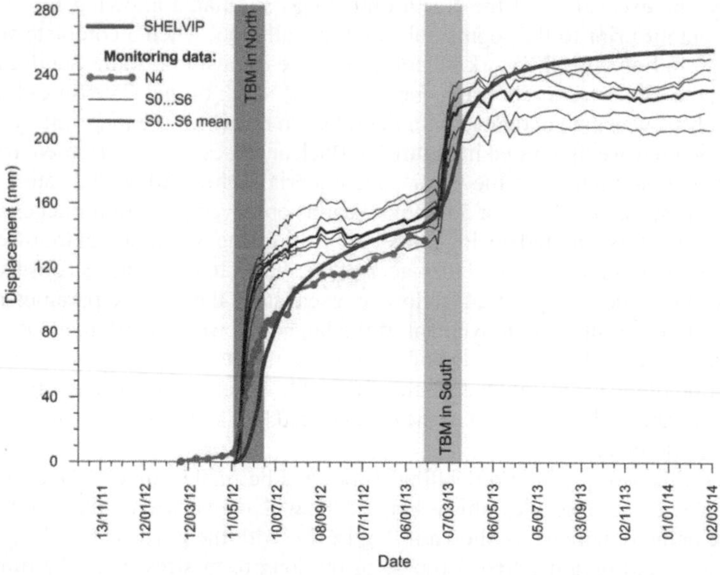

Figure 25. Computed and monitored displacements at the ground surface (Barla et al. 2015, Barla 2018).

With the above in mind, it is of interest to see the computed internal stresses in the tunnel linings some time following installation, e.g. when the doorstopper stress determinations previously discussed were performed. As an example, the computed principal stresses in the North tunnel are plotted in Figure 27. It is shown that the computed maximum principal stress in the segmental lining at the top right hinge and at the lower left hinge are in excess of the strength limit of 50 MPa for the concrete used.

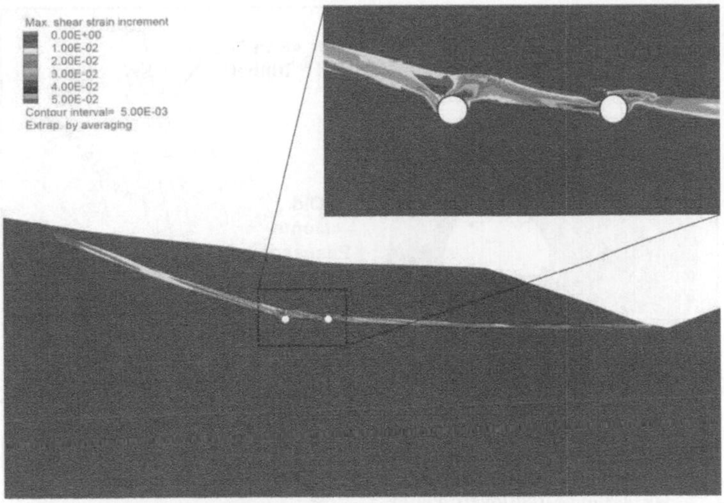

Figure 26. Maximum shear strain increment around the tunnels (Barla 2018).

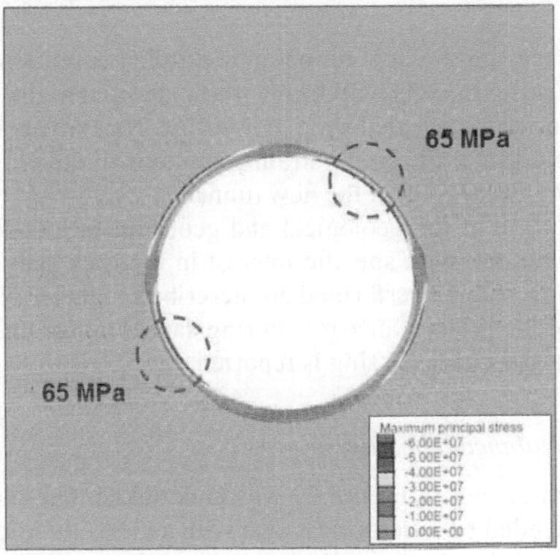

Figure 27. Maximum principal stress in the North tunnel lining.

Based on the results illustrated, the decision was reached to adopt appropriate reinforcement measures in both the tunnels along the 250 m length where the interaction with the deep-seated landslide was shown to occur. The stresses acting in the segmental linings were thought not to be acceptable for putting the tunnel into service and for an appropriate long term performance (Barla 2018).

5 CASE STUDY 3

5.1 *Preliminary remarks*

The “San Paolo” road tunnel was recently excavated in Italy under the old “Pasasco” twin tunnel, in service along the Ventimiglia-Genova Highway (A10), which was built in the 1960s. The new tunnel is at a distance of 11 to 15 m below the old tunnels as shown in Figure 28. Excavation took place by using a 13.6 m diameter Single Shield TBM through a paragneiss micaschist of poor to fair rock mass quality. The segmental lining of the new tunnel is 35 cm thick and has a 12.8 m final diameter.

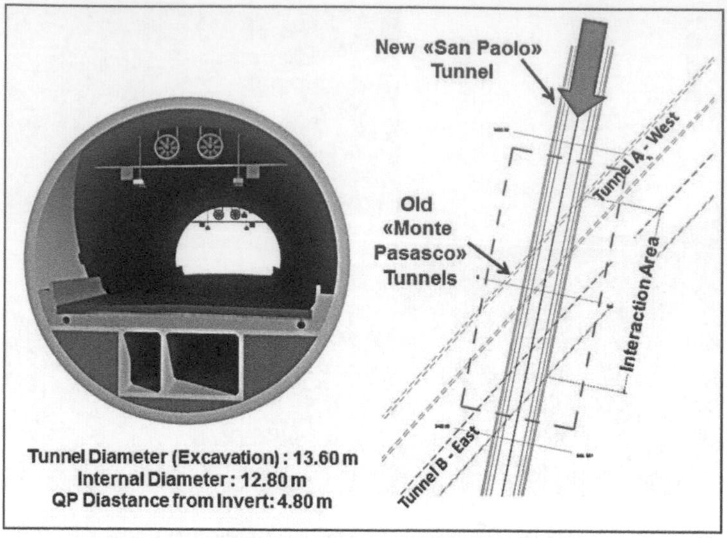

Figure 28. The new "San Paolo" tunnel under passing the old "Monte Pasasco" old tunnel.

The objective was to excavate the new tunnel in controlled conditions, i.e. without the need to stop service of the old, existing tunnels with heavy traffic day and night. In order to allow the work to be completed satisfactorily, a near real time monitoring system was implemented. This was to give a well documented picture of the stress-strain conditions in the old tunnels linings and in the surrounding ground during excavation of the new tunnel.

In the following, an outline of the geological and geotechnical conditions around the existing twin tunnel and the new tunnel, with specific interest in the rock slab in between, is given. The numerical modelling design studies performed are described together with the monitoring system adopted. Finally, a view of the performance monitoring data obtained during excavation of the new tunnel up to completion of the under passing is reported.

5.2 *Geological and geotechnical conditions*

The investigation program shown in Figure 29 was undertaken. On one side, the geological and geotechnical conditions needed be ascertained, also with a close attention to the characteristics of the linings of the old tunnels, given the very limited information available on them. The main investigations carried out comprised borehole drilling from the ground surface and from the old tunnels.

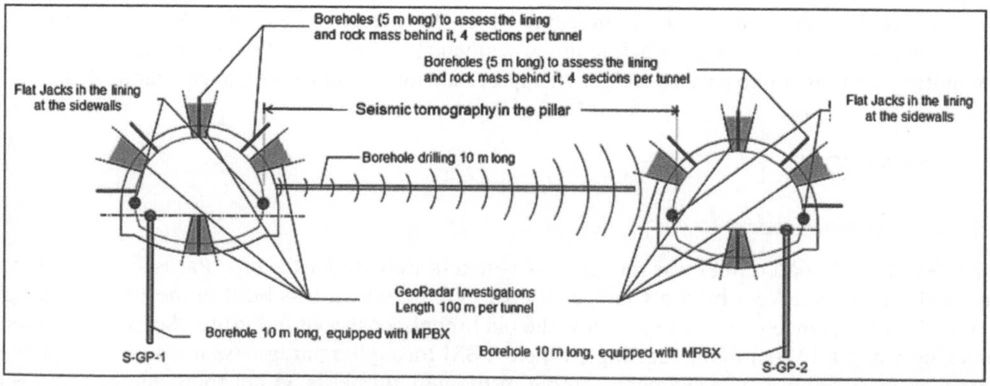

Figure 29. Schematic view of the investigations performed.

Also, seismic tomography of the intermediate pillar and cross-hole tomography of the rock below the old tunnels invert were performed.

In addition to pointing the attention on the final lining geometry (keep in mind that the old tunnels were excavated by using the conventional method with extensive use, at the time, of wood supports), also the state of stress in the lining was measured. With this in mind, the lining geometry was checked with TLS and GeoRadar investigations, including borehole drilling. For the state of stress in the lining, flat jack tests were performed.

It is of interest to highlight in Figure 30 the characteristics of the old tunnel linings with reference to its geometry, in particular the concrete thickness, the presence of voids and wood elements around the tunnel, with some uncertainty left regarding the presence/thickness of the invert. At the same time, as shown in Figure 31, the results of the seismic tomography of the intermediate pillar, in the interaction/intersection zone between the highway tunnels above and the new TBM tunnel below, give a P-wave velocity in the range 2000-2400 m/s.

Based on borehole drilling with full core recovery and laboratory testing, the rock mass conditions around the old tunnels (in particular in the zone below the invert) were investigated. The Geological Strength Index (GSI) values result to be in the range 40-50 (max) and down to 30-20 (min) locally. For appropriate reference, Table 3 gives the rock mass strength and deformability properties estimated for GSI = 45.

5.3 *Numerical modelling*

The interaction between the "San Paolo" tunnel and the old "Monte Pasasco" tunnels was analyzed with the Finite Element Method (FEM) in both 2D and 3D conditions. This was to: (1) represent the "conditions" of the old tunnels based on the investigations performed. (2) Analyze the mutual interaction, to determine the "stress-strain redistribution" in the ground, taking place during excavation of the new tunnel. (3) Make a forward prediction of the expected "scenarios" during excavation.

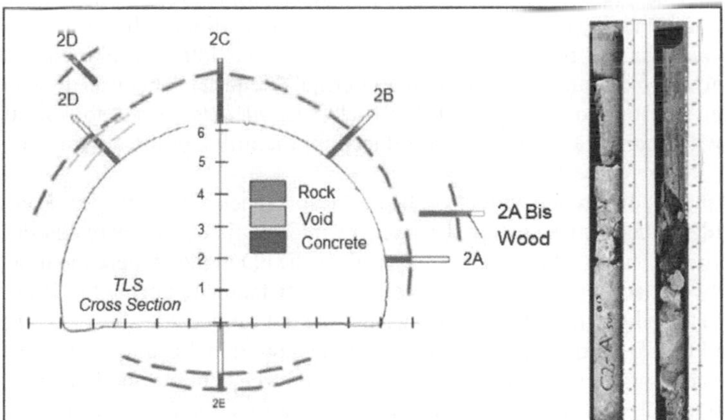

Figure 30. Typical conditions of the old tunnel lining.

Table 3. Geotechnical parameters.

	Parameters								
Paragneiss Micaschist	γ (kN/m^3)	σ_{ci} (MPa)	m_i (-)	m_b (-)	s (-)	σ_{cm} (MPa)	α (-)	E_m (MPa)	ν (-)
	25	50	12	1.03	0.001	6.6	0.058	4200	0.25

γ unit weight, σ_{ci} and σ_{cm} intact rock and rock mass rock uniaxial compressive strength, E_m deformation modulus, ν Poisson's ratio, m_i m_b, s, α parameters of the Hoek and Brown criterion for intact rock and rock mass respectively.

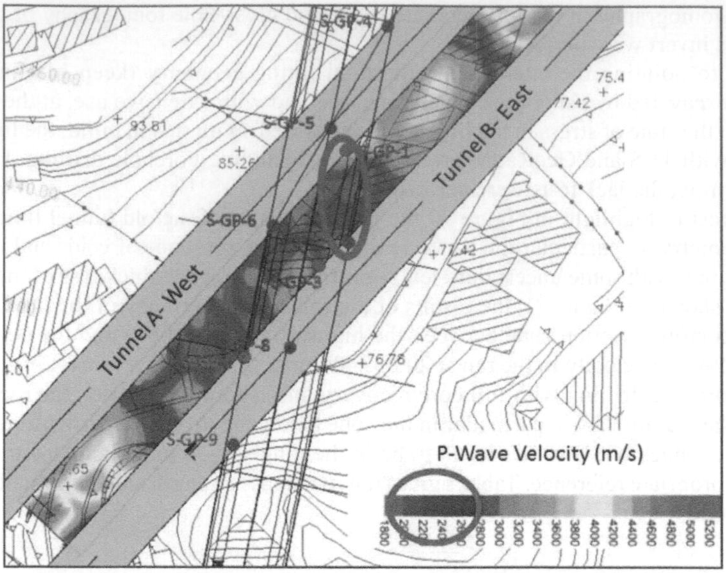

Figure 31. P-wave velocity in the pillar between the old "Monte Pasasco" East and West tunnels.

Essential in this process, which represents a thorough application of the Observational Method, was the adoption of an appropriate model of behavior to reproduce satisfactorily the interaction response between the existing tunnels and the new tunnel under it. The aim, in view of the monitoring program implemented, was the definition of "warning", "pre-alarm" and "alarm" levels of the monitored quantities to be adopted during the TBM excavation of the new tunnel.

The design analyses were carried out in 2D conditions with the RS2 computer code (Rocscience 2019). The Plaxis 3D code (Plaxis 2019) was used in 3D conditions. In both cases, based on the results of the investigations performed, a destressed zone was introduced to characterize the rock mass around the two old tunnels, including the intermediate pillar. This was done keeping in mind the results of the investigations performed. The rock mass was assumed to follows an elastic plastic brittle model, with realistic assumptions introduced for the lining of the old tunnels, i.e. in particular a very weak concrete invert.

With the 2D model, different positions of the new tunnel needed be considered with respect to the old tunnels, under either tunnel section or under the intermediate pillar between the East and West tunnels respectively. This is obviously not the case for the 3D model, where the new tunnel excavation under the two existing tunnels is simulated more satisfactorily during the TBM face advance.

The results obtained for the plastic zones around the old tunnels arc shown in Figures 32 and 33 with the 2D and the 3D models respectively, when the new tunnel is not yet present in the models. The advantage of 3D modelling with respect to 2D modelling is in the detailed analysis of the yield zones, which is possible during TBM excavation of the new tunnel. In all cases, when the new tunnel is not yet excavated, the extent of the plastic zones obtained with the 2D model compares reasonably well with that obtained with the 3D model.

Figures 34 and 35 show the plastic zones around the tunnels again with the 2D and the 3D models, however in this case with the new tunnel excavated completely under the existing tunnels (note that with the 2D model the plotting is for the case when the new tunnel is under the East tunnel). In these conditions, plastic zones extend around the old tunnels including the pillar between them. Under the assumptions introduced to simulate TBM excavation (with a controlled stress relief), no plastic conditions are present in the intermediate rock slab between them.

With the intent to better understand the interaction problem, with due attention to the observed conditions of the lining, not disregarding the uncertainties regarding the invert arch already mentioned, additional 2D analyses were carried out. Different conditions were considered such as: (1) a weak rock mass in the intermediate pillar of the two old tunnels. (2) A weak rock mass on one

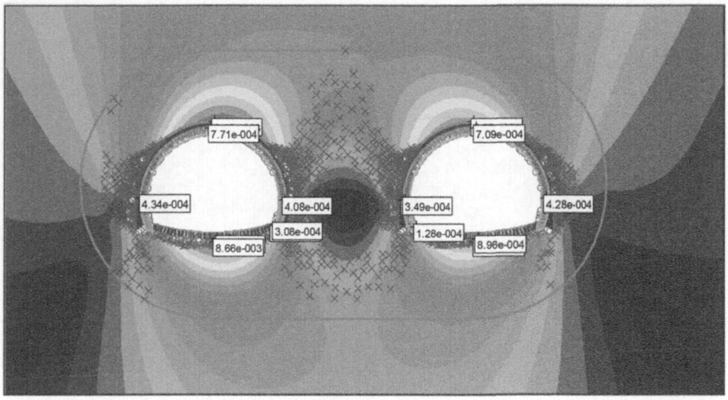

Figure 32. Plastic zones around the old tunnels. Initial conditions obtained with the 2D model.

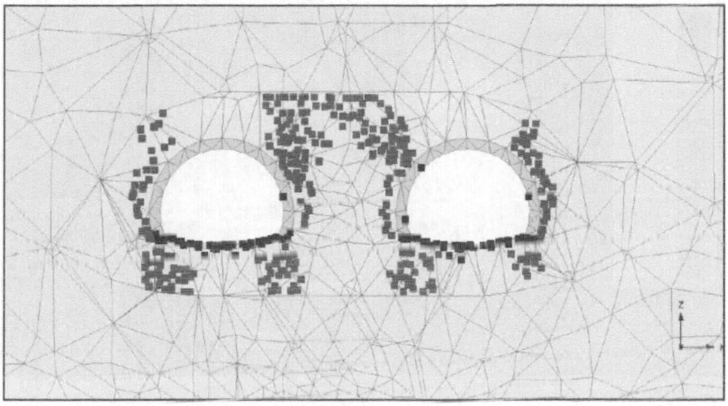

Figure 33. Plastic zones around the old tunnels. Initial conditions obtained with the 3D model.

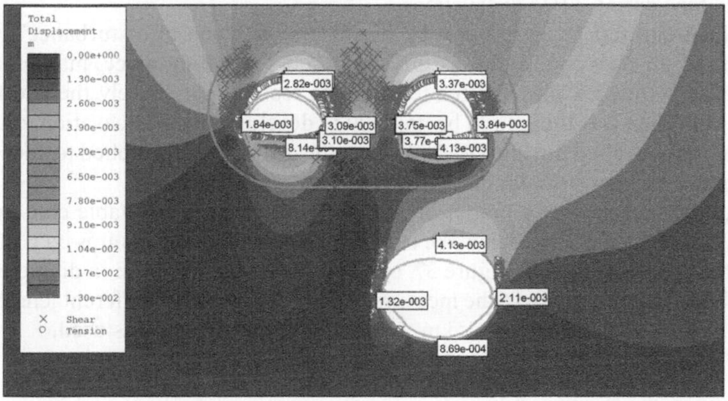

Figure 34. Plastic zones around the old tunnels with excavation of the new tunnel completed. 2D model.

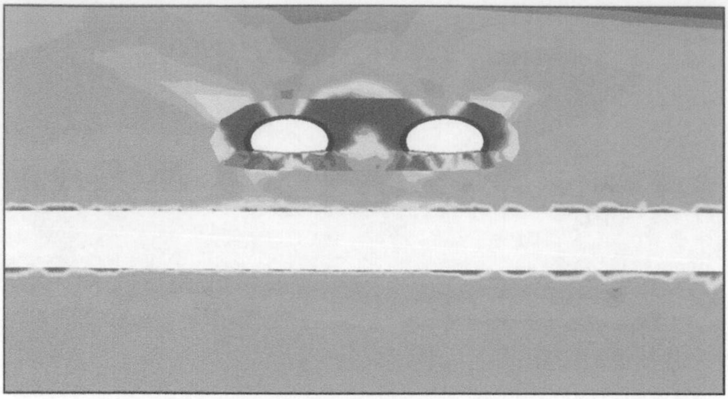

Figure 35. Plastic zones around the old tunnels with excavation of the new tunnel completed. 3D model.

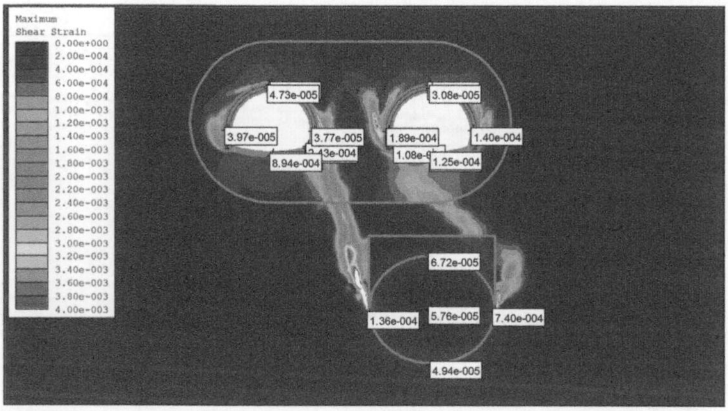

Figure 36. Maximum shear strain zones due to stress relief zone above the new tunnel. 2D model.

sidewall of the tunnels toward the pillar. (3) A possible over excavation at the crown of the new tunnel or at the face during TBM excavation.

Figure 36 illustrates this latter case (3) by introducing a "zone of disturbance", where a stress relief is permitted. In such conditions, as this stress relief is taking place, maximum shear strain zones do develop in the rock mass with a tendency to reach progressively the two tunnels above. Although this condition was thought to be unlikely to develop during excavation, given the strictly controlled conditions applied during TBM advance, the decision was reached to adopt the stabilization/reinforcement measures shown in Figure 37.

This decision was reached as a matter of caution, given the unacceptable consequences of any instability developing in the zone below the existing tunnels. These measures consisted in the creation of a grouted zone as shown in Figure 37, by using steel pipes (60 mm in diameter) with valves, "tube-a-manchette" (sleeve pipe), in the measure of 3 valves/meter, of different length. The longitudinal spacing of each ring shown was 1.5 m. Expansive cement grout was used.

5.4 *Monitoring methods and results*

With the understanding of the problem illustrated above and attention paid to the old tunnels response during TBM excavation of the new tunnel, the near real time monitoring system illustrated in Figure 38 was implemented. This comprises: (1) convergence/displacement monitoring in cross sections along the old tunnels by means of a RTS. (2) Multipoint extensometers installed in the rock between the old tunnels and the new tunnel. (3) Flat jack measurements in the old tunnels lining.

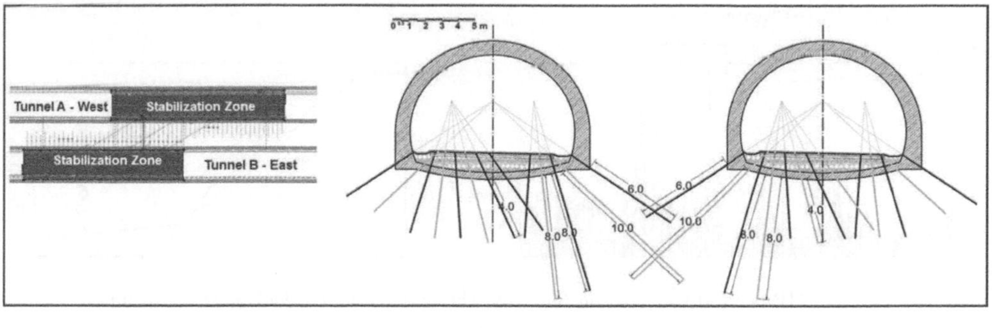

Figure 37. Measures adopted to create a stabilization/reinforcement zone above the new tunnel.

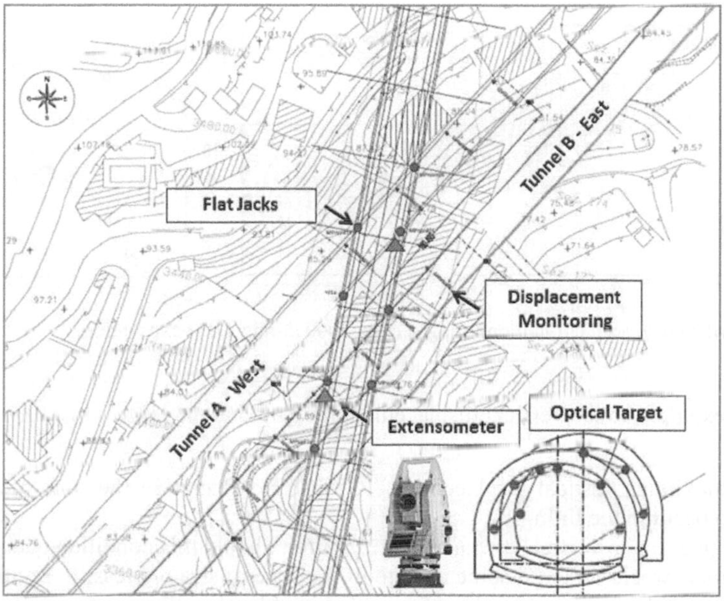

Figure 38. Schematic view of the monitoring system.

In addition, during the excavation of the new tunnel, performance monitoring of the TBM was carried out continuously by using a data acquisition system equipped within the TBM. The operational parameters of the TBM such as penetration per revolution, rotational speed, torque, advance rate, and machine thrust were obtained continuously. Also, geological/geotechnical mapping of the rock mass was performed systematically during face advance, generally when the lining segments were installed. The main objective was to check the rock mass conditions, possible over excavation, and stability conditions of the face.

As a follow-up of the design analyses performed and the results obtained, different "scenarios" were simulated, i.e. the most probable, anticipated conditions and likely variations with respect to the expected response of the old tunnels. In particular, reference was made to the displacement components x-y-z monitored by the RTS in the cross sections shown in Figure 38, which were positioned in view of the zone of interaction with the new tunnel.

Based on the above and having well defined the expected behavior of the old tunnels during the excavation of the new tunnel under them (this means that the expected response at the instruments (1), (2) and (3) is known), one is able to compare continuously the monitored values with the design values. The following warning levels were established as part of a safety management plan by

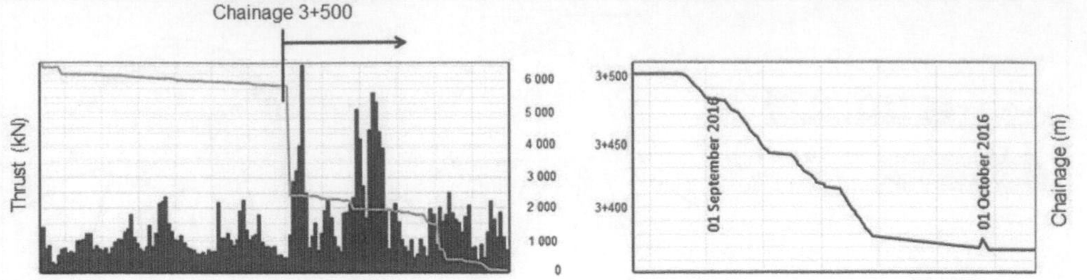

Figure 39. Typical TBM performance monitoring data. Thrust (left) and Face advance versus time

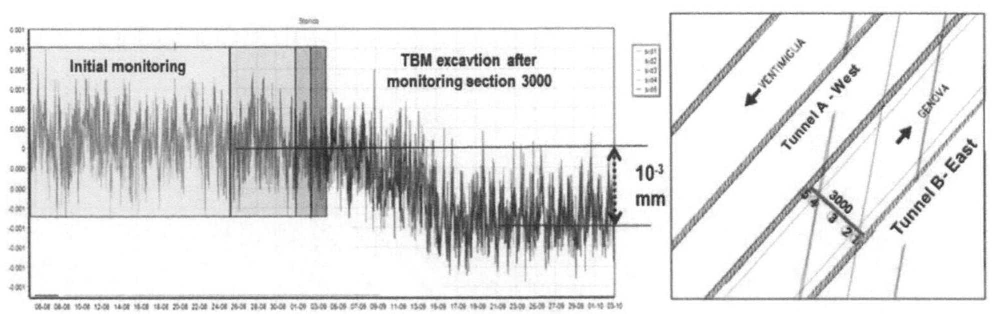

Figure 40. Vertical displacement in monitoring section 3000 during TBM excavation.

referring, for example, to the vertical displacement measured at the targets installed in the two old tunnels:

- Warning (maximum vertical displacement: 10 mm): No change in the state of strain/stress is expected to occur in the linings.
- Pre-Alarm (maximum vertical displacement: 25 mm): Should this condition occur, the lining of the old tunnels would undergo some change in the state of strain/stress, to require a first level strengthening of the lining, defined as "Strengthening Measure 1", prior to resuming TBM excavation in the new tunnel
- Alarm (maximum vertical displacement: 90 mm): Should this condition occur (with the Strengthening Measure 1 in place), the lining structure of the old tunnels would not comply with the required serviceability conditions. TBM excavation should stop immediately and "Strengthening Measures 2" would be required.

It is of interest to report in the following, for completeness, representative performance monitoring data obtained during the excavation of the new tunnel, which was completed successfully between September 1 and October 1 2016 as shown in the plots of Figure 39. Also shown in Figure 40 is the plot of the vertical displacement versus time at the crown of the East old tunnel. It is noted that the displacement monitored is negligible.

6 CONCLUDING REMARKS

In this lecture the "conventional" methods used for displacement monitoring in and around tunnels and on the ground surface (where needed) were first briefly described. Attention was in addition given, for example, to the instrumentation generally installed for determining strains and stresses in the tunnel support and in the reinforcement components. Also considered were "advanced" monitoring methods such as the Robotic Total Station (RTS), Interferometric Synthetic Aperture Radar (InSAR), and 3D Laser Scanning (TLS).

Then, attention was dedicated to the evaluation of the monitoring data by numerical methods, in particular for purpose of back analysis and assessment of the ground behavior and tunnel system response. The need to use such methods for obtaining realistic results, which can be carefully checked and reflect the in situ conditions, was pointed out. This is to be recognized also given that the development of new and improved monitoring tools is accompanied today by an increased ability to obtain and evaluate the ground characteristics.

All of this was discussed through three case studies of large size tunnels excavated with TBMs of different types, generally in poor ground conditions. Additional difficulties were considered such as: (1) tunnels in the near vicinity of the ground surface and under inhabited areas. (2) Tunnels excavated along a hilly terrain characterized by the presence of dormant deep-seated landslides, which were reactivated due to the tunnel excavation. (3) New tunnel under passing two existing old tunnels to be maintained in service.

In case studies (1) and (2), keeping with the use of "advanced" monitoring methods, a systematic InSAR analysis was carried out. In the first case, the method adopted let one describe the history of ground surface settlements in the city of Caltanissetta (Italy) due to tunnel excavation. In the second case, the reactivation of dormant deep-seated landslides could well be described, along a typical area of the Italian Apennines, during construction of the Sparvo tunnel, along a new section of the Milano-Napoli Highway (A1).

In both cases, "conventional" monitoring methods were used such as topographic measurements with TS. Note that in case (2) also placed on site, for displacements components monitoring were a number of RTSs, in addition to (e.g.) inclinometers installed along the slope. As previously described, the clear value added was the description with InSAR of the displacements occurred prior to tunnel excavation, particularly important in case (2) where a clear proof of the landslides reactivation was obtained.

With these monitoring data becoming available, associated with comprehensive geological and geotechnical studies for ground characterization, as mentioned in the lecture, the adoption of realistic numerical modelling became possible with significant knowledge being gained in both cases on a quantitative assessment of the influence of tunnel excavation. The results provided useful information for continuation of the work and long term maintenance of the tunnels.

In case study (1), 3D FDM modelling was carried out, associated with the information gained in a selected tunnel section. Settlements on the ground surface were obtained as the TBM was advancing in the tunnel, where, at the same time, the induced extrusions ahead of the face were being measured. The back analysis of the results obtained let one find out a likely motivation for the observed response, with appropriate, corrective actions being derived for a better control of the induced settlements on the ground surface.

In case study (2), 2D FDM modelling was in particular carried out to back analyse the observed interaction of the deep-seated landslide with the two tunnels excavated in the near vicinity. It is worth to remind here that this needed be done by using a time-dependent elasto-viscoplastic model which allowed to simulate the observed behavior, derived with monitoring data obtained with both "conventional" methods (inclinometers) and "advanced" methods (InSAR and RTS).

It is important to remind here that the back analysis performed in case (2) allowed also to obtain a quantitative estimate of the overstressing in the linings of the two tubes, along a tunnel length estimated to be 250 m. Although this problem was not addressed in the lecture, let remind here that the information gained, with the studies performed, shared light on the actions to be undertaken for putting the Sparvo tunnel into service, including the long term monitoring programme that was implemented and is now being used .

With case study (3), the attention moved to the nearly unique case of the TBM excavation of a road tunnel, the San Paolo tunnel, under an old twin tunnel (along the Ventimiglia-Genova Highway A10) which was built in the 1960s. The new tunnel is at a distance of 11 to 15 m below the old tunnels, which needed to be maintained in service, in near vicinity to it. The objective was to excavate the new tunnel in strictly controlled conditions. Also, in this case observations and near real time monitoring, associated with realistic numerical modelling, became the tools for completing the excavation of the new tunnel.

2D and 3D FEM modelling studies were carried out, associated with detailed geological and geotechnical investigations. The objective was the analysis of the interaction between the new tunnel and the old tunnels. In this case, the need was to predict the ground response above the new tunnel and the stress-strain redistribution in particular in the old tunnels linings. In addition, in view of the need to define a safety management plan, the modelling was finalized to establish the warning criteria/levels to be adopted.

With the 2D FEM model, different positions of the new tunnel needed be considered with respect to the old tunnels. This was obviously not the case for the 3D FEM model, where the new tunnel excavation was simulated. In view of the results obtained, a near real time monitoring system was implemented with RTS convergence/displacement monitoring in cross sections along the old tunnels, multipoint extensometers installed in the rock between the old tunnels and the new tunnel, flat jack measurements in the old tunnels lining. It is of interest to report in conclusion that the excavation of the new tunnel was completed successfully.

The aim of the lecture has been to show with case studies how the data, which can presently be obtained during tunnel excavation, with the many "conventional" and "advanced" monitoring methods and equipment presently available, can help in successful and safe construction. This has been done in the frame established with the observational method.

Emphasis has been placed on the use of numerical methods for purpose of back analysis and assessment of the ground behavior and system response, by paying attention to the need that the calculations performed lead to realistic results, which can be carefully checked and reflect the in situ behavior and tunnel response. This has been done, with the belief that this can presently be achieved also given the increased ability to obtain and evaluate the ground characteristics.

ACKNOWLEDGEMENTS

The invitation of the Organizing Committee to prepare this Keynote Lecture for the 14th International Congress of the International Society for Rock Mechanics and Rock Engineering, ISRM, held in Foz do Iguaçú, Brazil, offered the author the opportunity to describe in a single publication three case histories dealing with TBM excavation of large size tunnels.

The author would like to thank A. Perino for the help with the numerical analyses regarding in particular case studies 1 and 3, and both D. Debernardi and A. Perino for case study 2. In addition, to be thanked is F. Forlani for making available the results of the 3D analyses of case study 3. The help of F. Antolini for preparing Figures 16 and 17 introducing case study 2 is to be acknowledged.

Also Spea Ingegneria Europea SpA, Milano and Società Autostrade per l'Italia SpA, Roma for case studies 2 and 3, and of CMC SpA, Ravenna for case studies 1 and 3 are to be thanked.

REFERENCES

AGI (Associazione Geotecnica Italiana) 1977. *The Geotechnics of Structurally Complex Formations. Proceedings of Symposium held in Capri*, September 1977. Volumes 1 and 2.

Barla, G., Debernardi, D. & Perino, A. 2015. Lessons learned from deep-seated landslides activated by tunnel excavation. *Geomechanics and Tunnelling* 8(5): 394–401.

Barla, G., Tamburini, A., Del Conte, S. & Giannico C. 2016. InSAR monitoring of tunnel induced ground movements. *Geomechanics and Tunnelling* 9(1): 15–22.

Barla, G. 2018. Numerical modeling of deep-seated landslides interacting with man-made structures. *Journal of Rock Mechanics and Geotechnical Engineering* 10(6): 1020–1036.

Barla, G. 2019. Novel investigation methods of rock slopes and tunnels in view of realistic modelling in engineering applications. In *Geotechnical Challenges in Karst. International Conference, 8th Conference of Croatian Geotechnical Society*, Split/Omis, 11–13 April 2019: 13–27.

Debernardi, D. & Barla, G. 2009. New viscoplastic model for design analysis of tunnels in squeezing conditions. *Rock Mechanics and Rock Engineering* 42(2): 259–288.

Feng, Q. & Röshoff, K. 2015. A survey of 3D laser scanning techniques for application to Rock Mechanics and Rock Engineering. In R. Ulusay (Ed), *The ISRM Suggested Methods for Rock Characterization, Testing and Monitoring*: 265–293.

Ferretti, A., Prati, C., & Rocca, F. 2001. Permanent scatterers InSAR interferometry. *IEEE Trans. Geoscience and Remote Sensing* 39 (1): 8–20.

Ferretti, A., Fumagalli, A., Novali, F., Prati, C., Rocca, F. & Rucci, A. 2011: A new algorithm for processing interferometric data-stacks: SqueeSAR. *IEEE Trans. Geoscience and Remote Sensing* 49 (9): 3460–3470.

Huang, H.W., Zhang, D. M., & Ayyub, B.M. 2017. An integrated risk sensing system for geo-structural safety. *Journal of Rock Mechanics and Geotechnical Engineering* 10(6): 1020–1036.

Itasca, Inc. 2019a. FLAC (Fast Lagrangian Analysis of Continua). Itasca Consulting Group, Inc. (https://www.itascacg.com/software/flac).

Itasca, Inc. 2019b. FLAC3D (Fast Lagrangian Analysis of Continua in 3 Dimensions). Itasca Consulting Group, Inc. (https://www.itascacg.com/software/flac*3d).*

ITAtech. 2015. Guidelines for Remote Measurements Monitoring Systems. *ITAtech Report* 3 (2).

OGG, Austrian Society for Geomechanics 2014. *Geotechnical Monitoring in Conventional Tunnelling. Handbook.*

Plaxis. 2019. PLAXIS 3D (A powerful and user friendly finite element package intended for three-dimensional analysis of deformation and stability in geotechnical engineering and rock mechanics). (https://www.plaxis.com/product/plaxis-3d).

Rocscience. 2019. RS2 (Program for 2D Finite Element Analysis of Geotechnical Structures for Civil and Mining Applications). (https://www.rocscience.com/).

SGSS, 2018. Carta Inventario delle frane a scala 1:10000, edizione Giugno 2018, pubblicata dal Servizio Geologico, Sismico e dei Suoli della Regione Emilia-Romagna. (http://geo.regione.emilia-romagna.it/).

Steiner, P.H. 2007. Displacement measurements ahead of a tunnel face using the RH Extensometer. In *FMGM 2007: Seventh International Symposium on Field Measurements in Geomechanics.*

Xie, X. & Lu, X. 2017. Development of a 3D modeling algorithm for tunnel deformation monitoring based on terrestrial laser scanning. *Underground Space* 2: 16–29.

Rock Mechanics for Natural Resources and Infrastructure Development –
Fontoura, Rocca & Pavón Mendoza (Eds)
© 2020 ISRM, ISBN 978-0-367-42284-4

The geotechnical engineer in metalliferous open pit mines

John Read
John R Read Associates Pty Limited, Robina, Queensland, Australia

ABSTRACT: The open pit mine Geotechnical Engineer is responsible for creating slope designs that are expected by the owners, management, the workforce, the regulators, and the public domain to be stable for the life of the mine. Economic requirements are always implicit in the designs, but in today's world producing slope angles that ensure the workers in the pit are protected against death or injury have become additional moral and legal requirements. These factors require that the Geotechnical Engineer has a skill set that at a minimum: encompasses regional and site specific knowledge of the geological model; understands the spatial and temporal distribution of the structural defects that are likely to affect the stability of the pit slopes, together with the properties of the soil and rock materials in which the slope will be excavated. He also must have the ability to transform this knowledge into a soil and rock mechanics slope design framework that is understood clearly and can be implemented readily by the mine planners and pit operations professionals. The objective of this keynote is to critically examine how well these requirements are being met and the most effective approach to satisfy best practice beyond 2019.

1 INTRODUCTION

Reliable slopes are essential to open pit mine operations, at all scales and at every level of project development. Uncontrolled instabilities, that is, slope failures where displacement has reached a level where the intended function cannot be met, may have safety, economic or environmental consequences. The safety of personnel and equipment is of paramount importance, but any one of these consequences can effectively close the mine down.

To achieve reliable slopes the practitioners responsible for the design must possess the information and skills required to correctly assess the inherently variable properties and characteristics of the natural materials being dealt with. The responsibility for collecting, compiling and analysing this data will vary from company to company. Companies with strong in-house capabilities are likely to take responsibility for the conceptual and pre-feasibility project evaluations and for slope management in operating mines. Consultants are likely to have a role in the feasibility studies for large mines and for much of the work in smaller mines. In all cases they will be responsible for establishing robust designs, reducing uncertainty, reducing risk, identifying the preferred development option, and identifying opportunities.

Since the publication of the book *Guidelines for Open Pit Slope Design* (Read & Stacey, 2009), there has been a concerted effort by practitioners to prepare slope designs in a manner that enables mine owners, managers, and operators to fully understand the basis and shortcomings of the design. The slope design process followed, together with the responsibilities and expected performance of the geotechnical engineer (aka; engineering geologist, geological engineer) at all stages of the process, are critically examined under the following sections: the slope design process; required skills; skill shortages; and ensuring future best practice.

2 THE SLOPE DESIGN PROCESS

The slope design process, regardless of size or materials, is outlined in Figure 1. Slope designs prepared following this process must not only be technically sound, they must also address the broader context of the mining operation as a whole, taking into account safety, the equipment available to implement the designs, mining rates and the acceptable risk levels. They must also be presented in a way that will allow the mine executives, who are ultimately responsible for the project, and the operators, who implement the designs, to fully understand the basis of the designs.

Critical steps in Figure 1 are as follows.

2.1 *Preparing the geotechnical model*

The geotechnical model is the cornerstone of all slope design activities. The objective of the model is to provide the geotechnical engineer with the information required to correctly assess the inherently variable properties and characteristics of the natural materials being dealt with. It is comprised of the four components shown in Figure 1: the geological model; the structural model; the rock mass model; and the hydrogeological model.

The purpose of the geological model is to link the regional geology and the events that lead to the formation of the ore body to a mine-scale description of the setting, distribution, and nature of the overburden soils and rock types at the site, including the effects of mineralisation, alteration and weathering. The structural model describes the orientation and spatial distribution of the structural defects that are likely to influence the stability of the pit slopes. It must be configured in at least two overlays, one that shows the major structural features such as mine scale faults, folds and lithological boundaries that can be used to subdivide the mine into a select number of structural domains, and one

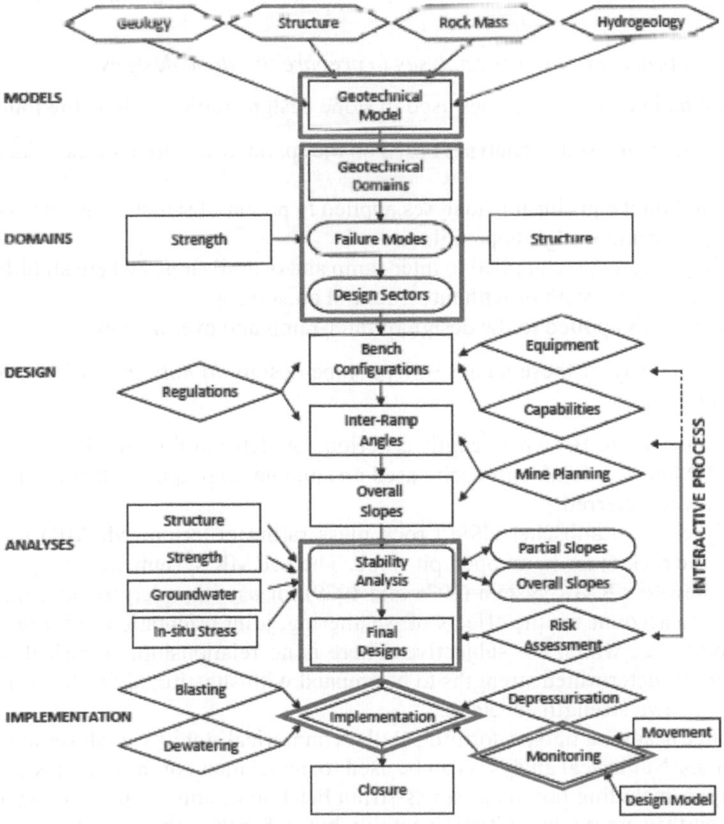

Figure 1. The slope design process, with critical steps outlined in red *(source, Read & Stacey, 2009).*

that shows the attributes of the more closely spaced fault and joint fabric that occur within each structural domain. The third model, the rock mass model, is erected to database the engineering properties of the rock mass for use in the stability analyses that will be used to prepare the slope designs at each stage of project development. The fourth and final model is the hydrogeological model, whose purpose is to present the surface water and groundwater aspects of the hydrogeological regime at the site.

2.2 *Recognising potential failure modes within mine scale geotechnical domains*

When the geotechnical model has been established the next step is to divide it into domains with similar lithological, structural and material properties, and then assess the potential for bench, inter-ramp and overall slope failures within each domain.

Slope failures occur in an open pit mine when displacement has reached a level where it is no longer safe to operate or the intended function cannot be met. In a metalliferous open pit they are most likely to occur in one of three ways.

1. Structurally controlled failures in strong rocks, where the rupture occurs only along the joints, bedding or faults. In this case, the strength and orientation of the structures are the most important parameter in assessing slope stability.
2. Failure with partial structural control, where the rupture occurs partly through the rock mass and partly along the structures. In this case, the strength of the rock mass and the strength and orientation of the structures are all important in assessing slope stability.
3. Failure with limited structural control in weak rocks, where the rupture occurs predominantly through the rock mass. In this case, the strength of the rock mass is the most important parameter in assessing slope stability.

When the most likely failure mechanism in each domain has been recognised the domains can be subdivided into design sectors and the task of preparing the operating slope designs commenced.

2.4 *Using the appropriate stability analyses to prepare the slope designs*

Historically, the main types of analyses used in slope design studies include the following.

- Kinematic planar and wedge analyses based on the spatial orientation of the joint fabric at bench scale.
- Kinematic and limit equilibrium analyses applied to potential structurally controlled bench and inter-ramp planar and wedge slope failures.
- Limit equilibrium analyses applied to inter-ramp and overall slopes where stability is controlled by rock mass strength, with or without structural anisotropy.
- Numerical analyses applied to the design of inter-ramp and overall slopes.

When performing any of these analyses, five slope design *do nots* (Read 2014, Read & Stacey 2017), are re-emphasised.

1. **Do not** use the Hoek-Brown strength criterion for determining weak (GSI$\leq$25) rock mass strengths. A Mohr-Coulomb approach based on suitable, high quality laboratory testing and/or back-analyses is preferred.
2. **Do not** rely on the Laubscher (1990) rock mass rating criterion (MRMR) to determine the strength of the rock mass in an open pit slope. The MRMR system was designed as an extension of Bieniawski's RMR system (1976 and 1979) for use in underground mining. It contains adjustments to account for the effects of weathering, joint orientation, mining-induced stress, blasting and water, which are subjective. There is no relationship, empirical or other, which enables MRMR determined strengths to be equated with/substituted for the GSI values as used in the Hoek-Brown strength envelope.
3. **Do not** plunge headlong into performing endless numerical stability analyses at the outset of the project studies. Numerical analyses can be used to model many of the complex conditions found in rock slopes, including nonlinear stress-strain behaviour, anisotropy, and changes in geometry. However, they are not the initiating activity but rather the penultimate activity in the design

process. They should not be performed until the failure mechanisms perceived for each design sector have been thoroughly evaluated using common sense and limit equilibrium methods, and a need for such numerical analyses has been demonstrated.

4. **Do not** rely on moment equilibrium solutions when utilising limit equilibrium methods of slices codes (e.g. Bishop's simplified method of analysis). Large scale jointed rock slopes do not form rigid bodies that obey the laws of moment equilibrium. Bishop's simplified method of slices is popular because it is quick and easy to use but, because it is rotational, it rarely approximates the type of geologically controlled failures that occur in large-scale slopes. For such slopes, use only force equilibrium methods (e.g. Janbu's simplified, Morgenstern & Price, Spencer). Although they will not consider deformations within the sliding body and will ignore out-of-slice forces, they will at least enable the modelling of irregular failure surfaces such as occur in large-scale slopes.
5. **Do not** confuse the phreatic and piezometric surfaces when calculating pore pressure forces. Phreatic surfaces represent the free ground water level within the slope whereas the piezometric surface represents the actual pressure head relative to a surface within the slope.

2.6 *Defining the implementation and monitoring requirements for the designs*

A monitoring program for assessing design performance and failure risk, and for aiding risk minimisation, should be established as soon as possible during the early stages of mining and maintained throughout the operating life of the open pit; in many cases it may be required beyond the closure of the pit.

The main objectives of the monitoring program can be summarised as follows.

- Maintaining safe operating conditions to protect personnel and equipment.
- Providing advance notice of zones of potentially unstable ground so that mine plans can be modified to minimise the impact of slope displacement.
- Providing geotechnical information for analysing any slope instability mechanism that develops, designing appropriate remedial action plans and conducting future slope design.
- Assessing the performance of the implemented slope design.

The golden rule when preparing a monitoring system to achieve these objectives is that all of the instruments installed, be they prisms or radar units for detecting slope movements or piezometers for measuring pore pressures and/or target pit dewatering levels, should be selected and placed to assist in answering a specific question.

3 REQUIRED SKILLS

The design process outlined may look straightforward, but putting it in place requires the application of skills unique to the geotechnical engineer, a sound knowledge of the ins and outs of open pit mining, and a great deal of common sense.

In summary, geotechnical engineering skills are underpinned by an understanding of five guiding principles outlined 20 years ago by Dr Fred Baynes (Baynes, 1999).

1. Geological knowledge (regional and site specific).
2. Spatial and temporal distribution of the geological attributes (the model).
3. Encoding of data in geotechnical language (engineering geological description systems).
4. Transformation into an engineering framework (application of soil and rock mechanics).
5. Communication of knowledge in cognisance of the project objectives and limitations due to any uncertainties attached to knowledge.

These principles remain as part and parcel of the studies for a degree in engineering geology, which usually will take an undergraduate student four years to complete. After graduation, young geotechnical engineers should work at least four years in an open pit, during which time they must demonstrably achieve competence in bench, inter-ramp and mine-scale mapping, core logging, slope design and implementation procedures, and slope monitoring. At least six to eight years' hands-on experience should be gained before a Senior Position is occupied and a minimum of 10 years' experience demonstrated before being employed as a Superintendent.

4 SKILL SHORTAGES

Today's open pit mining industry faces a global shortage of Graduate, Senior and Superintendent-level geotechnical engineers capable of performing the fundamental engineering geological tasks of outcrop mapping, core logging and preparing mine-scale geotechnical models. Many reasons are advanced to explain why this has happened and continues to happen. Cited frequently is the cyclical "boom and bust" nature of the commodities market, which has led to retrenchments and related cost-cutting manoeuvres as mining companies who had over-extended themselves during the "good" times struggled to survive in a re-ordered economic world.

Reasons both for and against can be advanced as to whether or not the boom and bust cycle has been the primary driver of the current shortfall in geotechnical capital. Another likely driver is the loss of the mining industry's social "licence to operate" during the last decade because of corporate's recurring investment write-downs, perceived inattention to environment concerns, and other examples of corporate misbehaviour. No matter which, there have been undeniable consequences. Re-trenched staff have moved on to alternative employment and have been lost to the industry, and would-be geotechnical engineers, seeing a lack of employment security in the industry, have looked elsewhere for a career. This waning interest is reflected in figures published recently by the Curtin University Western Australian School of Mines that show between 2013 and 2018 enrolments have plummeted from the mid three hundreds to less than one hundred (PWC, Curtin University, 2018).

The widespread introduction of 3D digital photogrammetric and laser imaging technologies for mapping structures at bench and inter-ramp scale has helped to redress this human resource and skills shortage, but is having unwanted consequences. Geological maps prepared by such techniques do provide reliable images and information concerning the orientation, frequency and spacing of structural discontinuities. However, they do not provide information on the persistence or roughness of the discontinuities, or of the nature of any infill materials, information that can only be obtained by ground-proofing.

Too often the co-dependent issues of time and under-resourcing lead to a map that has been prepared remotely in the office being committed to the database without adequate and, in some instances, any ground proofing having been performed. The consequences of this shortcoming are a diminished hands-on understanding and ownership of the geotechnical model, and a database lacking in quality and practical value. Increasingly, amongst some practitioners it has also led to a "smart technology mindset", where the emphasis is on office IT rather than field based activities. I agree with and strongly support the concept that IT expertise and smart technologies are a "must have" in the geotechnical toolbox of the future, but we must not allow the Internet of Things (IoT) and/or the Internet of Everything (IoE) to result in graduates who are dumbed-down "least-common-denominator" system managers rather than skilled, hands-on operators.

5 ENSURING FUTURE BEST PRACTICE

What can we do to alleviate the current shortfall in geotechnical capital and ensure that beyond 2019 the industry is supplied with geotechnical engineers who possess the skills and experience required to correctly assess the inherently variable properties of the materials they are dealing with, and create pit slope designs that will remain stable for the life of the mine? Four areas of endeavour stand out: education; training; research; and the appropriate use of smart technologies.

5.1 *Education*

Currently, many companies are using widespread television and newsprint public relations campaigns, for example, the BHP "Think Big" campaign in Australia, as a means of restoring their corporate face. However, if the industry is to remain sustainable, let alone grow, a substantial portion of the money now being spent on these public relations activities must instead be directed towards attracting earth science and mining engineering graduates into careers that are perceived as being financially and technically rewarding in an industry that is socially responsible and of national importance. Companics must again provide financial support to Universities, and publically work with Government Departments, Minerals Advisory Councils and Universities to develop Centres of Excellence that not only provide the required geotechnical skill base but can out-perform the competing demands for "Green" and IoT and IoE graduates. It would also help if hands-on geotechnical and mine development expertise was added to corporate investment decision making bodies that some view as being top heavy with financial engineers and political lobbyists.

5.2 *Training*

As already noted, quality field mapping, including mine-scale, inter-ramp, and bench-scale structural mapping, is an under-performing aspect of today's data gathering process. For those working in the field I consider this is best resolved by corrective skills training on site, not off-site (Read, 2018).

Off-site training and course work has a number of drawbacks. It reduces the site resources, the question always being, "can he/she be spared"? And the learning gained tends to benefit the individual not the site team, as it is often not transferred to the team members when the individual returns to the site.

On-site training brings the trainer to the site, which enables the training to be directed at and performed as part of the day-to-day activities; for example, inter-ramp and bench mapping, and core logging. This will not only enable the building of both individual and team skills, but will help build an understanding and ownership of the site geotechnical model. It will also build the confidence of the planning and operations groups in the skills of the geotechnical team.

5.3 *Research*

An excessive amount of university geotechnical engineering research and academic papers published at conferences such as this are devoted to the unproductive black arts of numerical modelling, expert systems, neural networks and statistical analyses, which are based on the virtual unreality that they can imitate real geological processes.

To help redirect this mindset, ongoing support is required for the upcoming third stage of the Large Open Pit (LOP) project, an industry-sponsored research and technology project that was initiated by CSIRO Australia in 2005 (LOPI) and continued in 2016 (LOPII) under the auspices of the University of Queensland. Funded by a diverse group of multinational mining companies, joint venture partners and individual mines representing a majority of the world's diamond and base metals production, the project's objective has been to develop and transfer skills to manage geotechnical risk in large open pits, address critical gaps in our knowledge and understanding of the relationships between the strength and deformability of rock masses, and the likely mechanisms of rock slope failures and landslides in large open pit mines (LOPII, 2016).

Significant contributions from LOPI include four in a series of five reference books on the state-of-practice related to the design and stability of large slopes associated with open pit mines: *Guidelines for Open Pit Slope Design* (Read and Stacey, 2009); *Guidelines for Evaluating Water in Pit Slope Stability* (Beale and Read, 2013); *Guidelines for Mine Waste Dump and Stockpile Design* (Hawley and Cunning, 2017); and *Guidelines for Open Pit Slope Design in Weak Rocks* (Martin and Stacey, 2018). The fifth book in the series, *Guidelines for Slope Performance Monitoring* (Sharon and Eberhardt, 2019) is expected to be published this year under the sponsorship of LOPII. It compiles the current and suggested monitoring practices for open pit operators, illustrating each topic with examples and case histories.

Ongoing LOPII research projects that will be continued in LOPIII include four fundamental aspects of open pit slope design: data uncertainty; quantifying mining induced rock mass damage; real time slope monitoring by integration of laser, radar, total station and 3D kinematic analysis; and groundwater flow, pore pressures and effective stress.

The Data Uncertainty project is focussed on improving the target levels of data confidence suggested in Table 8.1 (i.e. target levels of data confidence) of Chapter 8 of Read and Stacey (2009). Although the system outlined by Read and Stacey has gained widespread acceptance across the industry, it remains subjective, relying heavily on the judgement and opinion of the practitioner. The project task is to replace this subjectiveness with objective measures of the quantity and quality of data that can be confidently used by the slope design engineer, the owner, and the investor to help them assess the reliability of the slope design at the different stages of project development.

It is standard practice in stability analyses to account for rock mass damage resulting from stress redistribution and blasting during mining using the Hoek-Brown rock mass strength criterion D-factor. However, there are no clear guidelines for quantifying this damage. Similarly to the choice of the levels of data confidence in Table 8.1 of Chapter 8 of Read and Stacey mentioned above, the choice of the D-factor and its distribution behind the pit wall relies on the judgement and opinion of the practitioner, which often differ. Hence, the objective of the project is to match numerical analyses of hydro-mechanical changes and blast damage to quantify the spatial distribution of rock mass damage behind open pit walls.

Nowadays most large open pit mines have sophisticated systems in place for near real-time slope monitoring. Despite the availability of massive datasets of monitoring data, negative events associated with slope failures, having in some cases strong impact on mine safety and productivity, are still occurring in surface operations. The LOPII project has been supported by the Sponsors to develop a methodology to support mine site engineers in the daily task of interpreting monitoring data aimed at identifying potential failure events, mechanisms and setting up alarm thresholds. The project is based on collecting point cloud data from laser scans and then processing them in a semi-automatic way to characterise the main structural features, perform 3D kinematic analyses, and identify potential failure events.

Major advances in the understanding of groundwater flow, pore pressures and effective stress in pit slopes were presented in Chapter 6 of Read and Stacey (2009) and in Beale and Read (2013). Notwithstanding these advances, there remains across the mining industry a relatively poor understanding of groundwater flow and effective stress in a fractured medium and the significance of the coupled deformation/pore pressure response on pit slope stability. The LOPII project has been granted a unique opportunity to improve the industry knowledge of these topics utilising an extensive, comprehensive and world-class data set from the Rio Tinto Bingham Canyon Mine (BCM) drainage gallery project. The data comes from both the South Wall Drainage Gallery (SWDG) and the North Wall Drainage Gallery (NWDG) of the mine. The research initiated incorporates a phased programme of work to further our understanding of groundwater, pore pressure and effective stress in deep open pits.

5.4 *Appropriate use of smart technologies*

Smart Markers, enabling real time linkage of subsurface deformation with pore pressure changes (Wiszyk-Capehart, 2018) and UAVs, which allow for changes in perspective, eliminate shadows and enable the measuring of linear structures when mapping (Read, 2018), are examples of smart tools that have already become part and parcel of the geotechnical toolbox.

Smart data management tools now in use include the OSI-approved open source Python software that is freely available to access, interrogate and visualise the geotechnical data now routinely collected during project investigation, development and operation. A current example of the use of Python is the interrogation and visualisation of the comprehensive Rio Tinto Bingham Canyon Mine drainage gallery project data set being accessed by the LOPII project to further the understanding of groundwater flow, pore pressures and effective stress in deep open pits.

Artificial Intelligence (AI) and machine learning technologies now being used by many financial and commercial institutions to process vast quantities of data to find patterns and make predictions without being explicitly programmed to do so (the Economist, 2018) provide additional avenues for smart interrogation of the project database, particularly the mine planning database. Potentially, AI gap analysis, design performance, hazard inventory, and emergency response algorithms would all be additions to the toolbox that would greatly enable a safe and economic design for the bench, inter-ramp and overall slopes of the pit.

6 CONCLUSIONS

Currently, the supply of Graduate, Senior and Superintendent-level geotechnical engineers capable of performing the fundamental tasks of outcrop mapping, core logging and preparing mine-scale geotechnical models is out of step with the needs of the open pit mining industry. The boom and bust nature of the commodities market, which has led to large scale retrenchments, coupled with the loss of the industry's social "licence to operate" during the last decade because of perceived corporate misbehaviour are the most cited reasons for this loss of geotechnical capital. Re-trenched staff have moved on to alternative employment and have been lost to the industry, and would-be geotechnical engineers, seeing a lack of employment security in the industry, have looked elsewhere for a career.

Education, on-site training, applied research led by practitioners, and the appropriate use of smart technologies have been suggested as ways of ensuring that the industry is again fully supplied with appropriately skilled and experienced geotechnical engineers. However, to be successful, these activities must be strongly underpinned by the mining industry. As mentioned, many companies are using widespread television and newsprint public relations campaigns as a means of restoring their licence to operate. If the industry is to remain sustainable, let alone grow, it is imperative that a substantial portion of the money now being spent on these public relations activities is directed towards attracting earth science and mining engineering graduates into careers that are perceived as being financially and technically rewarding in an industry that is socially responsible and of national importance.

REFERENCES

Baynes, F.J. 1999. Engineering geological knowledge and quality. *Proceedings of the 8th ANZ Conference on Geomechanics, Hobart*. Institute of Engineers Australia, pp. 227–234.

Beale, G. and Read, J. 2013. Guidelines for Evaluating Water in Pit Slope Stability (eds., G. Beale & J. Read) CSIRO Publishing, Melbourne, Australia.

Bieniawski, Z.T. 1976. Rock mass classification in engineering. In *Exploration for Rock Engineering* (Ed. Z.T Bieniawski), V1, pp. 97–106, Balkema, Cape Town.

Bieniawski, Z.T. 1979. The geomechanics classification in rock engineering applications. In *Proceedings of the 4th Congress of International Society of Rock Mechanics*, Montreux, **2**, pp. 4–48. Balkema, Rotterdam.

Hawley, M. and Cunning, J. 2017. Guidelines for Mine Waste Dump and Stockpile Design (eds., M. Hawley & J. Cunning), CSIRO Publishing, Melbourne, Australia.

Hoek, E. and Brown, E.T. 2018. The Hoek-Brown failure criterion and GSI – 2018 edition. *Journal of Rock Mechanics and Engineering Geology*. August 10, 2018.

Laubscher, D. 1990. A geomechanics classification system for the rating of rock mass in mine design. *Journal of South African Institute of Mining and Metallurgy* **90** (10), pp. 279–293.

LOPII, 2016. The Large Open Pit Research Project, Stage II (www.lopproject.com).

Lorig, L., Stacey, P. and Read, J. 2009. Slope Design Methods. In *Guidelines for Open Pit Slope Design* (eds. J. Read & P. Stacey), Chapter 10, pp. 237–264, CSIRO Publishing, Melbourne, Australia.

Lorig, L. and Harthong, B. 2013. The lattice formulation and the *Slope Model* code. In *Guidelines for Evaluating Water in Pit Slope Stability*, Appendix 6, pp. 568–577, CSIRO Publishing, Melbourne, Australia.

Mars Ivars, I., Pierce, M.E., Darcel, D., Montes, J.R., Potyondy, D.O., Young, R.P. and Cundall, P.A. 2011. The synthetic rock mass approach for jointed rock mass modelling. *International Journal of Rock Mechanics and Mining Science* **48** (2), pp. 219–244.

Martin, D. and Stacey, P. 2018. Guidelines for Open Pit Slope Design in Weak Rocks (eds., D. Martin & P. Stacey), CSIRO Publishing, Melbourne, Australia.

PwC, Curtin University. 2018. Resurgent miners warned to lure talent. In *The Australian, 21 November, 2018*.

Read, J.R.L. 2014. Implementing a Reliable Slope Design. 5th ISRM Online Lecture, April 10, 2014.

Read, J.R.L. 2018. Developments in Open Pit Slope Design since 2009. *Keynote Lecture, XIV International Congress of Energy and Mineral Resources and Slope Stability*, Seville, Spain, April 10–13, 2018.

Read, J.R.L. and Stacey, P.F. 2009. Guidelines for Open Pit Slope Design (eds., J. Read & P. Stacey), CSIRO Publishing, Melbourne, Australia.

Read, J.R.L. and Stacey, P.F. 2017. Open Pit Slope Design. In *Rock Mechanics and Engineering* (ed., Xia Ting Feng), Volume 3, Chapter X. In Press.

Sharon, R. and Eberhardt, E. 2019. Guidelines for Slope Performance Monitoring (eds., R Sharon & E. Eberhardt), CSIRO Publishing, Melbourne, Australia.

The Economist. 2018. The Workplace of the Future. In *The Economist, March 28, 2018*.

Widzyk-Capehart, E. 2018. Mastering the moving rock; wireless in-ground slope stability monitoring. In *ACG Instrumentation and Slope Monitoring Workshop, XIV International Congress of Energy and Mineral Resources/Slope Stability*. Seville, Spain, April 10–13, 2018.

Rock Mechanics for Natural Resources and Infrastructure Development – Fontoura, Rocca & Pavón Mendoza (Eds)

In-situ computed tomography technique in geomechanical testing

Xiao Li, Shou-ding Li, Jian-ming He & Peng-fei He
Key Laboratory of Shale Gas and Geoengineering, Institute of Geology and Geophysics, Chinese Academy of Sciences, Beijing, China
College of Earth and Planetary Sciences, University of Chinese Academy of Sciences, Beijing, China

Rong-jian Shi
Institute of High Energy Physics, Chinese Academy of Sciences, Beijing, China

ABSTRACT: Predicting the occurrence and evolution of failure in geomaterials under the conditions of stress, temperature and pore pressure is a key aspect in many geomechanical applications, such as the exploitations of resources and energies and civil engineering. The geomechanical tests with in-situ CT tomography play an important role in improving the understanding of the mechanics and physics of deformation and failure on geomaterial specimens. This paper will provide a review of the applications of the in-situ CT scanning technique in the geomechanical tests. The review starts with the fundamental of the CT scanning technique, followed by a summary of the in-situ CT scanning in the geomechanical tests. Next, the limitations of the current in-situ CT scanning technique employed in the geomechanical tests are discussed briefly. Then, a geomechanics test system equipped with a linear accelerator CT scanner newly developed will be introduced. Some preliminary testing results are provided to show the capability of the geomechanics test system. The results show that the test system is capable of obtaining the complete stress-strain curve and the 3-D scanning images corresponding to the selected stress points for the full scale geomaterial specimens. Therefore, the test system is very useful to investigate the internal structure change in the deformation and failure process of the geomaterials under the uniaxial, triaxial, pore pressure, hydraulic fracturing and so on. Finally, the conclusions will be drawn with respect to the features and potential applications of the new geomechanics test system as well as the future works.

1 INTRODUCTION

With the demand for economic growth, social development, and environment protection, human beings' relying on the clean energy has been continuously increasing. Typical clean energy includes the unconventional natural gas represented by shale gas and the geothermal energy stored in the hot dry rock reservoirs. The exploitation of the shale gas and the geothermal energy faces the severe challenges. In China, the shale gas and hot dry rock reservoirs usually are buried in the depth over 3000 m. Because of the very low permeability of these reservoirs, horizontal drilling and hydraulic fracturing techniques are often employed to stimulate the reservoirs and increase the permeability manually. The occurrence, evolution, and distribution of the hydraulic fracturing induced fracture networks in the reservoirs have significant meaning in efficient and economic exploitation. Because of the inaccessibility of the deep underground reservoirs, the direct measurements of the stimulated fractures are very difficult to carry out in the field. The geophysical methods based indirect measurement are frequently used, e.g. micro-seismic monitoring of the hydraulic fracturing induced fractures. However, the field monitoring effect is not satisfactory due to the limitation of the micro-seismic monitoring accuracy. Therefore, it is very important to perform the laboratory scale tests and then investigate formation, evolution, and pattern of the fractures in different rocks induced by the high pressure fluid injection under the stress and temperature environment in the deep underground. The deformation and fracturing process as well as the internal structure change of the geomaterials under stress also play important roles in the mining, civil, and hydraulic engineering.

An appropriate tool is a prerequisite to achieving a goal successfully. Geomechanics test apparatus is the essential instrument to understand the deformation and fracturing behaviors of rocks. Development of the geomechanics testing facilities has experienced three phases since the 20th century (see Figure 1). Before the seventies of the 20th century, the pioneers of rock mechanics employed the material testing approach to study the rock behaviors. The first phase of the geomechanics testing facilities is so-called soft or conventional test machines. Using the conventional experimental techniques, the geomechanical parameters of rock speicemen, such as uniaxial compressive strength, Young's modulus, Poisson's ration, cohesion and internal friction angle, etc., can be obtained, which are essential for numerical simulation and safety design of the practical projects. However, because of the excessive energy stored in the loading system, the rock specimens usually collapse violently after reaching a certain peak value, i.e. compressive strength. The invention of the stiff testing machine with closed-loop servo-controlled system (Hudson, 1971) was regarded as the second phase of the geomechanics testing facilities, which make the complete stress-strain curve possible. The discovery of the post-peak behaviors of rocks is a milestone in the geomechanical testing techniques. The post-peak behaviors of rocks play very important roles in the pillar design of mining engineering and stability analyses of tunnels and slopes. Since the 21st century, the geomechanical researchers focus on not only the macro mechanical behaviors of rocks during the complete stress-strain process, but also the internal structure changes inside rocks during the process, i.e. formation and evolution of the cracks in the rocks. It is very important to achieve an full understanding of the micro-mechanism of the macro mechanical behaviors of rock deformation and fracturing. Hence, a wide range of experimental techniques have been applied over the years, which include optical and electronic microscopy, radiographic analysis, ultrasonic and acoustic techniques, multiple stress and strain local measurements, stereophotogrammetry and digital image analysis to name but a few.

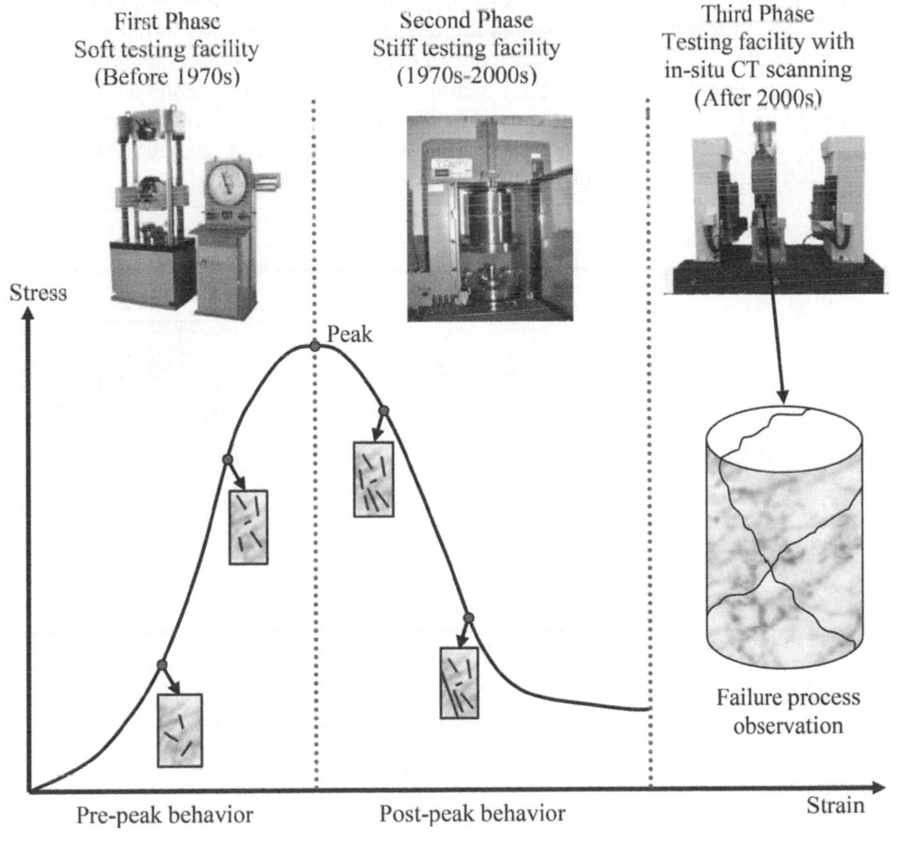

Figure 1. History of geomechanics testing facility.

In the recent decades, the non-destructive computed tomography (CT) technique has play an increasing important role in the geomechanical testing work because of its capability of capturing each individual fractures within the resolution. The geomechanics test system equipped with the powerful CT scanner is so called the third phase of the geomechanics testing facilities (see Figure 1). In the second section, the applications of the in-situ CT scanning in the geomechanical tests will be reviewed. Next, the limitations of the current in-situ CT scanning technique employed in the geomechanical tests are discussed briefly. In the third section, a newly developed geomechanics test system equipped with a linear accelerator CT scanner will be introduced. The fourth section will cover some preliminary ex-perimental results to show the capability of the new geomechanics test system. Finally, the conclusions will be drawn with respect to the features and potential applications of the new geomechanics test system as well as the future works.

2 REVIEW OF IN-SITU CT TECHNIQUE IN GEOMECHANICAL TESTS

2.1 *Fundamentals of the CT scanning technique*

The strong penetrability of X-ray leads to the application of the X-ray photographic technique after it was discovered by the German physicist Röntgen. However, this early technique cannot capture the structure information of the scanned object in the penetrating direction, until the American physicist Cormak proposed the theory of X-ray computed tomography, i.e. CT in 1963. A typical X-ray CT scanner is composed of the X-ray source, detectors, and controllers (as shown in Figure 2). The X-ray source radiates the X-ray beam towards the object to be scanned. Both fan beam and conical beam can be used for the X-ray source. After passing through the object, except the portion of the X-ray being absorbed, the residual X-ray is collected by the detectors to image. In order to obtain a sufficient number of projections of the scanned object at different angles for imaging, a medical CT (MCT) scanner usually adopts the scheme to rotate the X-ray source and detectors, whereas a industrial CT (ICT) scanner uses a positioner to rotate the objects (as shown in Figure 2).

In the early stage, applications of the CT scanning technique in the geomechanical tests is so called post-mortem CT scanning. The scanning objects were limited to the intact rock specimen before test and failed rock specimen after test. Some efforts have been reported using this method (Louis et al. 2006,Appoloni et al. 2007,Renard et al. 2009, Wennberg et al. 2009, Peng et al. 2011,

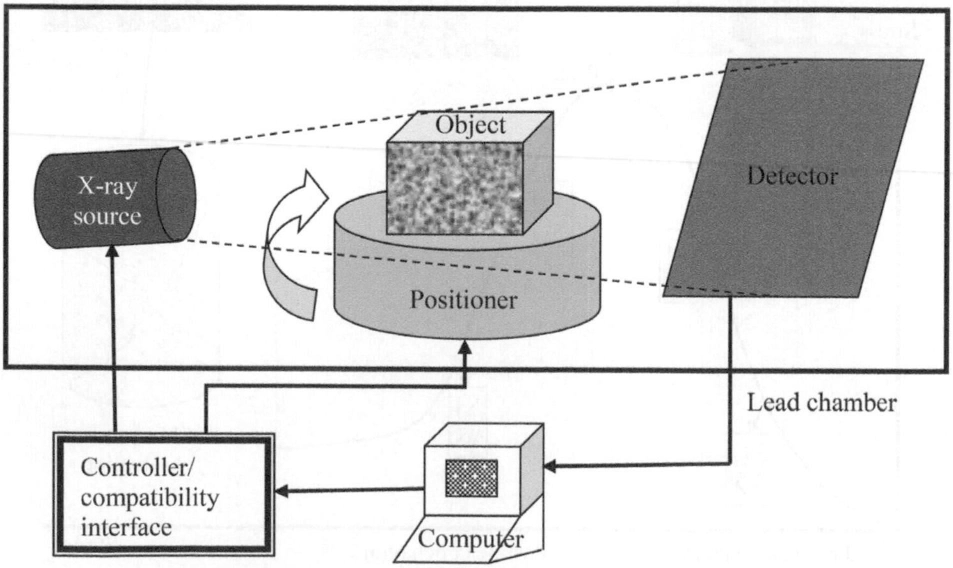

Figure 2. Components of a typical X-ray CT scanning unit (modified after Halverson 2008).

Morishita et al. 2012,Zhao et al. 2015, Machado et al. 2015, Arzilli et al.2016). Only the original structure and post-failure structure of the rock specimens can be obtained in this type of post-mortem CT scanning tests.

In order to observe the failure evolution process and fully understand the failure mechanisms of the geomaterials, the in-situ X-ray CT scanning experimental technique was proposed, where the "in-situ" means X-ray scanning at the same time as loading (Hall et al., 2010). The idea of in-situ CT scanning technique is combining a CT scanner and a geomechanics test apparatus.

The combinations are various depending on the different choices of the loading unit and CT scanners. In 1973, Dr. Hounsfield built the first MCT scanner. Soon after that, the application of the CT scanning technique rapidly extended to the industrial usage, which are separated into two extremes: high spatial resolution (micro-CT) and high penetrability (penetrate the rock specimens with considerable size). The micro-CT methodology mainly includes lab-based and synchrotron-based micro-CT (Cnudde & Boone, 2013). The high penetrability can be obtained using a linear particle accelerator (LINAC), which will be described in section 3. In the following, some research works based on the in-situ CT scanning technique that used the MCT and the ICT scanners will be reported, respectively.

2.2 *In-situ CT scanning technique in geomechanical tests using the MCT*

For the geomechanical test using a MCT scanner, the geomechanics test apparatus is usually placed in a medical spiral CT scanner. When performing the test, the geomechanics test apparatus and rock specimen stay still whereas the X-ray resource and detector are rotated to image. Vinegar et al. (1991) reported a series of triaxial compression tests on sandstone using an aluminum pressure vessel with MCT scanner. Desrues et al. (1996) designed an experimental setup for triaxial compression on sandy soil using a medical X-ray CT. The combination of a uniaxial loading unit and a helical MCT was reported by Yang et al. (1996, 1999). This facility was used to perform in-situ CT scanning tests on coal and sandstone specimens. Ge et al. (1999, 2002) independently designed a triaxial loading chamber to observe the rock failure process during the triaxial compression tests, fatigue tests, and loading-unloading tests. Some similar researches were also reported by Ren et al. (2000, 2001, 2004). Chen et al. (2000, 2005) conducted in-situ CT scanning tests on specimens with different non-persistent joints under uniaxial compression. An experimental work on rock specimen with single cracks was reported by Jian et al. (2002a, 2002b) using in-situ CT scanning. An undrained triaxial consolidation test on full weathered granite specimens was reported by Shang et al. (2004). Other research work based on the concept of in-situ MCT scanning approach are listed but not limited to the following references: Hirono et al. (2003), Zhou et al. (2008),Wennberg et al. (2009), Cao et al. (2011), Sun et al. (2016).

The energy of the X-ray source using MCT ranges from 120 to 140 kV. Most of the rock specimens reported are the cylindrical specimens with the diameter 50 mm and the height from 50 to 100 mm. Under the condition of the aforementioned X-ray energy level and the specimen sizes, the obtained spatial resolutions are approximately 0.3 mm. The axial loading capacity and maximum confining stress are 400 kN and 20 MPa. Other tests, such as fluid flow and freeze thawing tests, were also reported. Some achievements were obtained using this type of in-situ MCT scanning as follows: a) different specimens (soil, coal, concrete) were tested and initiation, propagation and coalescence of the cracks in the specimens were observed; b) the relationship between the CT value and damage variable was also established to quantitatively characterize the evolution process. The following drawbacks exist for the in-situ MCT scanning: a) limited by the direct use of the MCT scanner, the geomechanics test apparatus is quite simple with single loading mode; b) the penetrability and resolution of the MCT scanners can hardly satisfy the requirements.

2.3 *In-situ CT scanning technique in geomechanical tests using the ICT*

Because the MCT scanner adopted a relatively low X-ray energy (usually several hundreds kV), it is impossible to perform in-situ scanning on large size specimens. Therefore, the ICT scanner was later used in the geomechanical tests. Different from the in-situ MCT scanning, for the in-situ ICT scanning

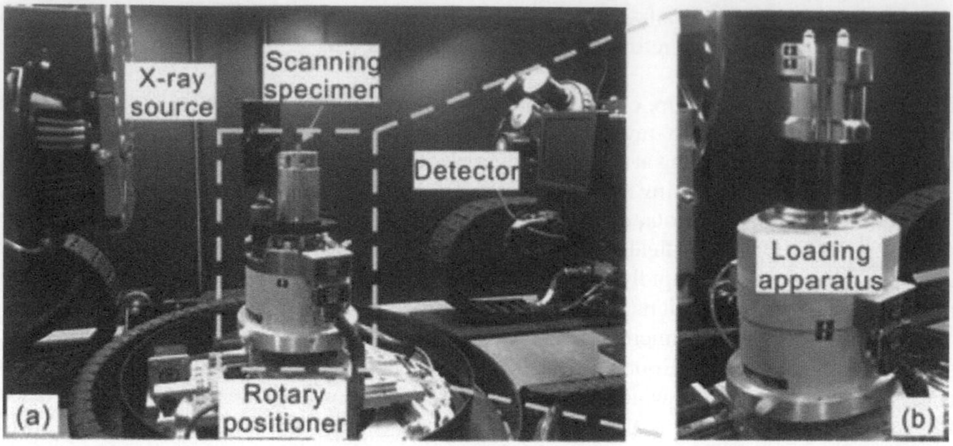

Figure 3. Schematics of in-situ ICT scanning technique: (a) X-ray CT scanner; (b) loading unit (modified after Duan et al. 2019).

the geomechanics test apparatus is mounted on a commercial positioner. When performing the test, the ICT scanner stays still whereas the loading apparatus and rock specimen are rotated together. Figure 3a provides a typical in-situ ICT scanner and the loading apparatus is shown in Figure 3b.

Otani et al. (2000, 2001 and 2002) designed a testing unit and this new apparatus was used in an ICT shield room to study the progressive failure with strain localization on clay/sand soil under uniaxial and triaxial compression. Watanabe et al. (2012) used the aforementioned unit to quantify the changes of the internal structure and deformation of the sandy soil specimens under the confining stress 50 kPa. Mukunoki et al. (2014) developed a bending testing apparatus to observe the inner cracking of clay with different water contents under punching or bending tests using the ICT scanner. Ju et al. (2017, 2018) incorporated the ICT (consisted of two separate scanners: GEIT 320 kV X-ray scanner and FXE 225 kV micro-focus X-ray scanner) with the servo-controlled triaxial loading technique to observe the interior fractures and permeability in coal samples at various loading stages under triaxial condition.

In order to obtain higher resolution of the CT images with time series, the lab-based micro-CT scanner was implemented in the in-situ CT scanning technique. Higo et al. (2006, 2011 and 2013) combined the triaxial compression apparatus with a lab-based micro-CT scanner to observe the internal structural evolution of sandstone with different saturation. Tisato et al. (2014 and 2015) invented an X-ray transparent high pressure chamber (a rotary shear apparatus) to work with the micro-CT scanner that available at Toronto University. Zhao et al. (2017) used this loading set-up inside the micro-CT scanner to directly observe the faulting of brittle rocks. Li et al. (2017) studied the progressive failure characteristics of Longmaxi shale under in-situ uniaxial tests using a loading device (DEBEN Micro test CT5000) integrated with the X-ray micro-CT (ZEISS Xradia 520) that are available at the Institute of Geology and Geophysics, Chinese Academy of Science, Beijing. Duan et al. (2018) proposed a damage value based on serial CT images to describe the internal structure evolution of shale specimen quantitatively. Subsequently, Wang et al. (2018), Yang et al. (2018), Zhou et al. (2018), Liu et al. (2018) and Duan et al. (2019) used this experimental setup to study the anisotropy and the cracking pattern of shale under uniaxial condition.

Other than the lab-based micro-CT scanner, the synchrotron-based micro-CT (e.g. European Synchrotron Radiation Facility, ESRF) was also used to perform in-situ ICT scanning tests. The X-ray beamline intensity is one trillion times of that used in the MCT. Viggiani et al. (2004) designed a loading apparatus including a small triaxial cell. Bésuelle (2006), Lenoir et al. (2007), Hall et al. (2010), Hasan et al. (2010), Viggiani et al. (2010) and Ando et al. (2011) operated this apparatus on the different energy beamlines at the ESRF in Grenoble to conduct the in-situ scanning to study the phenomena of strain localization in sandstone under triaxial compression. Some typical in-situ CT scanning setups using the ICT are summarized in the following Table 1.

Table 1. Some typical in-situ CT scanning setups using the ICT.

Research Institute	Testing condi-tion	Loading capacity	Sample Size	CT scanner	X-ray energy	Spatial resolution	References
Rock Engineering Laboratory at Kumamoto University, Japan	triaxial 400 KPa	1 kN	50×100 mm (cylinder, soil)	ICT	300/200/150 kV	0.2/0.3 mm	Otani et al., 2002; Watanabe et al., 2012
China University of Mining and Technology (Beijing)	triaxial 40 MPa	60 kN*	25×50 mm (cylinder, coal)	ICT	320/225 kV	150/50 μm	Ju et al., 2017, 2018
Laboratory 3S& ESRF, France	triaxial 1 MPa	7.5 kN	20×40mm (cylinder, soil)	synchrotron	50-70 keV	14 μm	Viggiani et al., 2004
School of Materials, University of Manchester, Manchester, UK	heating	N/A	3×3×7 mm^3 (clay)	synchrotron	20 keV	0.81 μm	Pilz et al., 2017
University of Texas, Austin University of Toronto, Canada	triaxial 30 MPa	not available	12×36mm (cylinder, rock)	Micro-CT	not available	25 μm	Tisato et al., 2014, 2015; Zhao, 2017
Kyoto University, Japan	triaxial 20 KPa	not available	φ 35×70 mm φ 50×100 mm (cylinder, sand)	Micro-CT	225 kV	5 μm	Higo et al., 2010, 2011,2013
Institute of Geology and Geophysics, Chinese Academy of Science, Beijing	uniaxial	5 kN	4×8 mm 5×10 mm (cylinder, shale)	Micro-CT	30-160 kV 90 kV	0.7 μm 11.27 μm	Li et al., 2017; Yang et al., 2018; Zhou et al., 2018

*The loading capacity is approximately calculated by the specimen size and maximum axial pressure provided in the reference.

2.4 *Digital image correlation method for in-situ CT imaging*

The further processing of the CT scanning images obtained from the geomechanical tests is very important in the in-situ CT scanning of geomechanical tests. With respect to the image processing for the CT scanning results, Digital Image Correlation (DIC) method is very promising. DIC is a mathematical method which essentially consists in recognizing the same material point on a pair of digital images of an object. A material point is assumed to be fully identified by its local pattern (e.g., the gray level distribution around the point in a black and white image). By optimizing an appropriate correlation function, DIC allows for determining for each point/pattern on the first image, the most likely location of such a point/pattern on the second image. By repeating this procedure for a number of points, a full displacement and deformation field can be obtained for the pair of images. Originally, DIC is a representative non-interferometric optical technique to provide full-field displacements by comparing the digital images of the specimen surface between the undeformed (or reference) and

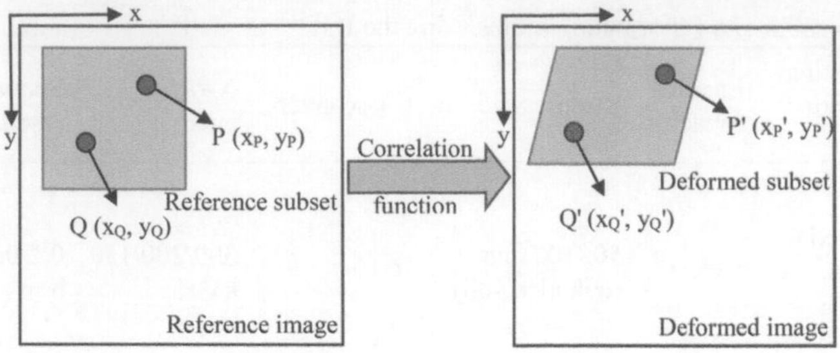

Figure 4. Schematic illustration of the DIC approach at 2-D level.

deformed states respectively (Pan et al. 2009), as shown in Figure 4. The application of DIC is to obtain the surface deformations of materials and structures subjected to various loadings. Because the surface of cylinder rock sample is of a curved surface and 3-D deformation occurs after loading, 3-D DIC was developed and the Digital Volume Correlation (DVC) method has been proposed (Pan et al., 2009). Through the comparison of couples of reconstructed 3D images of a specimen at two successive steps of loading, this allows to measure an incremental displacement field, from which a strain tensor field can be obtained. Beneficial from the in-situ CT scanning technique, DIC or DVC can be used to obtain the displacement and strain fields of a 2-D domain or a 3-D object at a certain loading point given a reference. From the processing results, the voxel rotations and translations can be detected, even the fractures beyond the scanning resolutions. DIC is often used to evaluate the onset and evolution of shear banding under the deviatoric loading on rock specimens.

2.5 *Limitations of current in-situ CT scanning technique*

Although the in-situ CT scanning technique has been applied in various aspects in geomechanical testing work and explained the failure process of some geomaterials under different loading conditions, the following problems exist in the aforementioned in-situ CT scanning facilities.

(a) The prevailing in-situ CT scanning tests are achieved through placing the triaxial compression apparatus on a commercial positioner, as shown in Figure 5a. The applied load on a rock specimen usually reacts on the triaxial cell wall. Hence the triaxial cell will suffer the tension force in the axial direction as well as the confining pressure. Generally, the thickness of the cell wall is limited to avoid excessive absorption of the X-ray. Therefore, when the load approaching the peak value of the rock specimen, the elastic strain energy stored in the cell wall will be suddenly released and cause a catastrophic failure occurring to the specimen. The post-peak behaviors of rock specimens cannot be obtained under this condition. The following Figure 6 provides a typical in-situ CT scanning test result (Li et al. 2017, Duan et al. 2018, 2019). The damage variable at different stress points were calculated from the CT scanning data. From Figure 6, the rock specimen experienced a sharp stress decrease shortly after the peak stress. Between the point 4 near the peak strength and the point 5 completely broken, no testing data was obtained because of the lost control of the loading unit. Therefore, this type of triaxial compressive apparatus with in-situ CT scanning still belongs to a soft test apparatus and cannot satisfy the stiffness requirement of capturing the post-peak behaviors.

(b) The limited cell wall thickness as mentioned also imposes restrictions on the axial loading capacity and maximum confining pressure of the triaxial cell. Otherwise, the significant tensile deformation will occur to the triaxial cell. For the destructive geomechanical tests, the limited loading capacity is not sufficient to break the rock specimen with large size, especially hard rock specimen. Therefore, the specimen size is generally limited to the cylindrical specimen with the diameter 25 mm for the prevailing in-situ CT scanning setups. However, for some newly emerging tests, e.g. hydraulic fracturing test, the diameter of the cylindrical specimen should be at least 50 – 100 mm, which is hardly satisfied by the triaxial compressive apparatus as shown in Figure 5a.

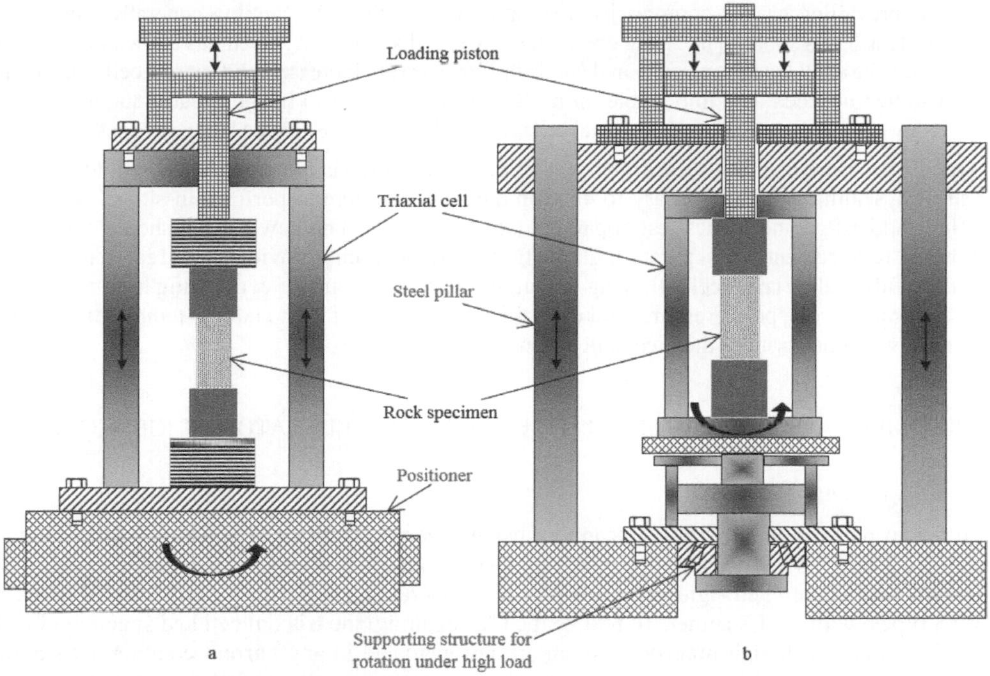

Figure 5. Schematic illustration of different reaction design (a) the current triaxial compressive apparatus with in-situ CT scanning (b) the newly designed triaxial compressive apparatus with in-situ CT scanning.

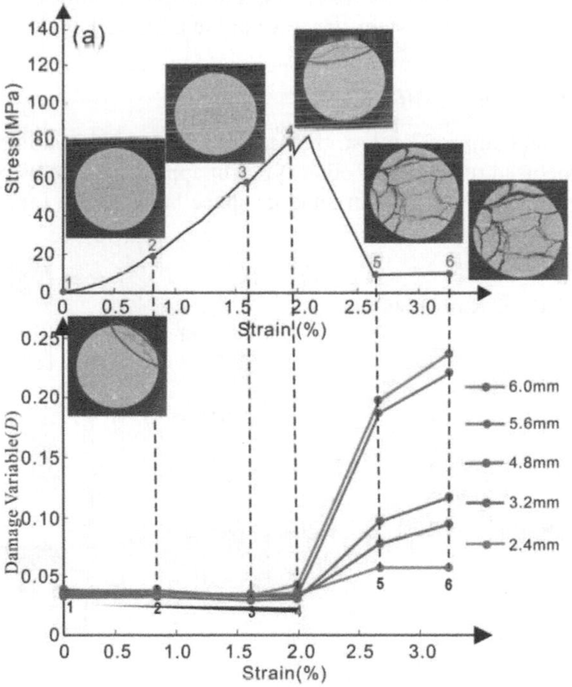

Figure 6. CT slices and damage evolution of a shale specimen at different stress points (from Duan et al. 2018).

(c) The prevailing triaxial compressive apparatus with in-situ CT scanning generally employ the MCT or ICT scanner. The X-ray energy level is less than 600 kV, which has a low penetrability. When the axial loading capacity and confining pressure are increased, the triaxial cell wall should also be thickened. It is impossible for a MCT or ICT scanner to image because the low energy X-ray can hardly penetrate the triaxial cell with thick wall as well as the large size rock specimen.

Due to the aforementioned the drawbacks of the prevailing triaxial compressive apparatus with in-situ CT scanning, it is necessary to develop a new test system to perform in-situ CT scanning on the standard geomechanics test apparatus (see Figure 5b). The new system should satisfy the following requirements: a) separate high stiffness loading frame providing the reaction force; b) standard full-scale triaxial cell allowing the large size rock specimen; c) applying high axial loading, confining stress, pore pressure, and even the environment of high and low temperature; d) CT scanning while performing the mechanical tests.

3 GEOMECHANICS TEST SYSTEM WITH LINEAR ACCELERATOR CT (GEOACT)

3.1 *Components of GEOACT*

In order to perform the in-situ CT scanning using a geomechanics test apparatus with standard loading capacity on a full scale specimen, our design concept is whether it is possible to equip the standard servo-controlled rigid geomechanics test apparatus (similar to the MTS, INSTRON, or GCTS types) with a CT scanner. To realize the CT scanning, the triaxial cell and specimen inside must be rotated with high precision, which are not available in any current commercial geomechanics test apparatus. Moreover, to penetrate the metal triaxial cell and full size specimen inside, it is necessary to use a LINAC CT with powerful penetrability, which a regular commercial ICT scanner does not possesses. Based on this idea, we developed the geomechanics test system with linear accelerator CT (GEOACT). The GEOACT is mainly composed of three parts: a LINAC X-ray source with high energy, high sensitivity detectors, and a geomechanics test apparatus with high precision rotating triaxial cell. The system structure of the GEOACT is shown in Figure 7.

3.2 *Operation principle of the LINAC CT*

During a standard triaxial compression test, the rock specimen is placed in a sealed metal triaxial chamber, just as the one used in a MTS or other types of apparatus. The triaxial cell is made of 30 mm thick high strength metal. The 70 mm annulus space between the inner wall and cylindrical

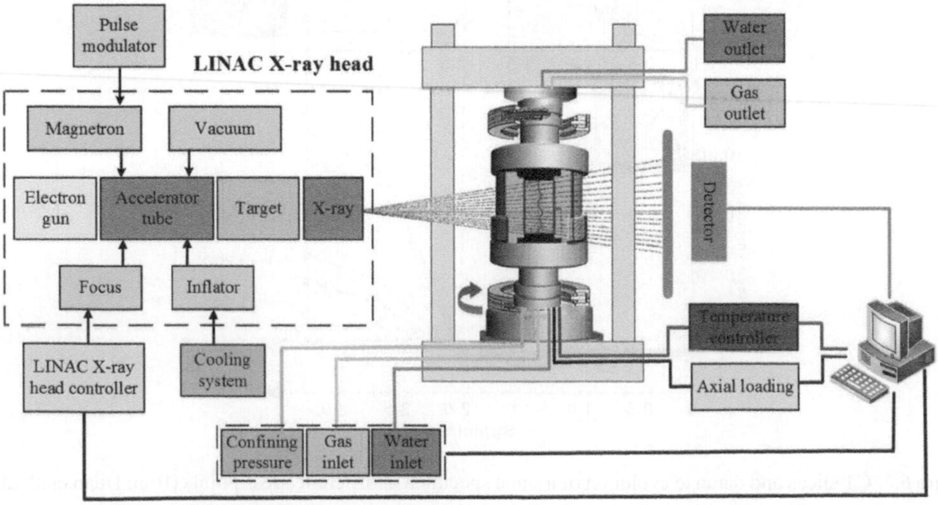

Figure 7. Structure of the geomechanics test system with linear accelerator CT (GEOACT).

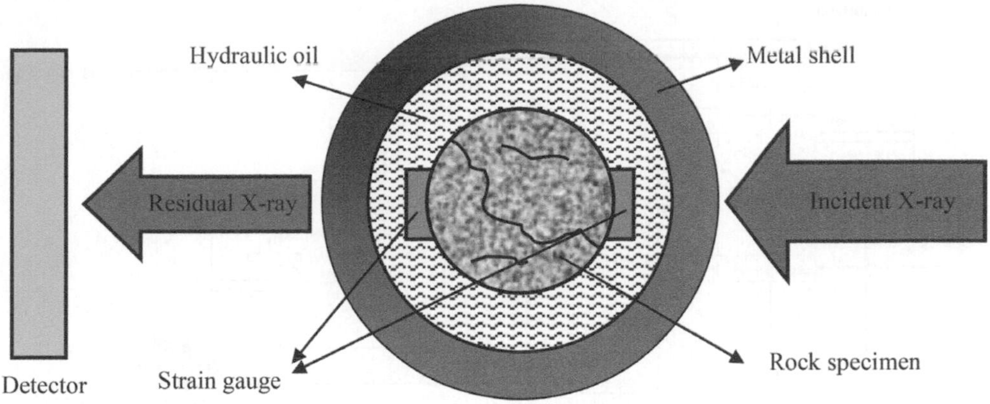

Figure 8. Schematic of the scanning objects.

specimen with the diameter 100 mm provides enough room for the axial and circumferential strain gauges. It is also filled with the hydraulic oil when performing the triaxial compression test, as shown in the following Figure 8. The X-ray must have enough power to penetrate the metal shell, hydraulic oil, and the cylindrical rock specimen. The total penetration thickness reaches 300 mm. This is far beyond the penetration capacity of the MCT or regular ICT scanners. To achieve a optimum balance between the penetrability and image resolution, a 6 MeV LINAC CT scanner was adopted after detailed calculation and design.

The X-ray generation principles for both the regular ICT and LINAC scanner are the same, i.e. bremsstrahlung or braking radiation induced by the high-velocity electrons hitting the target metal. The X-ray generated by the regular ICT scanner generally has a low energy level (less than 600 kV). Whereas, the LINAC has the capability of generating the X-ray at the energy level at least 1 MeV (equivalent to 1000 kV). The X-ray energy differences of the two types of X-ray generators are because of the different accelerating mechanisms. The regular ICT scanner adopts the electrostatic field accelerating, whereas the LINAC CT scanner uses microwave accelerating. The core components of the regular ICT and LINAC CT scanner are the X-ray tube and accelerator tube, respectively.

The LINAC basically accelerates the charged particles through the electromagnetic field. According to different microwaves, the linear accelerators are categorized as the traveling microwave and stationary microwave accelerators. The GEOACT adopted the stationary microwave accelerator design because of its compact structure. It has the following four components: LINAC X-ray head, modulator, water cooling unit, and controlling unit (as shown in the following Figure 9). The LINAC X-ray head used to generate the high energy X-ray is the core component of the stationary microwave LINAC, which is constituted of the following parts: side coupling stationary microwave accelerator tube, magnetron, waveguide, SF6 inflation, vacion vacuum, electron gun, target, shield, etc. The modulator or pulse modulator is the power supply of the magnetron, electron gun, and vacion pump in the LINAC. The water cooling unit is the auxiliary component, which provide the LINAC with the thermostatic cooling water. The controlling unit monitors the operation status of the accelerator and guarantees the safety and proper functioning of the LINAC.

The operation principle of the LINAC will be briefly described in the following: a) the pulse modulator transforms the main electric supply into the high voltage pulse; b) the high voltage pulse supplies power to the magnetron, so that the magnetron generates the radio-frequency (RF) microwave with a certain frequency; c) the RF microwave is led into the stationary microwave accelerator tube; d) the stationary microwave is obtained through the superposition of the forward and backward microwaves in the accelerator tube; e) the pulse modulator also supplies the electron gun with the high voltage pulse power; f) the electron gun ejects the electrons into the accelerator tube; g) the electrons are accelerated through absorbing energy from the stationary microwave until reaching the preset energy level; h) the accelerated electrons hit the target and generate the high energy X-ray; i) the shield and collimator blocks the X-ray radiating towards other directions and directs the X-ray towards the objects under detection.

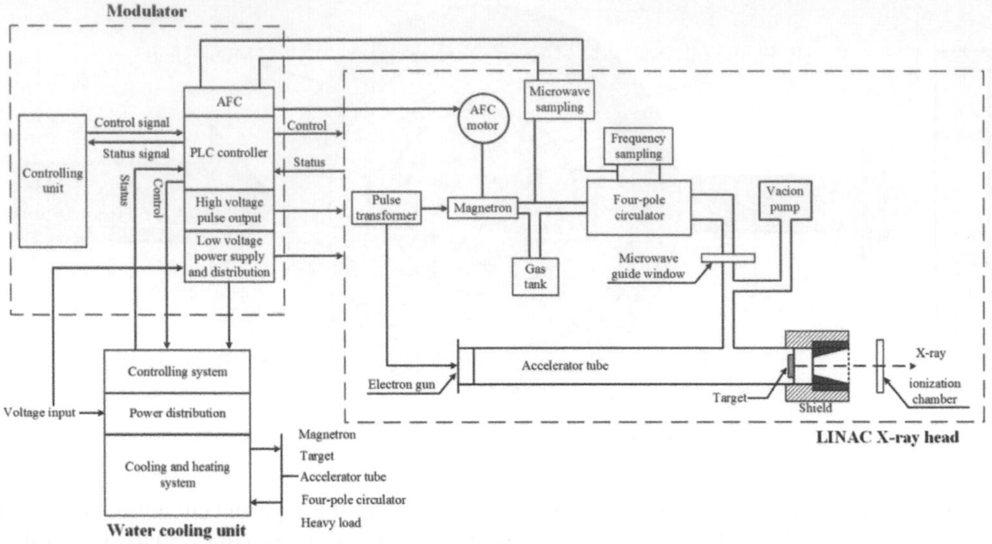

Figure 9. Structure of the stationary microwave LINAC.

3.3 *Geomechanics test apparatus with high precision rotating triaxial cell*

Benefit from the high penetrability of the high energy X-ray, a new geomechanics test apparatus has been invented. Other than the previous low stiffness apparatus with the reaction force applied on the triaxial cell wall, the reaction frame of the geomechanics test apparatus is composed of four stainless steel pillars with the diameter 300 mm. The loading frame not only can bear the axial force greater than 2000 kN, but also increase the system stiffness to 5.6 GN/m. The closed-loop servo-controlled technique is also incorporated to form a true servo-controlled rigid geomechanics test apparatus. Therefore, the complete stress-strain curves including the post-peak segment can be properly recorded. The triaxial chamber is made of high-strength metal material with the wall thickness 30 mm, which barely deforms bearing the confining pressure greater than 50 MPa. The triaxial chamber with the inner diameter 240 mm is capable of providing ample space for a cylindrical rock specimen (diameter 100 mm and height 200 mm) with the axial and circumferential strain gauges mounted (see Figure 11). Therefore, the axial and radial deformation of the rock specimen can be measured by the strain gauges through the tests. In the previous in-situ CT scanning of the geomechanical tests, the strain gauge cannot be mounted due to the narrow space inside the triaxial cell. The internal structure change in the rock sample during the complete loading process can be imaged by the CT scanning, which could reveal the micro-mechanism of the deformation and fracturing of the rock specimen.

In order to achieve a proper CT imaging, the CT scanner or the object to be scanned must be capable of rotating with high precision. For a MCT scanner, the rotatory part is the scanner whereas the object stays still. Because the LINAC CT scanner and detector occupy a considerable space, it is very difficult to realize a high precision rotation of the scanner itself. Moreover, the four steel pillars will block the X-ray if the scanner rotating design is adopted. Therefore, the LINAC CT scanner is fixed whereas the object is rotated in GEOACT. However, the challenge here is that a positioner cannot be used to rotate the whole test machine, just as the aforementioned design in Figure 5a. If so, the high precision positioner available cannot bear the test machine with the weight 40 tons and the steel pillars will block the X-ray as mentioned before. To solve this problem, a sophisticated rotating scheme is designed to rotate the triaxial cell and rock specimen inside only and keep the pillars still. The rotation accuracy of the triaxial cell is 0.003° while the undesired axial and radial deviation is controlled within 25 μm, which fully satisfy the requirement of CT imaging.

The GEOACT has three independent closed-loop servo-controlled systems to apply the axial load, confining pressure, and pore pressure simultaneously. The test machine and CT scanner are controlled by a unified software system.

Because the X-ray with high penetrability is harmful to any living thing exposed, the whole system is placed in a room enclosed by X-ray shielding structures. The operation center should also be located at a certain distance from the test room.

3.4 *Main specifics*

According to the aforementioned design, we invented the new geomechanics test system with linear accelerator CT (GEOACT, as shown in Figure 10). The GEOACT has the following characteristics and innovations:

(a) A rotation scheme is specially designed to achieve the high-precision rotation of the triaxial cell and rock specimen under 2000 kN axial force. The rotation accuracy, axial and radial deviation under testing can fully satisfy the requirement of CT imaging;
(b) It is the first time to manufacture a fully-functioning geomechanics test machine equipped with the LINAC CT scanner. Instead of the previous design of the reaction force applied on the triaxial cell wall, the vertical reaction frame is composed of four steel pillars. The stiffness of the test machine is drastically increased, so the post-peak behavior of rock specimens can now be obtained satisfactorily;
(c) A highly stable beam with high energy and small focal spot is obtained for the 6 MeV LINAC CT scanner, enabling the high-quality imaging of large-sized samples.

The main specifics of the GEOACT are listed in the following Table 2.

Figure 10. Panorama of the geomechanics test system with linear accelerator CT (GEOACT).

Table 2. Specifics of the GEOACT.

Parameter	Value
Maximum axial force	2000 kN
Maximum confining pressure	50 MPa
Maximum seepage pressure	50 MPa
Maximum gas pressure	50 MPa
Temperature range	-40-200°C
Maximum displacement of the piston	300 mm
Linear accelerator energy	6 MeV
X-ray dose rate	1000 cGy/min@1 m
Ability of the fracture detection	0.05 mm
Density resolution	1%
Density resolution	1%

4 PRELIMINARY EXPERIMENTAL RESULTS

The GEOACT is capable of performing in-situ CT imaging of uniaxial and triaxial compression tests or hydraulic fracturing tests on rock specimens with the diameter 50-100 mm and height 100-200 mm as well as post-mortem CT scanning of large size rock specimens (up to 0.6 m^3 cubic rock blocks), and also many other tests. In this section, some preliminary experimental results obtained using the GEOACT will be provided.

4.1 *Fracture aperture detection and imaging resolution test*

In order to show the fracture imaging quality and resolution of the GEOACT, a cylindrical granite specimen containing four artificial fractures with different apertures (0.5 mm, 0.4 mm, 0.1 mm, 0.05 mm) inside the triaxial cell was scanned. From Figure 11a, the GEOACT can detect the ideal fracture with a uniform aperture 0.05 mm in a cylindrical granite specimen with the diameter 100 mm. According to the section 2.2 and 2.3, the GEOACT has distinct superiority over the in-situ CT scanning setups using MCT or ICT scanners: a) when the specimen sizes are comparable, the GEOACT is capable of achieving a higher imaging resolution; b) the GEOACT can obtain the CT images of comparable resolution on the specimens with much larger size. Despite of the higher resolution with the in-situ CT scanning setups using the lab-based micro-CT scanners or the synchrotron-based micro-CT scanners, the sample sizes are not even comparable to the sample shown in Figure 11a. A cylindrical rock specimen mounted with the axial and circumferential strain gauges was also scanned. According to Figure 11b and c, the axial and circumferential strain gauges are perfectly visible from the CT images.

4.2 *Uniaxial compressive test in-situ tomography on the shale specimen*

A shale specimen with the diameter 50 mm and height 120 mm under uniaxial compression condition was scanned in-situ at some stress points in the complete stress-strain curve. The axial load was applied at the rate of 0.01 mm/min by the servo-control pattern of radial strain. The complete stress-strain curve and the CT images corresponding to these selected stress points are shown in Figure 12.

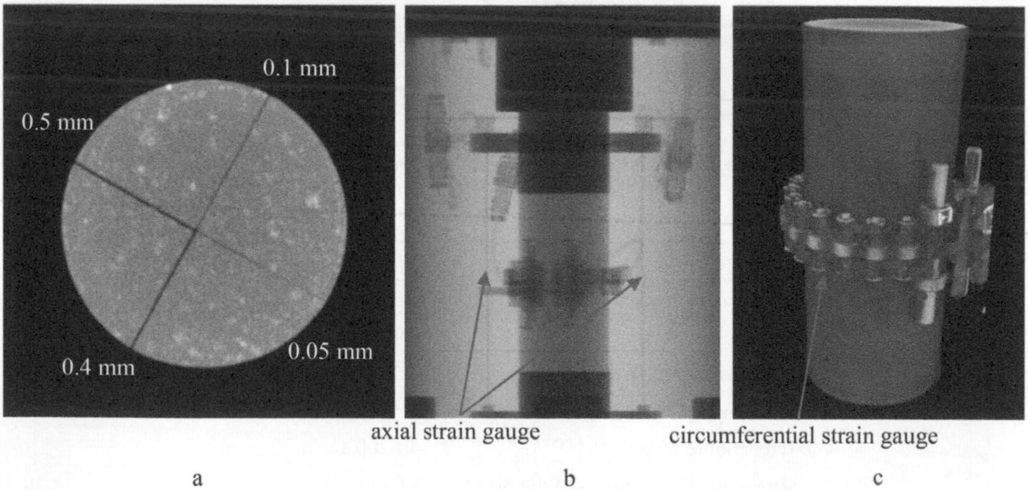

Figure 11. Different objects scanned by the GEOACT (a) a cylindrical granite specimen with the diameter 100 mm containing four artificial fractures with different apertures (b) axial strain gauge mounted on a cylindrical rock specimen (c) circumferential strain gauge mounted on a cylindrical rock specimen.

Comparing with Figure 6, the complete stress-strain curve can be obtained using the GEOACT as shown in Figure 12. Because the servo-controlled technique with the high frequency response is incorporated into the high stiffness loading frame, the stress points (4, 5, and 6 in Figure 12) in the post-peak region can be fixed as the scanning points. The fracture propagation can now be observed through the CT images at different loading steps in the whole deformation and failure process (including the post-peak behaviors), which is achieved for the first time.

4.3 *Triaxial compressive test in-situ tomography on the granite specimen*

In order to test the performance of the GEOACT under triaxial compression conditions, a cylindrical granite specimen with the diameter 50 mm and height 120 mm was subjected to the in-situ CT scanning test with the confining pressure 10 MPa. The axial load was applied at the rate 0.06 mm/min by the servo-control pattern of radial strain. The granite specimen yielded a peak strength 105.3 MPa. The following Figure 13 provides the complete stress-strain curve of the granite specimen under triaxial compression condition. One stress point in the pre-peak region and three stress points in the post-peak region were selected to scan the crack evolution in the granite specimen as shown in Figure 13.

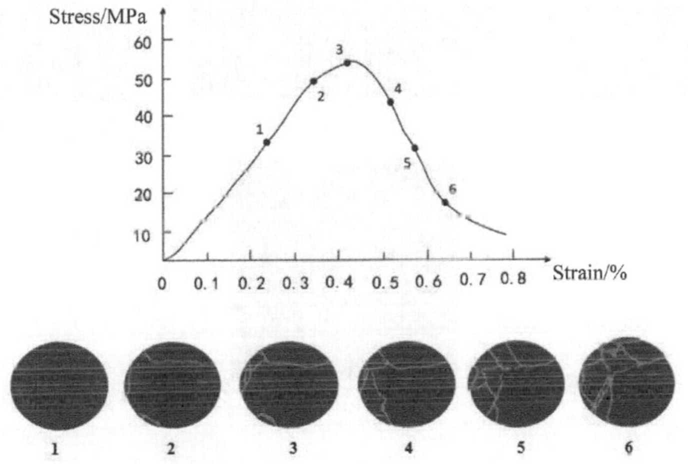

Figure 12. The complete stress strain curve of the shale specimen under uniaxial compression and CT images corresponding to some selected stress points.

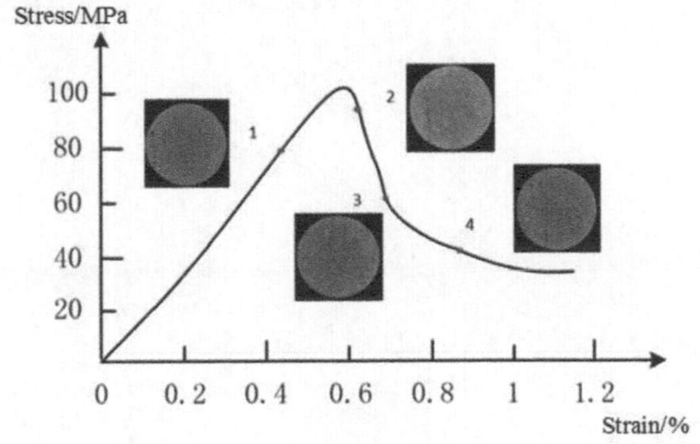

Figure 13. The complete stress-strain curve of the granite specimen under triaxial compression condition and CT images corresponding to some selected stress points.

The CT slices of the granite specimen at point 4 in Figure 13 were used to reconstruct the 3-D model. In the 3-D model, any desired cross section can be obtained through placing the corresponding cutting plane (as shown in Figure 14). The 3-D fracture network profile in the granite specimen can be also obtained (see Figure 15). On the basis of the 3-D model and fracture network profile, we can perform statistic analyses based on the reconstructed granite specimen model, e.g. the fracture aperture in the 3-D granite specimen model at any scanning points. Figure 16 shows the fracture aperture statistic characteristic of the scanning point 4 (see Figure 13) for the granite specimen. According to the fracture aperture statistic distribution, the mean value, maximum value, and standard deviation of the fracture aperture of the failed granite specimen are 1.43 mm, 0.74 mm, and 0.2mm, respectively. Statistical analyses can be also performed with respect to the orientation, connectivity, and other properties of cracks.

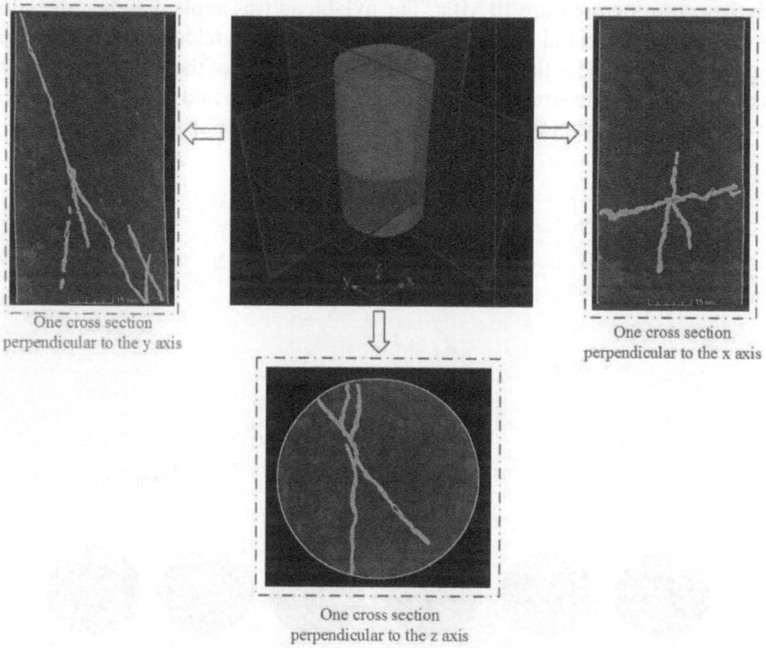

Figure 14. 3-D reconstruction of the granite specimen in the post-peak region and the cross sections perpendicular to each axis.

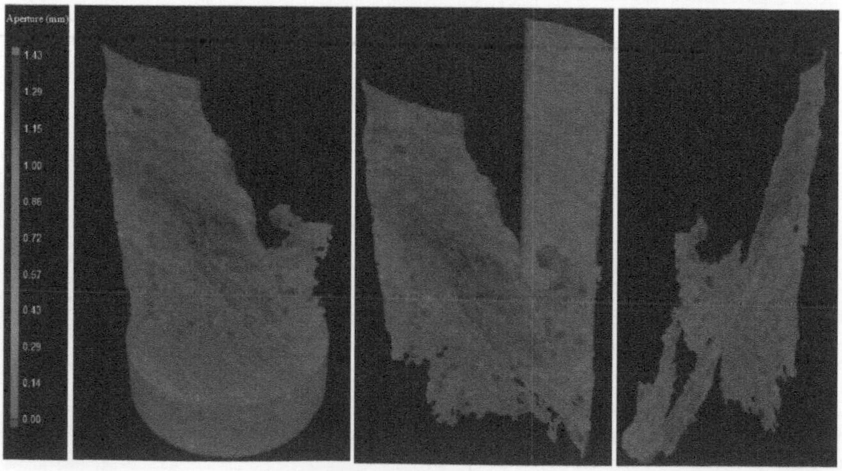

Figure 15. 3-D fracture network reconstruction in the granite specimen at the scanning point 4 (see Figure 13).

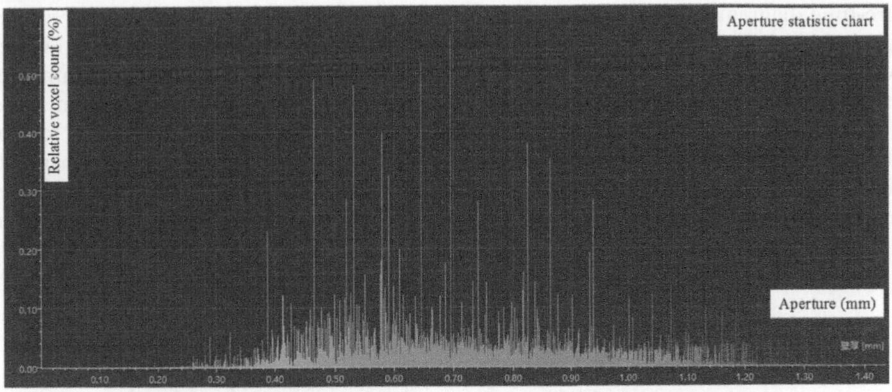

Figure 16. Fracture aperture statistic chart of the granite specimen at the scanning point 4 (see Figure 13).

4.4 *Hydraulic fracturing test in-situ tomography on the shale specimen*

The linear accelerator CT scanner is capable of penetrating rock specimen with large size. Therefore, the GEOACT can be used to perform the complex experiments which require large size specimens, e.g. hydraulic fracturing test. The specimen used in this test is a cylindrical shale specimen with the diameter 100 mm and height 200 mm. The natural density is 2.66 g/mm^3. The testing scheme is listed in the following Table 3.

The hydraulic fracturing test is shown in the schematic (see Figure 17). The confining pressure and axial load were applied until reaching the predeterministic values, then maintained the same through the whole testing process. Super critical CO2 was selected as the fracturing fluid in this test. Certainly, other types of fracturing fluid, such as water, can be also used in the tests. Through a hole of diameter 3 mm at the bottom of the specimen, the fluid pressure was injected until the specimen failed. The fluid

Table 3 Testing scheme of the hydraulic fracturing test of shale specimen

Confining pressure (MPa)	Axial pressure (MPa)	Fracturing fluid	Injection rate (ml/min)	Breakdown pressure (MPa)
30	35	Super critical CO_2	40	42.42

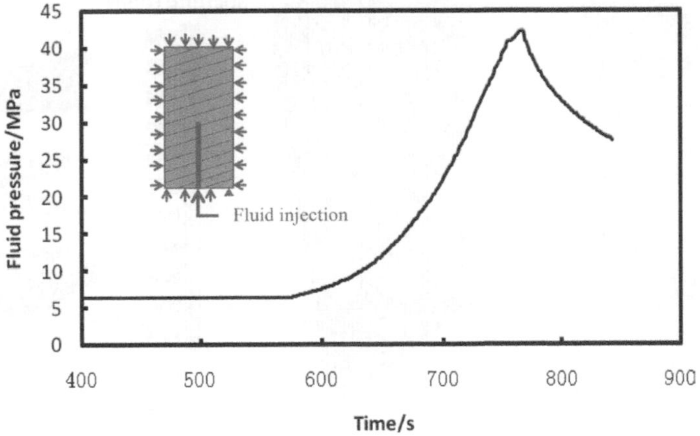

Figure 17. Fluid pressure-time diagram of the hydraulic fracturing test on the shale.

pressure-time diagram is shown Figure 17. The shale specimen yielded a breakdown pressure 42.42 MPa as listed in Table 3. The photographs of the failed shale specimen are provided in Figure 18. A sub-horizontal persistent fracture was induced by the super critical CO_2 fracturing in the middle of the shale specimen near the drill hole side. The 3-D reconstruction of the failed shale specimen is shown in Figure 19a. A vertical cutting plane parallel to the axis was placed in the 3-D model to show one half of the shale specimen in Figure 19b. According to Figure 19a and b, some fractures parallel to the shale bedding planes could be imaged by GEOACT, which are hardly detected by the naked eyes. The fracture aperture statistic analysis was also performed for the shale specimen as shown in Figure 19c.

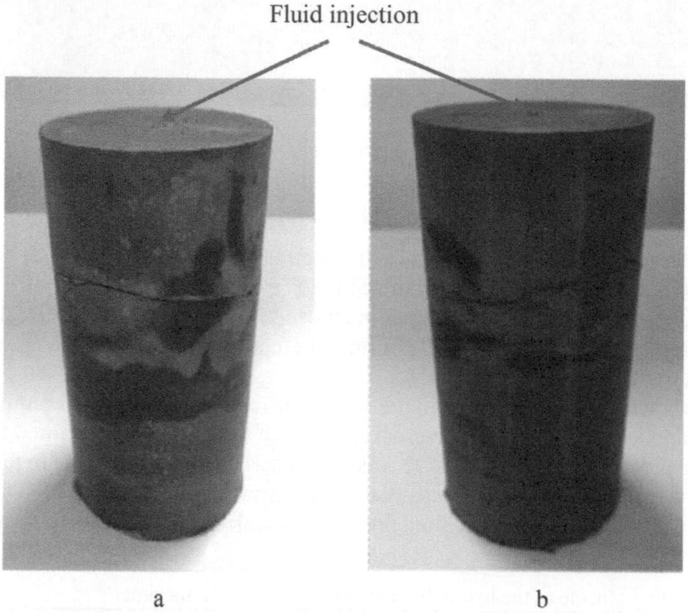

Figure 18. Photographs of the failed shale specimen in the hydraulic fracturing test a) front b) back.

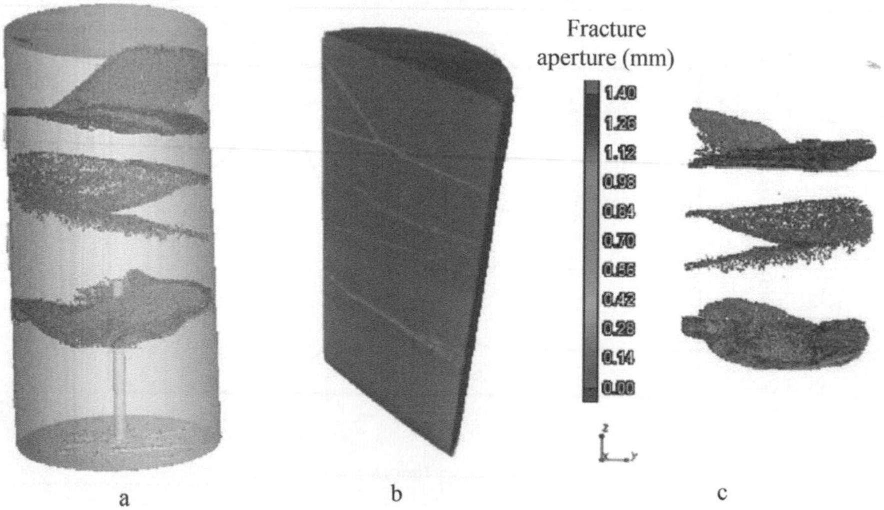

Figure 19. Image analyses of the failed shale specimen (a) 3-D fracture network reconstruction (b) fracture traces on the cutting plane parallel to the axis (c) fracture aperture contour of the failed shale.

4.5 *In-situ CT test and DIC application on the soil and rock mixture specimen*

The mixture of soil and rock, so called the soil and rock mixture (SRM) or gravelly soil, is frequently encountered in the slope engineering. The mechanical behaviors of SRM, especially onset and evolution of the internal fracture under loading, play an important role in the slope stability calculations. Benefit from the significant density difference between the soil and rock materials, CT scanning technique is especially useful to capture the internal gravel structure in the SRM specimen and examine the influence of the internal structure on its mechanical behaviors. A cylindrical SRM specimen with the diameter 50 mm and height 100 mm was scanned and tested by the GEOACT. The complete stress-strain curve and CT images at some selected stress levels are shown in Figure 20.

The fractures within the CT scanning resolutions can be directly imaged as shown in Figure 20. However, some tiny fractures beyond the CT scanning resolutions cannot be imaged. The DIC method can be used to obtain the tiny fractures beyond the CT scanning resolutions from the strain field in the specimens. The fracturing process of the geomaterials is accompanied by the variation of strain field. If the variation of strain filed can be obtained, the prediction of the geomaterial failure will be possible. To obtain the strain field, the displacement field should be obtained first. In this test, the DIC method is used to calculate the displacement field from the CT images.

The CT images of the original SRM specimen corresponding to the stress point O in Figure 20 was selected as the reference images. The total number of pixels in each CT slice is 1024 × 1024. The size of each pixel is 0.129 mm. Generally, the images contain millions of pixels. Therefore, it is impossible to correlate the reference and deformed images by running the matching algorithm for each individual pixel. To increase the practicability of the matching algorithm, a moving square (50 × 50 pixels) having the pixel to be matched in the center was used to correlate the reference and deformed images. The correlation function used in this research is the sum squared difference (SSD) expressed in Equation (1). When a particular square in the deformed image yields a minimum SSD value with respect to the reference square, the two center pixels are match successfully.

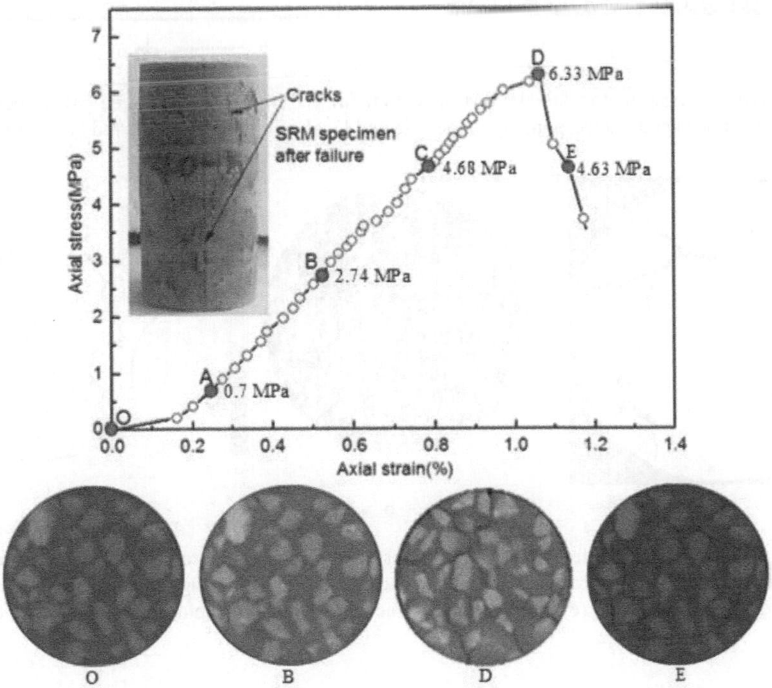

Figure 20. The complete stress strain curve of the SRM specimen and CT images corresponding to some selected stress points.

$$C_{SSD} = \sum_{x=-M}^{M} \sum_{y=-M}^{M} \left[f(x,y) - g(x',y') \right]^2 \quad (1)$$

The displacement field in the SRM specimen was successfully achieved using the SSD based DIC approach. Figure 21 provides the calculation result using the DIC approach. A local area defined by the red box is selected in the CT image of failed SRM specimen (corresponding to stress point E in Figure 20) as shown in Figure 21a. Although no visible deformation can be observed in the region from the CT images, the displacement vector diagram obtained using the DIC approach in Figure 21b displays evident deformation occurring in the selected area. Further than that, the fractures were extracted based on the displacement field. We can make the comparison between the fractures extracted directly from the CT image in Figure 22a and that obtained by the DIC method in Figure 22c. Through this technique, some tiny cracks can be detected in the center area, which cannot be captured by CT image alone due to the resolution limitation. Therefore, combining the CT scanning and DIC approach may be a useful tool to detect the tiny cracks beyond the scanning resolution, which is important to analyze the geomaterial fracturing process.

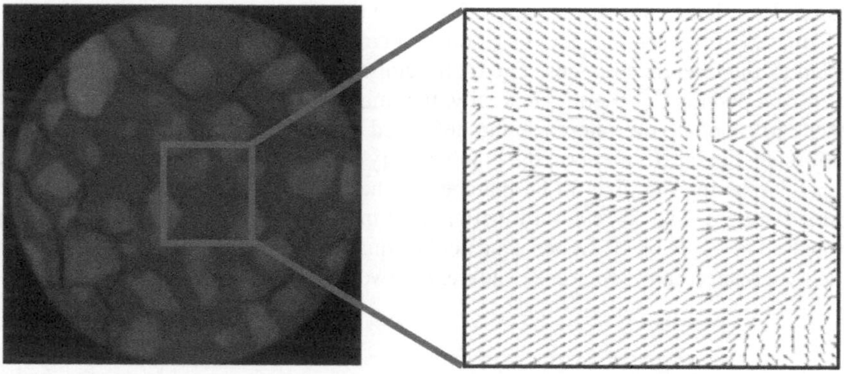

a b

Figure 21. DIC analysis of the failed SRM specimen (a) CT image of the failed SRM specimen (b) the displacement vector diagram corresponding to the selected local area defined by the red box in the CT image.

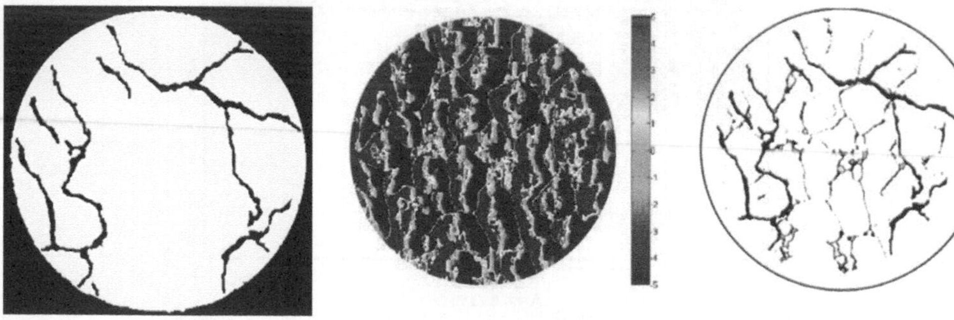

a b c

Figure 22. Comparison of the fractures in the failed SRM specimen extracted from the CT image directly and using the DIC approach (a) fractures directly extracted from the CT image (b) displacement contour of the SRM specimen along x direction (c) tiny fractures extracted through the displacement filed obtained by the DIC approach.

5 CONCLUSIONS

The full scale geomechanical tests with in-situ CT scanning are very useful to deep understanding the micro-mechanisms of mechanical behaviors of the geomaterials. This paper reviewed the application of the in-situ CT scanning technique in the geomechanical testing work firstly. Then a newly developed geomechanics test system with linear accelerator CT (GEOACT) was introduced. Some preliminary tests on geomaterial samples were conducted using GEOACT. The experimental results were shown that GEOACT has the following features:

(a) The developed linear particle accelerator (LINAC) CT scanner has fairly strong penetrability and high resolutions. It is capable of penetrating the metal triaxial cell with the outer diameter 300 mm and wall thickness 30 mm as well as the cylindrical rock specimens with the diamter 100 mm to image. The fracture aperture as low as 50 μm in a rock specimen can be detected;
(b) The triaxial cell with the function of high precision rotation is capable of performing geomechanical tests on a full scale rock specimen with the diameter 50 - 100 mm and height 100 -200 mm;
(c) The test apparatus has sufficient axial loading capacity (2000 kN) and high machine stiffness (5.6 GN/m). Three separate high frequency response closed-loop servo-controlled units for axial loading, confining pressure, and pore pressure are employed in the system. The complete stress-strain curve on rock sample can be obtained using the GEOACT;
(d) It is capable of capturing the formation and propagation of the interior fractures as well as the evolution of displacement field in rock samples;
(e) It is capable of performing the standard geomechanical tests, such as uniaxial compression, triaxial compression, pore pressure, hydraulic fracturing, and Brazilian tests, etc. By appropriate loading apparatus, some physical simulation tests with small size may also be conducted using the LINAC CT scanner with high penetrability.

The potential practical applications of the GEOACT could be as follows: a) new energy exploitation, such as shale gas, coalbed methane, enhanced geothermal resource, and natural gas hydrate; b) new geological engineering, such as underground storage of oil and gas, geological disposal of highly radioactive nuclear waste and underground CO_2 storage; c) important engineering constructions of mining engineering, hydraulic engineering, transportation engineering, civil engineering, and environmental protection.

The GEOACT introduced in this paper is the first prototype recently manufactured. Some technical performances are still under improvement. For example, how to reduce the scanning time duration as much as possible, how to improve the imaging resolution further while maintaining the high penetrability of the X-ray beam, how to develop advanced processing approaches (e.g. DIC).

ACKNOWLEDGEMENTS

This study is funded by the Chinese National Major Research and Development Program of Scientific Instruments (Contract No. 41227901), the Strategic Priority Research Program of the Chinese Academy of Sciences (Contract Nos. XDB10030301 and XDB10030304). We thank Zheng Bo, Wu Yan-fang, Duan Yong-ting, and Mao Tian-qiao for their contributions during this work.

REFERENCES

Ando, E., Hall, S.A., Viggiani, G., Desrues, J., Besuelle, P. 2011. Grain-scale experimental investigation of localized deformation in sand: a discrete particle tracking approach. *Acta Geotechnica* 7 (1): 1–13.

Appoloni, C.R., Fernandes, C.P. & Rodrigues, C.R.O. 2007. X-ray microtomography study of a sandstone reservoir rock. *Nuclear Instruments and Methods in Physics Research* 580(A): 629–632.

Arzilli, F., Cilona, A., Mancini, L. & Tondi, E. 2016. Using synchrotron X-ray microtomography to characterize the pore network of reservoir rocks: A case study on carbonates. *Advances in Water Resources* 95: 254–263.

Besuelle, P. 2006.X-ray Micro CT for Studying Strain Localization in Clay Rocks under Triaxial Compression. *In Advances in X-Ray Tomography for Geomaterials, 2006.*

Cao, G.Z., Qiang, Y. & Li, F.2011. Real-Time Observations of Fracturing Processes of Brittle Rock in Compression by X-Ray Computed Tomography. *Advanced Materials Research* 361–363: 171–178.

Chen, Y.S., Li, N., Li, A.G., Pu, Y.B. & Liao, Q.R. 2000. Analysis on Meso-Damage Process of Non-Interpenetraed Jointed Media by Using CT. *Chinese Journal of Rock Mechanics and Engineering* 19(6): 702–706. (In Chinese)

Chen, Y.S., Li, N., Han, X., Pu, Y.B. & Liao, Q.R.2005. Research on Crack Developing Process In Non-Interpenetrated Crack Media by Using CT. *Chinese Journal of Rock Mechanics and Engineering* 24(15):2665–2671. (In Chinese)

Cnudde, V. & Boone, M.N. 2013. High-resolution X-ray computed tomography in geosciences: A review of the current technology and applications. *Earth-Sci. Rev.* 123: 1–17.

Desrues, J., Chambon, R., Mokni, M. & Mazerolle, F. 1996. Void ratio evolution inside shear bands in triaxial sand specimens studied by computed tomography. *Geotechnique* 46 (3): 529–546.

Duan, Y.T., Li, X., He, J.M., Li, S.D. & Zhou, R.Q. 2018. Quantitative analysis of meso-damage evolution for shale under in situ uniaxial compression conditions. *Environ. Earth Sci.* 77(4):154.

Duan, Y.T., Li, X., Zheng, B., He, J.M. & Hao J. 2019. Cracking Evolution and Failure Characteristics of Longmaxi Shale Under Uniaxial Compression Using Real-Time Computed Tomography Scanning. *Rock Mech. Rock Eng.* DOI 10.1007/s00603–019–01765–0.

Ge, X.R., Ren, J.X., Pu, Y.B., Ma, W. & Zhu, Y.L.1999. Real-in-time CT triaxial testing study of meso-damage evolution law of coal. *Chinese Journal of Rock Mechanics and Engineering* 18(5): 497–502. (In Chinese).

Ge, X.R. Ren, J.X., & Pu, Y.B.2002. Real-time CT test of damage failure process of jointed rock sample in unloading confining pressure. *Rock and Soil Mechanics* 23(5): 575–578. (In Chinese)

Hall, S.A.; Bornert, M.; Desrues, J.; Pannier, Y.; Lenoir, N.; Viggiani, G.& Bésuelle, P. 2010. Discrete and continuum analysis of localized deformation in sand using X-ray μCT and volumetric digital image correlation. *Géotechnique* 60: 315–322.

Halverson, C. 2008. Characterization of geomaterials with X-ray computed tomography (X-ray CT). *Retrospective Theses and Dissertations*. Paper 15309.

Hasan, A. & Alshibli, K.A. 2010. Experimental assessment of 3D particle- to-particle interaction within sheared sand using synchrotron micro-tomography. *Geotechnique* 60(5): 369–379.

Higo, Y., Oka, F., Kodaka, T. & Kimoto, S. 2006. Three-dimensional strain localization of water-saturated clay and numerical simulation using an elasto-viscoplastic model. *Philosophical Magazine: Structure and Properties of Condensed Matter* 86 (21–22), 3205–3240.

Higo, Y., Oka, F., Kimoto, S., Sanagawa, T., Sawada, M., Sato, T. et al. 2010. Visualization of Strain Localization and Microstructures in Soils during Deformation Using Microfocus X-ray CT. Advances in Computed Tomography for Geomaterials, March 1-3, 2010, GeoX2010, ISTE Ltd. *In: Alshibli*, K.A., *Reed*, A.H. *(Eds.), Proceedings of the 3rd International Conference on X-ray CT for Geomaterials. John Wiley & Sons, Inc., New Orleans*, *Louisiana*, *USA* pp. 43–51.

Higo, Y., Oka, F., Sanagawa, S. & Matsushima, Y. 2011. Study of Strain Localization and Microstructural Changes in Partially Saturated Sand during Triaxial Tests using Microfocus X-Ray CT. *Soils Found.* 1(51):95–111.

Higo, Y., Oka, F., Sato, T., Matsushima, Y. & Kimoto, S. 2013. Investigation of localized deformation in partially saturated sand under triaxial compression using microfocus X-ray CT with digital image correlation. *Soils Found*. 53(2): 181–198.

Hirono, T., Takahashi, M., & Nakashima, S. 2003. In situ visualization of fluid flow image within deformed rock by X-ray CT. *Eng. Geol*. 70(1): 37–46.

Hudson, J.A. 1971. Effect of time on the mechanical behaviour of failed rock. *Nature* 232(5307): 185–186.

Jian, J., Zhu, W.S., Li, S.C. & Qi, L.F. 2002a. Real-Time CT Testing on Damage Propagation Law of Jointed Rock-like Material under Uniaxial Compact. *Chinese Journal of Rock Mechanics and Engineering* 21(6): 2115–2120. (In Chinese)

Jian, J., Zhu, W.S., Li, S.C., Pu, Y.B. & Qi, L.F. 2002b. Real-Time CT Test of Hydraulic Fracture Process for Jointed Rockmasses. *Chinese Journal of Rock Mechanics and Engineering* 21(11): 1655–1662. (In Chinese).

Ju, Y., Zhang, Q.G., Zheng, J.T., Wang, J.G., Chang, C. & Gao, F. 2017. Experimental study on CH4 permeability and its dependence on interior fracture networks of fractured coal under different excavation stress paths. *Fuel* 202: 483–493.

Ju, Y., Xi, C.D., Zhang, Y., Mao, L.T., Gao, F. & Xie, H.P. 2018. Laboratory In Situ CT Observation of the Evolution of 3D Fracture Networks in Coal Subjected to Confining Pressures and Axial Compressive Loads: A Novel Approach. *Rock Mech. Rock Eng.* 51(11): 3361–3375.

Lenoir, N., Bornert, J., Desrues, J., Besuelle, P. & Viggiani, G. 2007. Volumetric digital image correlation applied to X-ray micro-tomography images from triaxial compression tests on argillaceous rock. *Strain* 43(3): 193–205.

Li, X., Duan, Y.T., Li, S.D. & Zhou, R.Q. 2017. Study on the progressive failure characteristics of Longmaxi shale under uniaxial compression conditions by X-ray micro-computed tomography. *Energies* 10: 303–315.

Liu, S.X., Wang, Z.X. & Zhang, L.Y. 2018. Experimental study on the cracking process of layered shale using X-ray microCT. *Energy Exploration & Exploitation* 36(2): 297–313.

Louis, L., Wong, T.F., Baud, P. & Tembe, S. 2006. Imaging strain localization by X-ray computed tomography: discrete compaction bands in Diemelstadt sandstone. *J. Struct. Geol.* 28(5): 762–775.

Machado, A.C., Oliveira, T.J.L., Cruz, F.B., Lopes, R.T. & Lima, I. 2015. X-ray microtomoraphy of hydrochloric acid propagation in carbonate rocks. *Appl.Radiat. Isot.* 96: 129–134.

Morishita, R., Yoshida, T., Higo, Y., Oka, F. & Kimoto, S. 2012. Observation of microstructures of unsaturated Toyoura sand with different water retention states using microfocus X-ray CT. *In: Proceedings of the 47th Annual Meeting of Japanese Geotechnical Society. Hachinohe, 14–16 July, 349 (DVD-ROM)* pp. 695–696.

Mukunoki, T., Nakano, T., Otani, J. & Gourc J.P., 2014. Study of cracking process of clay cap barrier in landfill using X-ray CT. *Applied Clay Science* 101: 558–566.

Otani, J., Mukunoki, T. & Obara, Y. 2000. Characterization of failure in sand under triaxial compression using an industrial X-ray CT scanner. *Soils Found.* 40 (2): 111–118.

Otani, J., Miyamoto, K. & Mukunoki T. 2001 Visualization of interaction behaviors between soils and reinforcement using X-ray CT. *Proceedings of the International Symposium on Earth Reinforcement, Fukuoka, Japan.* 1: 117–120.

Otani, J., Mukunoki, T. & Obara, Y. 2002. Characterization of failure in sand under triaxial compression using an industrial X-ray CT scanner. *Int. J. Phys. Model. Geotech.* 2(1): 15–22.

Pan, B., Qian, K., Xie, H.M. & Asundi, A. 2009. Two-dimensional digital image correlation for in-plane displacement and strain measurement: a review. *Measurement Science and Technology* 20(6): 62001–0.

Peng, R.D., Yang, Y.C., Ju, Y., Mao, L.T. & Yang, Y.M. 2011. Computation of fractal dimension of rock pores based on gray CT images. *Chinese Science Bulletin* 56 (31): 3346–3357.

Pilz, F.F., Dowey, P.J., Fauchille, A., Courtois, L., Bay, B., Ma, L., Taylor, K.G., Mecklenburgh, J. & Lee, P.D. 2017. Synchrotron tomographic quantification of strain and fracture during simulated thermal maturation of an organic-rich shale, UK Kimmeridge Clay. *Journal of Geophysical Research: Solid Earth* 122(4): 2553–2564.

Ren, J.X., Ge, X.R., Pu, Y.B. 2000. CT Real-time Analysis of Meso-damage Propagation Law of the Whole Process of Rock Failure. *Journal of Xi'an Highway University* 20(2): 12–16 (In Chinese)

Ren, J.X., Ge, X.R. & Yang, G.S. 2001. CT Real-time testing on damage propagation microscopic mechanism of rock under uniaxial compression. *Rock and Soil Mechanics* 2(22): 130–133. (In Chinese)

Ren J.X. & Ge X.R. 2004. Computerized Tomography Examination of Damage Tests on Rocks under Triaxial Compression. *Rock Mechanics and Rock Engineering* 37 (1): 83–93.

Renard, F., Bernard, D., Desrues, J. & Ougier-Simonin, A. 2009. 3D imaging of fracture propagation using synchrotron X-ray microtomography. *Earth and Planetary Science Letters* 286(1–2): 0–291.

Shang, Y.J., Wang, S.J., Yue, Z.Q., Tan, G.H. & Zhao, J.J. 2004. Triaxial Test of Undisturbed Completely Decomposed Granite under CT Monitoring. *Chinese Journal of Rock Mechanics and Engineering* 23(3): 365–371. (In Chinese)

Sun, W., Wu, A.X., Hou, K.P., Yang, Y., Liu, L. & Wen, Y.M. 2016. Real-time observation of meso-fracture process in backfill body during mine subsidence using X-ray CT under uniaxial compressive conditions. *Constr. Build. Materi.* 113: 153–162.

Tisato, N.B., Quintal, S., Chapman, C., Madonna, S., Subramaniyan, M., Frehner, E., Saenger, H.& Grasselli, G. 2014. Seismic attenuation in partially saturated rocks: Recent advances and future directions. *The Leading Edge* 33(6): 640–646.

Tisato, N., Zhao, Q., Biryukov, A. & Grasselli, G. 2015. Experimental rock deformation under μCT: two new apparatuses. *Geoscience new horizons, Geoconvention 2015, Telus Convention Center, 4-8 May 2015, Calgary, AB Canada.*

Viggiani, G., Lenoir, N., Bésuelle, P., Michiel, M.D., Marello, S., Desrues, J. & Kretzschmer, M. 2004. X-ray microtomography for studying localized deformation in fine-grained geomaterials under triaxial compression. *CR Mecanique* 332(10): 819–826.

Viggiani, G., Besuelle, P., Hall, S.A.& Desrues, J. 2010. Sand deformation at the grain scale quantified through X-ray imaging. Advances in Computed Tomography for Geomaterials. In: Alshibli, K.A., Reed, A.H. (Eds.), *Proceedings of the 3rd International Conference on X-ray CT for Geomaterials, March1-3, 2010, USA, GeoX2010, ISTE Ltd. John Wiley & Sons, Inc., New Orleans, Louisiana* pp.43–51.

Vinegar, H.J., Dewaal, J.A. & Wellington, S.L. 1991. CT Studies of Brittle Failure in Castlegate Sandstone. *International Journal of Rock Mechanics and Geomechanics Abstracts* 28(5): 441–448.

Wang, Y., Li, C.H., Hao, J. & Zhou, R.Q. 2018. X-ray micro-tomography for investigation of meso-structural changes and crack evolution in Longmaxi formation shale during compressive deformation. *J Petrol Sci Eng.* 164: 278–288.

Watanabe, Y., Lenoir, N, Otani, J. & Nakai, T. 2012. Displacement in sand under triaxial compression by tracking soil particles on X-ray CT data. *Soils Found.* 52(2): 312–320.

Wennberg, O.P., Rennan, L. & Basquet, R. 2009. Computed tomography scan imaging of natural open fractures in a porous rock; geometry and fluid flow. *Geophysical Prospecting* 57(2): 239–249.

Yang, G.S., Xie, D.Y., Zhang, C.Q. & Pu, Y.B.1996. CT detection of the damage characteristics in coal rock. *Mechanics in Engineering* 18(2): 19–21. (In Chinese)

Yang, G.S., Xie, D.Y., Zhang, C.Q. & Pu, Y.B. 1999. CT analysis of mechanical properties of rock damage propagation. *Chinese Journal of Rock Mechanics and Engineering* 18(3): 250–254. (In Chinese).

Yang, B.C., Xue, L. & Zhang, K. 2018. X-ray micro-computed tomography study of the propagation of cracks in shale during uniaxial compression. *Environ. Earth Sci.* 77(18):.652.

Zhao, B., Wang, J., Coop, M.R., Viggiani, G. & Jiang, M. 2015. An investigation of single sand particle fracture using X-ray micro-tomography. *Géotechnique* 65(8): 625–641.

Zhao, Q., Tisato, N. & Grasselli, G. 2017. Rotary shear experiments under X-ray micro-computed tomography. *Rev. Sci. Instrum.* 88(1): 015110.

Zhou, M.Y., Zhang, Y. F., Zhou, R. Q., Hao, J. & Yang, J. J. 2018. Mechanical Property Measurements and Fracture Propagation Analysis of Longmaxi Shale by Micro-CT Uniaxial Compression. *Energies* 11(6): 1409–1426.

Zhou, X.P., Zhang, Y.X. & Ha, Q.L. 2008. Real-time computerized tomography (CT) experiments on limestone damage evolution during unloading. *Theor. Appl. Fract. Mec.* 50(1): 49–56.

Rock Mechanics for Natural Resources and Infrastructure Development – Fontoura, Rocca & Pavón Mendoza (Eds)
© 2020 ISRM, ISBN 978-0-367-42284-4

Advances in numerical approaches for coupled problems in rock mechanics and recent developments in related fields

Y. Ohnishi
Kyoto University (Prof. Emeritus), Kyoto, Japan

ABSTRACT: Natural rock masses consist of solid and fluid materials. In rock mechanics and engineering, researches have made an effort to understand rock mass behavior under different mechanical and hydraulic conditions. After invention of powerful method FEM, it was introduced into rock mechanics to analyze coupling behavior of rock masses.

Coupling analyses at first have been developed in continuous media and then extended to discontinuous fractured rock masses. Major contribution to solve coupled problems came from the DECOVALEX project which deal with nuclear waste repository.

Couple between solid rock and water was extended to broader area. Fully coupled thermal-hydro-mechanical analysis becomes common in research area. In addition, chemical and time effect is under consideration. Theoretical and experimental researches are actively performed worldwide. How the coupling problems were tackled is shown in the paper.

Numerical methods to handle the coupled problems keep on evolving. In addition to FEM, discrete methods such as DEM, DDA etc. have been utilized to interpret complex natural rock mass behavior. Recent new development for fluid modeling is so called particle methods (MPS, SPH etc.). They are used to explore the physical and mechanical effects of water on rock mass. Some numerical examples are shown in the paper.

1 INTRODUCTION

In the underground, fluids are always present; most of the time, water is saturating the existing geological materials. These fluids are in general moving because of pressure gradients imposed by a complex function of the topography and of infiltration. The flow of water in rock mass is considered to cover a wide range from the ground surface to the deep underground in a broad sense. The study of rock mass infiltration flow, is extremely diverse not only in civil engineering and resource engineering but also in science, agriculture and earth sciences, and forms a typical interdisciplinary area. For example, there is a wide range of social interests such as general groundwater flow, unsaturated flow, tunnel, dam, slope stability with rain, land subsidence, movement of materials and energy, thermal convection simulation, earthquake engineering (liquefaction) and earthquake prediction etc. There is a need for this research to solve some practical problems.

Speaking of rock mass and water in the field of civil engineering, dams are the first to be mentioned. Dams are intended to store water. However, the large-scale dam failure that motivated to create the foundation of rock engineering all related to rock mass failure caused by water, in other words, stress-flow coupling effect.

Groundwater surveys and research are also required to solve the following problems. In fuel underground storage and storage, water sealing, oil containment by water, prevention of gas leakage, groundwater salinization when the cavern is near the shoreline, air leakage prevention in underground storage of compressed air, radioactive waste in underground disposal, heat transfer, advection and diffusion of radionuclides, the amount of spring water, water quality, and underground nuclear power plants may cause problems such as leakage of nuclide during accident and advection and diffusion of pollutants. In addition, there are heat transfer problems as effective use of heat energy and heat transfer problems related to geothermal power generation and CO2 underground storage sequestration.

Considering the relationship between rock structures and water, understanding of hydraulic characteristics of rock, grasping of hydrogeological characteristics classified according to geological characteristics, regional hydrological effects on groundwater and surface water during and after excavation.

In addition to the study of the rock mass, numerous issues such as the stability of the rock mass, the amount of impounding water flow related to the flow of water around the tunnel, etc.

The major problem of rock groundwater is its modeling and analysis method. The analysis method changes greatly depending on how to model the rock and the fluid present in it. In these cases, if the rock is a continuum or an equivalent continuum including discontinuities, the determination of the rock mass permeability coefficient is the top priority. Similarly, there is also a double porosity model in which the continuity between a continuum and a discontinuity is included, in which the hydraulic conductivity is important. A model focusing only on the network of discontinuities is also available. Analytically, determination of water permeability by inverse analysis is also used as a powerful tool.

The basic equations used for analysis regardless of the soil ground, rock mass or water are such that materials that are a continuum (in a broad sense) must be satisfied in the thermodynamic process. These basic equations are well known as a continuous equation (mass conservation law), an equation of motion (momentum conservation law), and an energy balance equation (the first law of thermodynamics) (Bear, 1972).

In the above three basic equations, assuming that the body force and the heat source strength are known, an equation concerning time and position (space) holds. Besides this, the Clausius-Duhem inequality, i.e., the second law of thermodynamics, is added as a constraint. In order to solve the equations of these thermodynamic systems, it is necessary to construct equations which further define the physical and mechanical properties of the continuum. Considering from this basic formula, even if the object of analysis is rock mass, even if it is soil ground, if it is modeled as a continuum, the water existing in it will be the same (water modeled similarly). If so, analysis methods can be used without any distinction between rock and soil ground. It is also clear that the discussion of whether it is continuous or discontinuous is about the geometrical properties of the rock, and it does not mean that the basic equation becomes discontinuous.

In rock mechanics and engineering, researches have made an effort to understand rock mass behavior under different mechanical and hydraulic conditions. Complex rock engineering problems are solved first analytically, and with the advance of digital computer many numerical analysis methods were proposed and applied to actual engineering problems.

Due to the influences of water effect on rock mass behavior, stress-flow coupling was first recognized in consolidation process and the same idea was used to analyze the liquefaction phenomenon. Then the method has been used to analyze slope stability problems with heavy rain.

Coupling analyses have been developed at first in continuous media and then extended to discontinuous jointed rock masses. Major contribution in solving coupled problems came from DECOVALEX project which deal with nuclear waste repository in deep geological formation. DECOVALEX is an acronym for "Development of Coupled Models and their Validation against Experiments." The DECOVALEX project is a unique international research collaboration for advancing the understanding and mathematical modeling of coupled thermo-hydro-mechanical (THM) and thermo-hydro-chemical (THC) processes in geological systems.

After starting the first DECOVALEX project, importance of understanding the coupled process in geological engineering was broadly recognized. Couple between solid rock and water was extended to broader area. Fully coupled thermal-hydro-mechanical analysis with multiple fractures becomes common in research areas. In addition, chemical effect, like weathering, mineral dissolution etc. in geological processes is under consideration. It means that very complex interaction between solid rock (stress or strain), water, air, heat, chemical reaction is coupled with time. Theoretical and experimental researches are actively performed worldwide now.

Numerical methods to handle the coupled problems keep on evolving. In addition to FEM (mainly continuous), discrete methods such as DEM, DDA, etc. have been utilized to interpret complex natural discontinuous rock mass behavior.

Recent new development for fluid modeling is so called particle methods (MPS, SPH etc.). The particle method expresses a fluid-like continuum as a collection of many particles, analyzes its behavior based on the Navier-Stokes equation, and can express dynamic fluid motion well (Koshizuka & Oka, 1996). They are used to explore the physical and mechanical effects of water on rock mass. A new attempt in analysis methods is coupled analysis of particle method, discrete element method (DEM) (Cundall, 1971) and discontinuous deformation method (DDA) (Shi & Goodman, 1989). The coupling phenomenon that occurs when the groundwater particles flow through the fracture space in rock masses. By tracking, it is possible to understand in detail the mechanism of rock movement or failure by water infiltration. Interaction analyses between solid (use discrete models) and water particles reveal and visualize what is happening inside and outside of rock masses. These related methods are considered to be applicable to geotechnical engineering such as slope failure and liquefaction, and further development is expected.

2 COUPLING PHENOMENA AND COUPLED PROCESS

2.1 *Introduction*

Rock and soil ground are composed of solid and fluid. Interaction of these substances with one another makes the work of the earth. Their actions can be roughly resolved into four components: mechanics, hydraulics and hydrology, heat and temperature, and chemical reactions, each of which interacts with each other.

From the past, academic fields have been developed to explore the action of these elements, and they have been called soil/rock mechanics, hydraulics/hydrology, heat conduction, and geochemistry respectively. However, in the real world, these elements are intertwined with each other and have been regarded as complex phenomena, but sufficient academic development is obtained to construct a theoretical composition that actually understands such phenomena.

However, as a result of research and development for many years, it is coming to a stage where it can be practically applied in engineering. Along the way, it can be said that a coupled analysis in which two to four elements are combined has developed and reached the current stage. The need for a proper performance assessment of a nuclear waste repository has opened up new areas of THMC coupling research in many geoscience disciplines, including geology, geochemistry, geophysics, geomechanics, and hydrogeology. Specially, the International project, so called "DECOVALERX" (explained later) made great contribution in developing innovative tools for the study of THMC interaction phenomena. Note that most of these models have been developed for the design of waste disposal sites.

There are the various types of possible coupling among the four effects (T, H, M, C). Table 1 indicates the stronger couplings (shown as double lines) and weaker ones (shown as single lines). Such a distinction is based on current knowledge and may have to be modified as new data and information emerge.

Table 1 shows six 2-way couplings as follows, some of which are relatively well known (Tsang, 1987),

1. T = C, includes chemical thermodynamics as well as all temperature effects on the equilibrium constants and kinetics of chemical reactions.
2. T = H, corresponds to known buoyancy effects, resulting in convection. These are well studied in porous media, especially in the context of geothermal systems. However more work needs to be done on such effects in sparsely fractured porous systems.
3. T = M, represents the well-known thermo-mechanical effects, although more work should be done in this area to evaluate the impact of existing fractures and joints.
4. H = C, includes chemical transport in water-saturated fractured porous rocks. Interaction between fluids, the rock matrix, and fracture-lining materials must be taken into account.
5. H = M, is also of importance for fractured media. Here, changes in pore pressures may affect fracture apertures, which, in turn, critically control fluid flow rates.
6. C - M, which is a weak type of coupling. Stress corrosion is an example. Table 1 also shows four 3-way couplings.

7. T = H = C includes chemical reactions and heat and mass transport in rock masses. A chemical reaction could be mineral precipitation (or dissolution), which changes the rock permeabilities to fluid flow. Such a process has been extensively studied in hydrothermal systems.
8. T = M - C coupling, may be thermo-mechanical deformation of rock masses, where the mechanical strengths of the rock or joints may change because of thermochemical effects.
9. T = H = M describes the thermally induced hydro-mechanical behavior of fractured rocks, where both heat and fluid pressure may change the fracture apertures with the result that buoyant flow may be strongly affected.
10. M = H = C, an example of which is the case where chemical transport is affected by changes in hydraulic properties of fractured rocks due to hydromechanical effects.
    The last type of coupling shown in Table 1 is the 4-way coupling.
11. T = H = M - C, all four effects are interrelated. Here one may be studying the chemical reaction and mass transport under both thermal and hydraulic loading. The thermal effects may cause mechanical changes in fracture apertures and, hence, fracture permeability. Chemical precipitation and dissolution may also change the permeability of the rock mass. These two effects join together to influence the fluid flow and chemical transport.

General interaction phenomena in rock engineering is summarized by Hudson (1997) as an interaction matrix as shown in Figure 1. The interactive elements in Figure 2 are self-evident from the sketches. Hudson (1977) stated that interpretation of the diagrams represents each element within

Table 1. Types of coupled process (C.F. Tsang, 1987).

No.	Type	Example
1.	T = C	**phase changes**
2.	T = H	**buoyancy flow**
3.	T = M	**thermally induced fractures**
4.	H = C	**solution and precipitation**
5.	H = M	**hydraulic fracturing**
6.	C–M	**stress corrosion**
7.	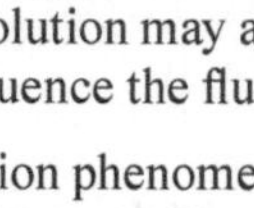	**chemical reactions and transport in hydrothermal systems**
8.	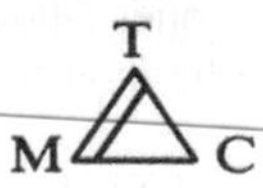	**thermomechanical effects with change of mechanical strengths due to thermochemical transformation**
9.		**thermally induced hydromechanical behavior of fractured rocks**
10.	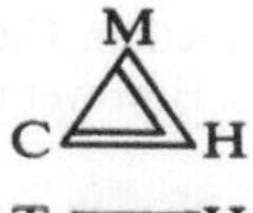	**hydromechanical effects (in fractures) that may influence chemical transport**
11.		**chemical reactions and transport in fractures under thermal and hydraulic loading**

Note: T = Thermal, M = Mechanical, H = Hydrological, C = Chemical. A single line indicates weak coupling; a double line indicates strong coupling.

the structure of the total interaction matrix must be encouraged. The technique can be used for all interactions, so it is helpful not only to understand the underlying structure of such matrices, but also to be able to create new matrices oneself.

The interaction matrix is the basic device used in rock engineering systems. If the state variables are conceptual in nature, the off-diagonal interaction can be assessed using a semi-quantitative method of coding. If the state variables are physical variables, a new modeling technique known as the fully-coupled model can be used. The physical mechanisms linking the variables are identified for each off-diagonal box and relation quantified (Hudson & Harrison, 1997).

2.2 *Modeling of rock mass*

In the conventional mechanical analysis of rock mass, the mechanics of a continuum based on elasticity and plasticity has been used. In this case, analysis is performed under the assumption that the rock mass is a continuum of isotropic or anisotropic locally made of homogeneous material. On the other hand, if you look at the original appearance of the rock, it is natural to think that the rock is a discontinuous body, because there are many cracks, fractures and joints.

The simplest method of understanding the rock mass as a discontinuous body is the Rock Characterization, which has been accumulated over the past, and based on this, the mechanical characteristics and hydraulic characteristics considering the discontinuity are determined. That is, the characteristics of the rock are judged by comprehensive engineering judgment based on the experience of the expert engineers. In this case, an engineering constant as an equivalent continuum is usually determined in consideration of the influence of discontinuity (simple equivalent continuum model). Ideally it would be possible to estimate mechanical properties directly from the assessment of rock mass, but that mechanical constants can only be used as an indicator. Therefore, at present, there is not much ambiguity in modeling and property determination of rock mass.

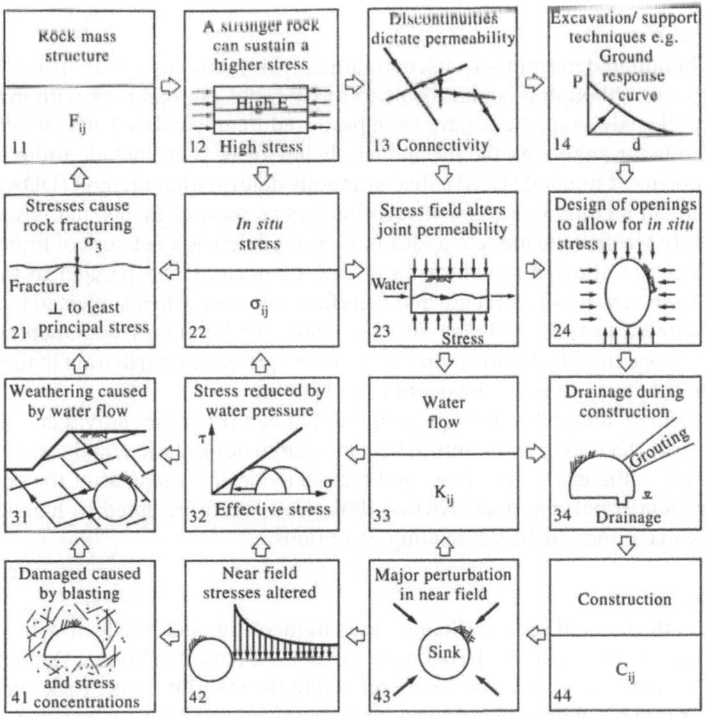

Figure 1. Rock mechanics-rock engineering interaction matrix (Hudson & Harrison, 1997).

On the other hand, as an analytical approach from another view point, an attempt is made to grasp the geometrical shape of the rock discontinuity and use it for the rock mass evaluation. There are two methods for evaluating geometric shapes such as size, direction, and position of discontinuities (Priest, 1993) included in rock mass, treating them as deterministic discontinuities and treating them as equivalent continuum (Long et.al., 1982). The former is a method in which analysis is attempted on the basis of the existence of all discontinuities, but this is not possible in reality, and idealized modeling is performed after some statistical processing (Priest & Hudson, 1981). The latter is a method of statistically processing survey data on discontinuities obtained from actual rock mass, and replacing discontinuous rock with equivalent continuum and analyzing them as homogeneous or anisotropic rock masses. Continuum based analysis methods are known as "crack tensor" (Oda, 1986), "damage tensor", etc.

In any case, in order to realistically model rock mass, accurate in-situ information is required. In addition, new models are being developed one after another, and in the case of three-dimensional models such as 3-dimensional disk models, en-echolen models. In general, the discrete models discussed above are used in stability assessment for many structures of rock masses with a limited number of discontinuities and blocks. Discrete models require a complete understanding of the fracture network. Such an understanding would be extremely expensive, if even possible, to develop. Dershowitz models and their improved models are presented (Dershowitz, Herbert & Long), and they become the basis of Discrete Fracture Network (DFN) analysis.

The spatial discretization method of three-dimensional DFN models for FEM-based methods is still a challenge. Many 3D mesh generation methods were developed for discretizing the three-dimensional fractured media embedded with DFNs into conforming tetrahedral elements (Mustapha et.al., 2011), which can reconstruct fracture's geometrical characteristics and provide matching nodes on fracture surfaces. This kind of mesh is efficient and effective for numerical simulations that physical fields are continuous across fractures, such as pressure field, thermal field and concentration field.

2.3 *Mechanics for rock mass*

2.3.1 *Introduction*

The method of handling deterministic discontinuities with methodological difficulties is still making steady progress. Although it is impossible to finely represent real rock with many discontinuities as it is, several analysis methods have been proposed in an idealized and simplified form. That is, typical methods for analyzing the mechanical behavior of rock include finite element method (FEM), distinct element method (DEM), discontinuous deformation method (DDA), etc.

The finite element method represents discontinuities as joint elements and is solved using continuum analysis. In the discrete models, a rock mass is modeled as a network of finite planar hydraulic conduits which are allowed to deform in directions normal and parallel to their planes. The discontinuities have zero tensile strength and therefore can open when subjected to tensile stresses. Also, they are allowed to shear and their shear strength can be mobilized. Opening and sliding of discontinuities change their hydraulic properties and the pressure distribution in the rock mass. The new pressure distribution creates more opening and sliding.

The distinct element method is a unique method that can track the individual movements of the rock that constitutes the rock masses until it becomes large deformation, and can know the collapse of the cavern wall or the excavated slope, and the behavior and shape after the collapse. Furthermore, DDA and numerical manifold (NMM) (Shi, 1991) were invented to handle discontinuous behavior of rock mass under different loading conditions.

2.3.2 *Rock mechanics*

Mechanics forms the basis of rock mechanics and engineering, and has dealt with stress and deformation phenomena under various external forces with boundary conditions. Human life affects the rock masses in various ways, and it is closely related to the stability of the rock mass as well as the external force caused by natural phenomena, and stress-deformation analysis has been conducted for its evaluation. The theory of elasticity and plasticity in continuum analysis, and the elastic modulus, deformation coefficient, and failure condition (c and f) used in it are the basis of rock mass analysis.

Understanding the rock mass as a continuum and porous body and constructing the theory has been the basis of research for many years based on the continuum mechanics as the foundation of rock mechanics. However, focusing on the fact that there are innumerable discontinuities in the rock mass, the mechanics of discontinuities taking into account the existence of discontinuities are emerging along with the progress of numerical analysis methods, it has been recognized as a major pillar of mechanical analysis research.

2.4 *Modeling for groundwater flow analysis in rock mass*

Anyone knows that the soil and rock materials are involved in fluids, in particular water (groundwater). Water flows in the void space of the rock mass and becomes underground water. Macroscopically, rains fall down to the ground and penetrates into the ground. The water becomes groundwater passes through the rock mass, but there is also water that leach out on the way and pour into the river, and those penetrate deeply and become the source of hot springs due to the heat of the earth magma. During this process, interaction between water and heat occurs. Figure 2 shows the movement (circulation) of water in the subsurface area.

The theory of groundwater movement in geological materials is mathematically studied for many years, assuming that the medium is porous, and theoretical solutions are presented under various boundary conditions. The most commonly used is the theory of well. The porous continuum is a medium for analyzing the behavior of groundwater, but with the progress of research, analysis taking into consideration the discontinuity of rock mass has been developed along with the development of numerical analysis methods. The analysis by the finite element method incorporating joint elements has greatly expanded the field of groundwater flow in rock masses. For this reason, modeling of rock mass taking into consideration discontinuities is performed in various ways, and sophisticated models are presented. The representative is known as a DFN (Discrete Fracture Network) model.

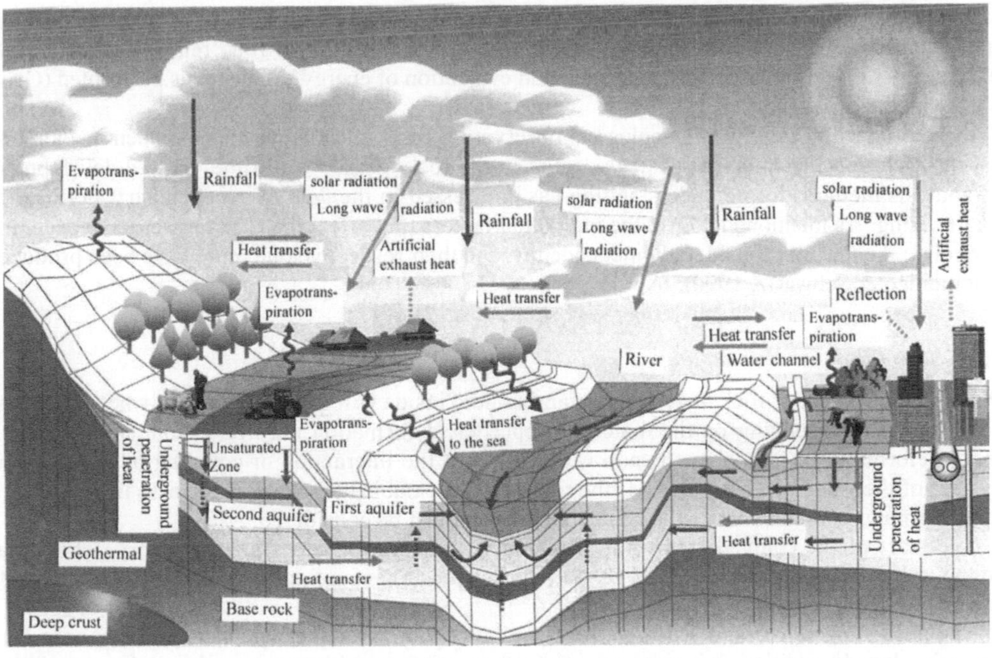

Figure 2. Water circulation and heat transfer in subsurface area.

Uncertainty in the geometric and hydraulic properties compromises the accuracy of seepage simulation in rock masses. Fracture networks are commonly described by several statistical characteristics, including orientation, trace length, aperture, and location. These statistical characteristics are represented with distribution models, whose parameters are evaluated from fracture outcrops shown on excavation surfaces or the scanline and sampling windows (Priest & Hudson, 1981). Unlike discrete element models, equivalent continuum models can be used for heavily jointed rock masses. The jointed rock mass is replaced by an equivalent continuum with anisotropic deformability and flow properties. The permeability coefficients and the deformability properties of the equivalent continuum depend on the spacing, orientation and aperture of the different joint sets but not on their exact location.

In the case of a dam, there are grouting performance, piping handling, groundwater fluctuation and behavior of the rock mass, understanding laminar flow/turbulence state and leakage amount, and grouting method. In the tunnel, there are estimation of inflow water volume, influence on surrounding groundwater, prediction of sudden flooding water, collapse of face and side wall by inflow water, etc. Underground openings and radioactive waste geological disposal further add to the infiltration behavior in the low permeable rock mass. For waste, it is also important to understand the water circulation characteristics in the long-term period.

For the slope stability problem, the situation is slightly different, and the relationship with soil engineering becomes strong, and changes in permeability due to weathering and alteration, piping, and the behavior of pore water pressure are discussed. Also, when the rock is near the ground surface or near the excavated surface, unsaturation problems occur. In this case, it is basic to evaluate the unsaturation property of the rock mass, but it is important to estimate the rainfall as boundary condition and the inflow from the surrounding ground.

In some cases, the fluid in the rock mass is not limited to water. Various materials are sent into the rock mass and interact with water. For grouting, cement suspension, air for compressed air storage in the rock mass, polymer solution for water harvesting of oil extraction, geothermal development, vaper due to water phase change for waste, carbon dioxide (CO2) for sequestration. Also, since water transports heat efficiently, its effect also needs to be considered.

2.5 *Heat and temperature effects*

Conventional rock mechanics and engineering have rarely dealt with heat and temperature issues individually. As an example, analysis of temperature distribution in rock related to geothermal power generation is regarded as important, and evaluation of energy efficiency is performed (Gringarten et.al., 1995).

In addition, it is clear in the geological observation cases such as weathering phenomenon and rock dissolution that chemical change occurs in the rock masses with time, but the development of analysis methods for its understanding has progressed in the field of geochemistry, not in rock engineering. As mentioned above, the individual factors that affect the rock masses have been historically taken up and considered independently, and their respective outcomes have been produced (Domenico & Schwartz, 1990).

2.6 *Stress-flow (M-H) coupling*

Permeable ground and rock materials containing fluid inside will generally deform when pore pressure changes. Depending on the material properties, deformation may vary from small cases to large deformations. This phenomenon is the result of the interaction or coupling of the dynamic system of the water and rock mass. If the rock is a deformable body (as it is the case in practice), if its deformability is noticeable, then a coupled action must be considered.

The start of the coupled stress-flow analysis is well-known consolidation theory, which was founded by Terzaghi, but it is Biot (1941) that tackled this coupling problem and systematically developed the theory.

The governing equation consists of the equation of motion of seepage water and the mass conservation law. In addition, it is necessary to further satisfy equilibrium equation and compatibility condition regarding the stress and the deformation of the ground. And when considering the movement of the underground water and the deformation of the ground at the same time, there must be a relationship showing the interaction between the two. This is the principle of the effective stress of Terzaghi used for saturated soil. It was Biot that integrated these relations and derived simultaneous partial differential equations with unknown displacements of the structural framework and pore water pressure, and his general theory is three-dimensional consolidation theory. Derivation of basic equations are presented now in ordinary geotechnical engineering textbooks and published papers.

In the rock, water mechanically works as pore pressure in stress, so the problem is considered using the principle of effective stress. Civil engineering works often are controlled by existence of water, and this is the effect of coupling. Collapse or over deformation due to flooding during excavation, fluctuation of fill dam bank and slope due to water level change, land subsidence due to excessive flooding, occurrence of micro earthquake motion due to injection are directly related themes.

The procedure is exactly the same as in the case of rock mechanics analysis. However, in mechanical rock discontinuity analysis, the problem is the transferring and resistance of the fracture force, but in the seepage analysis, importance is placed on the fracture connectivity and water permeability. In such a case, the idealized finite element method, in which discontinuous surfaces are represented by parallel plates, is often used, but with the development of 3-dimensional geometric shape modeling methods for rocks, new numerical methods have been developed and used such as DDA, DEM..

Based on discontinuous deformation analysis (DDA), Jing et al. (2001) simulated the fully coupled fluid flow and stress deformation in fractured hard rocks. As an implicit method, it allows for setting a large time step in the dynamic analysis and provides unconditional stability. Discrete element method (DEM) as an explicit method was also used to estimate the local stress state and study the effect of stress ratio on the hydraulic properties of fractured rocks. The hybrid finite-discrete element method is becoming an attractive way to solve the coupled mechanical-hydraulic problems of complex fracture geometries. These researches are limited in either 2D problems or 3D problems with simple fracture networks. More contributions are needed in the numerical simulation of complex three dimensional discrete fracture networks.

In the flow analysis, it is necessary to consider the modeling of the fluid existing in the rock mass other than the water, in the case of saturated water, laminar or turbulent, and in the unsaturated state, multiphase flow of water and air, and also in some cases, multiphase flow with oil or salt water, etc.

As in an example for stress-flow problem, Malpasset dam, France, uplifting pressure was applied to the fault zone of the rock mass foundation during reservoir filling, and the non-uniform lifting of the rock mass twisted the arch dam, resulting in a catastrophe in 1959. The cause of the accident was unknown for a long time, but it was revealed that after about 10 years, the stress of the fault and the water passing through it had a complicated entanglement and had an adverse effect. Subsequently, the disaster at Vajont Dam in Italy in 1963 was not a problem of the dam itself, but because the slopes of the landslides around the reservoir collapsed in large scale by rising reservoir water levels, and the full water in the reservoir was expelled. The failure of the U.S. Teton Dam in 1976 was caused by piping in base rock. Here too, the coupled phenomenon of stress and water pressure at the core part of the dam is a cause of failure.

Slope failure can be seen in natural and man-made areas, with heavy rains or water rise in the slopes. The collapse of the slope causes a huge human lives loss and property damage once it occurs. From the perspective of disaster prevention, it is urgently necessary to clarify the relationship between the ground water flow and the stability of rock slopes.

2.7 *Themo-Mechanical (T-M), Thero-Hydraulic (T-H) and Thermo-Hydro-Mechanical (THM) Coupling*

It is well known that the effects of heat on rock masses are important. When a high temperature or low temperature heat source exists in the rock, temperature changes affect the behavior of the rock as well as external load and water pressure changes, stress and deformation of the rock structure, groundwater flow, and heat transfer phenomena. With the increasing interest in heat transfer phenomena underground, such as geothermal power generation, thermal storage utilization in underground aquifers (Gringarten et.al., 1975), and underground disposal of radioactive wastes, there is a growing need to consider these interactions.

Heretofore, coupled analysis taking these interactions into consideration has mainly been conducted by targeting two elements. The coupling of heat and stress/deformation has been studied in many related fields including solid mechanics, and the coupling of heat and flow has been studied in the fields of fluid mechanics and hydraulics, such as thermal diffusion and thermal conduction.

The coupling of rock, seepage water, and heat is closely related to the development of geothermal power generation, which is one of the topics of natural energy. When extracting the heat from the ground, the surrounding rock mass is affected and deformed due to temperature drop (Gringarten et.al., 1975). When the temperature of the hot rock decreases, the thermal efficiency decreases and it becomes necessary to abandon the pumping well. The examination of performance of the well is important as a coupled problem of rock, seepage water and heat (THM), and it is used in evaluation associated with geothermal energy development. The coupled heat-flow water problem, which is simpler case, targets the relationship between the temperature change and the groundwater in the urban area where the heat island phenomenon occurs, which also leads to the environmental problem and greenhouse effect.

Three elements, such as stress-deformation behavior, groundwater flow, and heat transfer phenomena (MHT coupling), relate to each other and bring extremely complex behavior to the rock. In order to analyze these interactions, it is necessary to solve the stress equilibrium equation, the water continuity equation, and the energy conservation law at the same time. Such research has emerged from the study of geothermal development, and the opening of that research is the theoretical examination result by Bear and Carapcioglu (1981). On the other hand, under various conditions, it has been recognized that these elements are intertwined, and studies have been made in consideration of the interaction between two to four factors.

Advances in numerical analysis have made it possible to analyze the problem of the interaction between pollutants and groundwater, and since they have strongly stimulated interest in groundwater environmental problems. In order to solve these problems, analytical methods based on advection and diffusion equations have been developed one after another, and their application to practical use has advanced, and they play a part in solving environmental problems. This issue can be said to be chemical coupling because the behavior of chemical substances is strongly linked to groundwater.

2.8 *DECOVALEX project for THMC coupling problems*

Emplacement of nuclear waste in a repository in geologic media causes a number of physical processes to be intensified in the surrounding rock mass due to the decay heat from the waste. The four main processes of concern are thermal (T), hydrological (H), mechanical (M) and chemical (C). Interactions or coupling between these heat-driven processes must be taken into account in modeling the performance of the repository for such modeling to be meaningful and reliable.

DECOVALEX (DEvelopment of COupled models and their VALidation against EXperiments) brings together organizations from industry, government and academia to tackle some of the most technically challenging problems in the geological disposal of]radioactive waste. The DECOVALEX project is typically conducted in separate 3-4 years project phases. Each phase features a small number (typically three to six) modeling test cases of importance to radioactive waste disposal. The idea of DECOVALEX for the international co-operative project came in among the discussion guided around year of 1991 by Dr. Chin-Fu Tsang and Prof. O. Stephansson with support of Swedish waste research groups for preparation in the various international rock mechanics meetings.

The DECOVALEX project is to support the development of mathematical models of coupled THMC processes in fractured geological media for potential nuclear fuel waste repositories. The capability of modelling coupled phenomena is of particular importance to the safety assessment of geologic disposal of radioactive waste materials. Different mathematical models and computer codes have been developed by different national research teams and these are used to study the so-called bench mark test and test case problems developed within this project.

During the first stage (May 1992 to March 1995), called DECOVALEX I, the main objective was to develop computer codes for coupled T-H-M processes and their verification against small-scale laboratory or field experiments. In the second stage, called DECOVALEX II, the main objective was to further develop and verify the computer codes developed in DECOVALEX I against two large-scale field tests with multiple prediction-calibration cycles, the pump test at the Sellafield, UK, with a hypothetical shaft excavation and the in-situ THM experiment at the Kamaishi Mine, Japan (Kurikami, et.al., 2003). The DECOVALEX III project is the third phase of the project series and was run through the period of 1999-2003. It was initiated with two main objectives. The first is the further verification of computer codes by simulating two additional large scale in-situ experiments: the FEBEX T-H-M experiment performed in Grimsel, Switzerland, and the drift scale heater test at Yucca Mountain, Nevada, USA (Min et.al., 2004).

During the project time six workshops and many task force meetings were held and numerous reports were generated reporting progresses in research. Near the end of each project an international conference on coupled THMC processes in geological systems was held. (For example, "GeoProc 2003" at KTH, Sweden by Stephansson et al., 2003) (Koyama, et.al., 2008). In addition, International Conference, so called "CouFrac", on Coupled Processes in Fractured Geological Media: Observation, Modeling, and Application was initiated by Dr. Rutqvist (LBL in Berkeley, USA).

The DECOVALEX project is still on going. The recent DECOVALEX-2019 (2016-2019) will end this year, but it will be continuing for the next step. As a product of DECOVALEX project, a large number of high-quality papers produced in the wide fields of numerical modelling of nuclear waste disposal, oil/gas reservoirs, geothermal energy extraction, geological systems, coal mining, geotechnical engineering, environmental engineering and, fundamental researches about coupled THMC processes in different geo-materials and geo-systems. The international meeting is the first of such academic gatherings in the specific field of coupled THMC processes in geo-systems.

The full development of THMC modelling is still in progress and do not come to the end. Although the geosphere is a system of fully coupled processes, this does not directly imply that all existing coupled mechanisms must be represented numerically. Modelling is conducted for specific purposes and the required confidence level should be considered. It is necessary to match the confidence level with the modelling objective. Coupled THMC modelling has to incorporate uncertainties, in the conceptual model and in data. Assessing data uncertainty is important when judging the need to model coupled processes, but also the confidence in the prediction need to be assessed. Even if comparing THMC models with real data is a fundamentally difficult problem, it is not impossible.

3 THMC COUPLING RESEARCH EXAMPLES FOR NUCLEAR WASTE REPOSITORY

Many countries that operate nuclear power plants are concerned of nuclear waste repository put very strong force on evaluation of adequate disposal sites. For safety management THMC coupling phenomena in rock masses of repository site must be examined carefully. In DECOVALEX and also in other meetings, very cooperative works are in progress. some new ideas came in the research works and three topicss of them are introduced here in this paper.

3.1 *Nuclear waste repository site evaluation at coastal undersea area in Japan*

Japanese government and associated agencies (JAEA (Japan Atomic Energy Agency) et. al. (2019)) are trying to find a candidate site for nuclear waste repository and coastal undersea areas are examined in recent years. As for the evaluation of groundwater environment for coastal area, evaluation methods combining actual geological environment investigation using boreholes and numerical analysis etc. have been used so far. In addition, a nuclide migration evaluation method for coastal areas is being developed in Nuclear Power Environment Organization, Japan. In these cases, using a numerical analysis code such as SEAWAT for the evaluation of the groundwater environment, a numerical model representing the geological formation as an equivalent porous media is used and the codes has been applied in the project. On the other hand, in foreign disposal site candidates for crystalline rocks such as Forsmark (Sweden) and Orkiloto (Finland), the movement of salinity associated with groundwater flow in a fracture and migration of salinity by diffusion for the rock matrix in contact with the fracture groundwater environment has been evaluated using a numerical model (SKB, 2011; Posiva, 2013; Yoshida et.al., 2014).

The migration behavior of solutes using a double porosity model with such fractures and matrix parts may be more complicated and slower than in the case of a single porosity model assuming a porous medium. For example, the movement of solute in the fracture is advection-dispersed movement and its velocity is relatively fast, but in the matrix part the diffusion is relatively slow and the porosity is large, then capacity to hold the solute increases. Therefore, it is conceivable that the retention of salinity in the matrix reduces the influence of movement of shoreline due to seagoing and retreating on the fluctuation of salinity distribution in groundwater and suppresses the fluctuation of groundwater environmental conditions used for nuclide transfer evaluation.

From the above point of view, when evaluating the groundwater environment change due to sea level change in a very long geologic time, uplift and erosion, targeting the fractured media in the coastal area, in addition to the groundwater flow in the fracture and the movement due to advection dispersion of salinity, groundwater environment evaluation method must take into consideration the effect that the rock matrix part in contact with the fracture will retain the salt content by diffusion. Numerical evaluation has been done by using two different scales: regional scale of a few tens of kilometers, and the block scale (area of about 200 m x 200 m x 200 m) of around two tunnels. The following figure (Figure 3) shows the representative fracture network model used for computation.

As a result, for fractured media in the coastal area, we have developed a method to quantitatively evaluate the influence of sea level change, uplift and erosion on the change of groundwater environment around the disposal site. Information about temporal changes of groundwater environment such as surrounding salinity and groundwater flow velocity were extracted. By using the extracted information for setting the boundary conditions of the fracture network model for the block scale around the disposal site, the evaluation results of changes in the groundwater environment around the disposal site can be treated as nuclide in the block scale. An analysis method has been developed that can be reflected in the evaluation of transition. Obtained results of analysis has been evaluated in view of performance assessment for nuclear waste repository.

3.2 *Evaluation of chemical effect in THMC long-term coupling*

The research team of Yasuhara and Elsworth (2006) developed coupled THMC models and procedures for water flow experiments in rock fracturing under various confinement pressure and temperature conditions to investigate the long-term evolution of permeability in sedimentary rocks

Figure 3. A representative 3D fracture network model in 200m x 200m x 200m.

(Yasuhara et al., 2016). The experiments track the progression of fracture permeability mediated by the binding process. A conceptual model is also presented for reproducing experimental measurements by describing chemical and mechanical processes such as pressure solutions.

The model solves heat transfer, groundwater flow, induced stress fluctuations, geochemical reactions (i.e. free surface dissolution/precipitation and pressure dissolution), and their interactions. In order to test the validity of the developed model in expressing the experimentally observed behavior, they examine the changes in porosity and silica concentration in quartz aggregates. This prediction is very in line with experimental observations. Subsequently, applying the validated model, the influence of pressure dissolution on the evolution of the permeability of sedimentary rock where the radioactive waste disposal site is constructed is examined considering the expected temperature, flow and stress conditions (Figure 4). According to predictions, it has been confirmed that the process of pressure dissolution causes the transmission of radioactive materials to be delayed, especially at locations close to the excavated tunnel, by one order of magnitude less than the initial value (Figure 5).

The developed model is direct and relatively simple because some interactions between the THMC processes are omitted, but it is somewhat accurate for long-term changes in permeability under any stress and temperature conditions. In general, uncertainty is important when considering coupling processes to solve complex nonlinear problems. Therefore, it is always important to consider whether you need a complex and fully coupled model to achieve a sufficiently high accuracy. In addition, coupled THMC numerical models are presented to predict long-term changes in rock permeability.

This THMC model is extended to predict fluid flow and mass transfer behavior of rock fractures. In addition, this model has succeeded in reproducing the actual changes in hydraulic aperture and effluent element concentration in mudstone fractures. From the analysis results obtained in this study, the present model is effective in evaluating the evolution of fluid flow and mass transfer

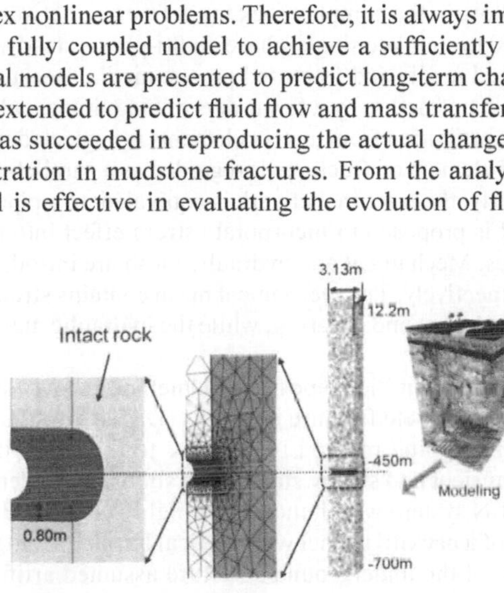

Figure 4. Target area for computation (after Yasuhara et al., 2016).

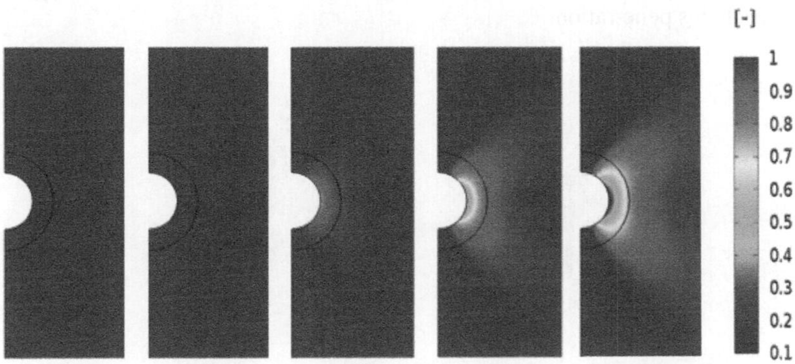

Figure 5. Change in normalized permeability with time in the range of 10° to 10^4 years under the PS condition. The permeability decreases with time (Yasuhara et al., 2016).

behavior within rock fractures that can enhance the coupled thermodynamic-hydraulic-mechanical-chemical conditions. However, some unpredictable behavior, such as evolution in the concentrations of Na and K elements for granite experiments and that of Al elements for mudstone experiments, still exists at this stage. Therefore, detailed investigation of reaction transport behavior by geochemical process is necessary.

3.3 *Influence of fracture network and coupling effect*

Ma et.al (2019a, 2019b) proposed a new approach based on the unified pipe-network method (UPM) (Ren et.al., 2017) to simulate nuclide migration in highly fractured rock masses and to solve the flow problems in three-dimensional highly discontinuous media. This method explicitly represents fracture networks with interconnected nodes and pipes in a spatial domain and applies a set of universal constitutive models to both fractures and rock matrices. The fundamental idea in this model is using geometry equivalence and universal constitutive models to simplify complex 3-dimensional problems in one-dimensional ones. UPM modeling demonstrates its capability in dealing with reactive mass transport problems in complicated fracture-matrix system under sophisticated boundary conditions. Then, the fluid flow in the complex rock mass fracture network can be simulated.

It has been effectively used to simulate single-phase, multi-phase, steady, transient and multi-physics problems in both fractured media and fractured porous media. Because the permeability of the rock matrix is often smaller than fracture by a few orders of magnitude, only fluid flow in fractures is considered. The hydraulic contribution from matrix elements is neglected in the hydraulic simulation stage. Every edge of triangle element on fractures is regarded as a parallel-plate pipe with a hydraulic transmissivity. Mathematically, the fracture network is equivalent to a pipe network.

A mesh mapping method is proposed to incorporate stress effect into the hydraulic analysis for highly fractured rock masses. Mechanical and hydraulic mesh are introduced for the hydraulic and mechanical simulations, respectively. The mechanical mesh contains strong discontinuities for capturing the fracture opening, closure and shearing, while the hydraulic mesh only consists of merged triangles on fractures.

Using finite element method and unified pipe network method (UPM), stress effects on the equivalent permeability of complex discrete fracture networks (DFNs) are studied. The numerical result demonstrates that the mesh mapping method is effective to the stress-dependent analysis of 3D complicated DFNs. Both transient and steady states were simulated to demonstrates the applicability of UPM to a complex DFN system with hundreds of highly connected fractures.

As an example, a section of a backfill tunnel with vertical burial hole in a repository was modeled the geometric configuration of the underground structure assumed artificially. One realization of fracture networks (See Figure 6) was generated based on the distribution models from the imaginary site statistics. By using these fracture network data, UPM enabled forecasting of the ultimate depth of a contaminant propagation in explicitly represented fracture networks and predicting the time taken to reach this penetration.

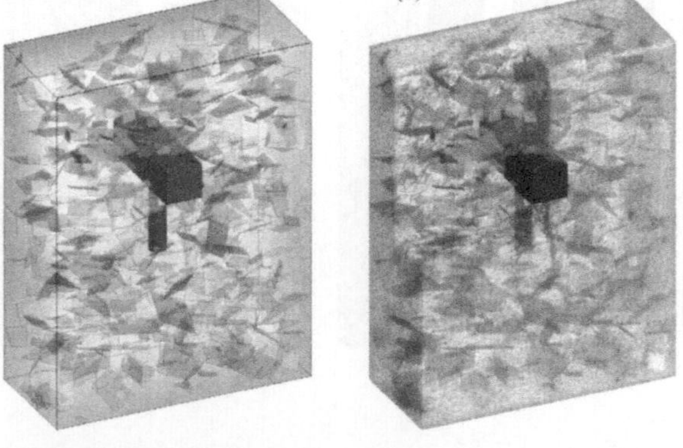

Figure 6. Simulation domain with fracture networks: (a) geometric configuration of fracture networks; (b) discretization of simulation domain.

4 SOLID AND WATER COUPLING ANALYSIS BY NMM-DDA AND MPS METHODS

4.1 *Scope of the NMM-DDA & MPS coupling analysis*

The disaster due to earthquake and heavy rain has occurred frequently. The analytic method also has been applied to the estimation of the hazards and damage. However, it is not always successful. One of the reasons is considered that continuum method such as FEM is applied to analyze the phenomenon with large movement and a breakdown occurring in slope disasters.

Slope failure is often caused by heavy rain fall and it crashes into small pieces with water. Then, use of discontinuous DDA and particle MPS analysis is proposed to solve a complex solid and fluid interaction problem, especially rock and water mixed debris flow.

In the form of limit analysis with c and F066 for slope stability, water pressures are often imposed in the equation. Stability of slope is evaluated by a factor of safety for a slope with this method. However, the limit analysis cannot analyze the time-dependent process of slope failure, then in rock mechanics fields numerical discrete methods such as DEM or DDA were often introduced. Usual slope stability analysis for discontinuous media only considers movement of soils and rocks and in these cases water is taken into account implicitly. Interaction (coupling) between geo-material and water is very complex but important to understand the behavior of rock and soil mass.

Recently a few people started to solve a complex problem of moving blocks with moving water. Here, we will use discrete blocks (DDA) and show how fluid goes through the void and intact blocks (particle method). Then they are coupled in the calculation to account for the mutual interaction. Discontinuous Deformation Analysis (DDA) and Distinct Element Method (DEM) has been used as analyses methods of the slope disasters (Wu et. al., 2004). However, these methods cannot handle the mixed body of fluid and solid material appropriately. It is known that the water explicitly affects stability of rock slope.

In the past decades, a particle method such as MPS (Moving Particle Semi-implicit) (Koshizuka & Oka, 1996) and SPH (Smoothed Particle Hydrodynamics) (Lucy, 1977), which is one of mesh free methods, has been used to analyze water movement. The meshless MPS method has been developed within the Lagrangian framework. In MPS, the fully Navier–Stoke equation can be solved without using computational grids or meshes, and MPS is suitable for simulating motion and fragmentation of fluid.

The particle method is simple but powerful method to solve the problem of fluid mechanics. We introduce the coupling analysis by a discontinuous DDA and a particle MPS method to solve complex solid and fluid interaction problems such as collapse of natural landslide dam and debris flow (Miki et al., 2017). The proposed method shows the potential ability for the application in landslide, debris flow and other related fields.

Numerical Manifold Method (NMM) (Shi, 1991) can simulate both continuous and discontinuous deformation of blocks with contact and separation. However, the rigid body rotation of blocks, which is one of the typical behaviors for rock slope failure, cannot be treated properly because NMM does not deal with the rigid body rotation in explicit form. DDA and NMM-DDA can be applied in case by case (Ohnishi et. al., 2016).

To analyze the coupling between the water flow and the falling blocks of rock, this study introduces the equivalent external forces for the falling blocks of rock caused by the water flow. The velocities are calculated for the fluid flow based on MPS. MPS analyzes the part of fluid alone and DDA analyzes the part of solid alone to set the initial state of the analysis. After the results is superimposed, the coupling analysis is carried out.

4.2 *Analysis methods*

4.2.1 *Discontinuous deformation Analysis (DDA) with disc element*

This study analyzes a debris flow with DDA which gets stable solution based on the implicit method. The basic formula of DDA is an equation of motion including a solid contact based on Hamilton's principle. The equation of motion shown in Eq. (1) is formulated by the principle of least work.

$$M\ddot{u} + C\dot{u} + Ku = F\left(\dot{u} = \frac{du}{dt}, \ddot{u} = \frac{d^2u}{dt^2}\right) \tag{1}$$

Where, M : Mass matrix, C : Dumping matrix (= ηM), η : viscous coefficient,
K : Stiffness matrix, F : External force vector, t : time,
u : displacement at the block center of the gravity, $\dot{u}$: velocity, $\ddot{u}$: acceleration

The original DDA was invented by Shi (1989) and uses convex polygon blocks. It can simulate complex fractured rock masses and results of the computation were recognized to be very effective. However, because of complex contact points problems, the original DDA had difficulty to handle large number of blocks. Ohnishi and Miki (1996) introduced a simple rigid disk (circular) element (block) to speed up the solution process. In this paper, we assume that debris flow consists of various size of disks (circular) elements (blocks) and water. This can be extended easily to a 3D problem using a ball element. Then we can handle many blocks for debris flow simulation.

For a rigid disc element, displacement vector (u, v) can be expressed as follows,

$$\begin{pmatrix} u \\ v \end{pmatrix} = \begin{pmatrix} 1 & 0 & -(y-y_0) \\ 0 & 1 & x-x_0 \end{pmatrix} \times \begin{Bmatrix} u_0 \\ v_0 \\ r_0 \end{Bmatrix} = [T_i] \times [D_i] \tag{2}$$

Where, (u_0, v_0) is a rigid displacement vector at the center in the equation, r_0 is rotation, (x_0, y_0) is a center of the disc. The total energy of the whole system is shown in the following Eq. (3).

$$\Pi_{sys} = \sum_i \Pi_i^B + \sum_i \sum_j \Pi_{i,j}^{PL} \tag{3}$$

The first term in the right-hand side shows the potential energy for each block and the second term shows the potential energy for contact between block i and block j respectively. The equation of motion for the block system is formulized by the Hamilton's principle. Displacement vector D and the external force F can be obtained by minimizing the total potential. And the equation of motion is solved by introducing time integration of Newmark β and γ method. This procedure is same for a rigid block or an elastic block.

In this study, solid part is analyzed by NMM-DDA (Miki et. al., 2010), which is coupling analysis of NMM and DDA, can deal with large movement of blocks with deformation and rigid body rotation. NMM-DDA are formulated in similar way as is shown in the procedure for the disc element. The total displacement (u, v) at the point (x, y) inside i-th DDA block can be calculated by:

$$\begin{pmatrix} u \\ v \end{pmatrix} = \left[T_i^d(x,y)\right]\left[D_i^d\right] \tag{4}$$

$$\left[T_i^d\right] = \begin{pmatrix} 1 & 0 & -(y-y_0) & (x-x_0) & 0 & \dfrac{(y-y_0)}{2} \\ 0 & 1 & (x-x_0) & 0 & (y-y_0) & \dfrac{(x-x_0)}{2} \end{pmatrix} \tag{5}$$

$$\left[D_i^d\right] = \begin{pmatrix} u_0 & v_0 & r_0 & \varepsilon_x & \varepsilon_y & \gamma_{xy} \end{pmatrix}^T \tag{6}$$

where $[T^d_i]$ is the DDA block deformation matrix (displacement function) for i-th block, (x_0, y_0) is the location of gravity center of block i, (u_0, v_0) in Eq. (6) is the rigid body transformation, r_0 is the rigid body rotation of the block at the gravity center (x_0, y_0), and ε_x, ε_y, γ_{xy} are the normal (in the x- and y-directions) and shear strains of the block, respectively. For the NMM, assuming that the shape of the cover mesh is triangle and the coordinates of three nodes of the triangle, C_1, C_2 and C_3

are (x_1, y_1), (x_2, y_2) and (x_3, y_3), respectively, the displacement (u, v) at the point (x, y) in the element defined by C_1, C_2 and C_3 can be given by:

$$\begin{pmatrix} u \\ v \end{pmatrix} = \left[T_j^m(x,y)\right]\left[D_j^m\right] \tag{7}$$

$$\left[D_j^m\right] = \begin{pmatrix} u_1 & v_1 & u_2 & v_2 & u_3 & v_3 \end{pmatrix}^T \tag{8}$$

where $[T^m_j]$ is the element deformation matrix (displacement function) for j-th element and (u_k, v_k) (k=1, 2, 3) in Eq. (6) means the displacements at the triangle nodes C_k. The deformation matrix $[T^m_j]$ can be extended as following equation.

$$\left[T_j^m(x,y)\right] = \begin{pmatrix} f_1 & 0 & f_2 & 0 & f_3 & 0 \\ 0 & f_1 & 0 & f_2 & 0 & f_3 \end{pmatrix} \tag{9}$$

where, $\begin{pmatrix} f_1 & f_2 & f_3 \end{pmatrix} =$

$$(1 \times y)\frac{\begin{pmatrix} x_2y_3 - x_3y_2 & x_3y_1 - x_1y_3 & x_1y_2 - x_2y_1 \\ y_2 - y_3 & y_3 - y_1 & y_1 - y_2 \\ x_3 - x_2 & x_1 - x_3 & x_2 - x_1 \end{pmatrix}}{\begin{vmatrix} 1 & x_1 & y_1 \\ 1 & x_2 & y_2 \\ 1 & x_3 & y_3 \end{vmatrix}} \tag{10}$$

The total potential energy Π_{sys} of the block system, which includes the DDA blocks and NMM elements, can be expressed as the following equation.

$$\Pi_{sys} = \Pi^m_{sys} + \Pi^d_{sys} + \sum_{B,i} \sum_{E,j} \Pi^{PL}_{i,j} \tag{11}$$

In Eq. (11), the first and second term on the right side are the potential energy for the DDA blocks and NMM elements, respectively. The last term on the right side of Eq. (9) represents the potential energy for the contacts between DDA block i and NMM element j. The matrices and vector in kinematic equations based on Hamilton's principle (Eq. (1)), can be also obtained by minimizing the potential energy expressed as Eq. (11). However, the potential energy for DDA part, Π^d_{sys} and NMM part, Π^m_{sys} are minimized with respect to the block displacement $[D^d]$ and displacement $[D^m]$, respectively.

4.2.2 *MPS (Moving Particle Simulation) method*

Discretization of continuum medium (usually fluid) by particle method is categorized into roughly two groups. Both SPH (Smoothed Particle Hydrodynamics) and MPS (Moving Particle Simulation) are used in the field of fluid dynamics and geomechanics.

The MPS method is a computational method which is suitable for simulating fragmentation of incompressible fluids. It is a macroscopic, deterministic particle method (Lagrangian mesh-free method). It solves the following Navier-Stoke equation which is fundamental equation for water.

$$\frac{Du}{Dt} = -\frac{1}{\rho}\nabla P + \frac{\mu}{\rho}\nabla^2 u + F \tag{12}$$

u: velocity, t: time interval, D/Dt: Lagrangean operator, ρ: density, P: pressure,
μ: viscosity, F: external force, g: gravity vector, ∇: gradient, ∇^2: Laplace operator

First term on the right side of Equation (10) means pressure gradient, and the second term means viscosity. The external force term and viscosity term are solved explicitly, and the pressure gradient term is solved implicitly. The gradient and Laplacian model for i-th particle are expressed as:

$$\langle \nabla \phi \rangle_i = \frac{d}{n^0} \sum_{j \neq i} \left[\frac{\phi_j - \phi_i}{\left| \mathbf{r}_j - \mathbf{r}_i \right|^2} \left(\mathbf{r}_j - \mathbf{r}_i \right) w\left(\left| \mathbf{r}_j - \mathbf{r}_i \right| \right) \right] \tag{13}$$

$$\nabla^2 \phi_i = \frac{2d}{n^0 \lambda^0} \sum_{j \neq i} \left(\phi_j - \phi_i \right) w\left(\left| r_j - r_i \right| \right)$$

$$\lambda^0 = \frac{\sum_{j \neq i} \left| r_j^0 - r_i^0 \right|^2 w\left(\left| r_j - r_i \right| \right)}{\sum_{j \neq i} w\left(\left| r_j - r_i \right| \right)} \tag{14}$$

where ϕ is the variable, r_i and r_j are the position vector of i-th and j-th particle, respectively. d is the number of the spatial dimension, and n° is the initial particle number density. The function w is the weighted function and expresses the degree of influence to the particle j that exists inside the influential radius r_e of the particle i as given by Equation (15).

$$\begin{aligned} w\left(\left| r_j - r_i \right| \right) &= \frac{r_e}{\left| r_j - r_i \right|} \quad \left(0 \leq r < r_e \right) \\ w\left(\left| r_j - r_i \right| \right) &= 0 \quad \left(r_e \leq r \right) \end{aligned} \tag{15}$$

In the process of MPS analysis, viscosity term and external force term are solved explicitly and pressure gradient term is solved implicitly. Therefore, this method is classified as a semi-explicit solution technique.

Then this study pursues the movement of water changing the boundary conditions every moment by MPS method. In the computer simulation of hydrodynamics by particle method, space and time are discretized. The discretization of the space means a fluid is set of multiple particles. One particle is regarded as a lump of the fluid having a size. Then, the discretization of the time means the time interval is short enough so that the water particles move gradually and the position of the water particle is calculated at each time interval.

4.3 *Method of coupling between DAA and MPS analyses*

4.3.1 *Procedure of the analysis (Flow chart)*

Procedure of the analysis (descriptive flow chart) is shown in Figure 7. In the initial state setting of the analysis, MPS analyzes the part of fluid alone and DDA analyzes the part of solid alone. At the beginning MPS analysis does not consider the existence of the DDA block and DDA analysis does not consider the existence of the MPS particles. After each analysis is carried out, DDA blocks and MPS particles are processed not to overlap as prepare for coupling analysis. The analysis by coupling fluid and solid is calculated step by step (at each time step) by repeating the MPS analysis and DDA analysis. Therefore, at this moment the fluid and solid is not fully coupled in the analysis.

4.3.2 *Modeling DDA block And NMM element*

In an actual joint between sound rock mass, flow water does not go through a solid rock because permeability of rock is usually very low comparing to soil or void of rock mass. Therefore, water particles in MPS are analyzed in conditions that they do not enter the interior of solid parts, such as DDA block and NMM element. To realize this condition in the analysis, wall particles of MPS,

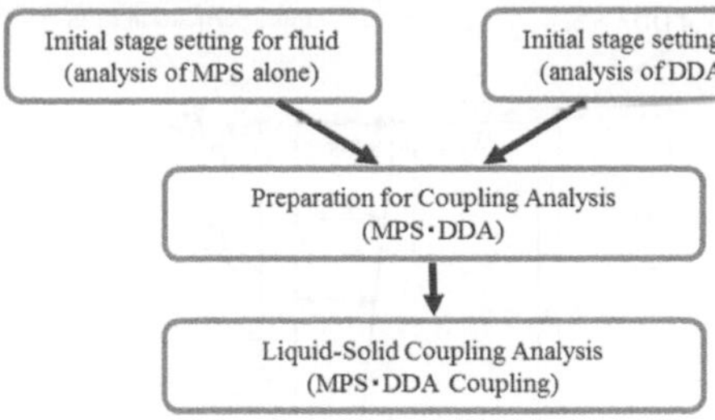

Figure 7. The flow chart for the MPS-DAA coupling analysis.

which are used for setting impermeable condition in MPS, are arranged inside the DDA block as shown Figure 8. The wall particles arranged inside the block act as an impermeable wall because particles of MPS avoid closing to each other in the principle of the analysis. At the time of analysis, the wall particles move with the movement of the block in the NMM-DDA analysis.

On the other hand, in the above mention method, water particles cannot go through joint and void between blocks or elements under contact. It means that water cannot flow along the joint and no water pressure act along the joint. To avoid this condition, the wall particles of MPS are arranged inside the DDA block to enable the flow of water particles (an imaginary hydraulic aperture is introduced) as shown in Figure 8. For NMM elements, the wall particles of MPS are arranged inside the joint loop, which is the periphery of NMM elements. Water pressure and water flow along the joints between DDA blocks, NMM elements, and DDA block and NMM element are performed by above mentioned method.

4.3.3 *Force acting from the water particles to DDA block and NMM elements*

Acting force from the MPS water particles to the DDA block and/or NMM elements is calculated by water pressure. Water pressure is caused by the water particles located in the range of influence to the DDA block (See Figure 9). Water pressure acts on periphery of the block and a direction of the force is normal to the periphery of the block. For the DDA block, water pressure act along the edge of the block as line distributed force. Assuming that the water pressure acting on 2 points (x_1, y_1) and (x_2, y_2) along the edge of DDA block is $F_1 (f_{x1}, f_{y1})$ and $F_2 (f_{x2}, f_{y2})$ as shown in Figure10, the sub-matrices for the line distributed force is expressed by Eq. (14). The distance between points (x_1, y_1) & (x_2, y_2), and (x_0, y_0) is the location of gravity center of DDA block i. Water pressure acting on the DDA block is derived by assembling Eq. (14) into global stiffness matrix. For the NMM element, water pressure act along the wall particle periphery inside the joint loop of NMM elements as line distributed fore. The sub-matrices for the line distributed force in NMM element is easily derived by the same procedure as that of DDA block.

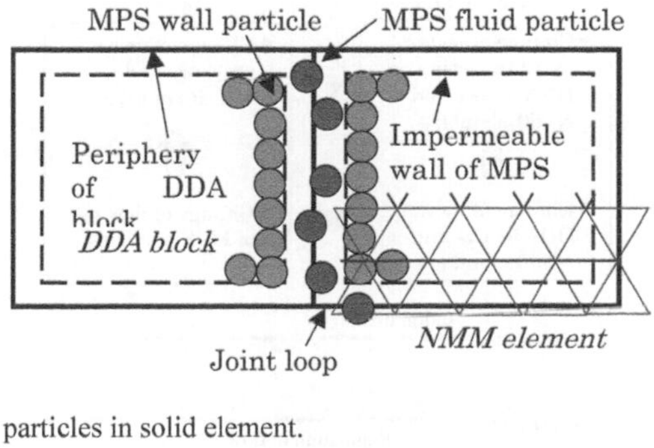

Figure 8. Concept of wall particles in solid element.

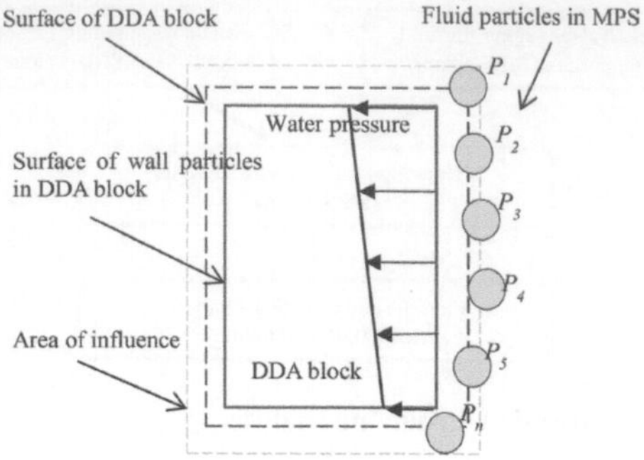

Figure 9. Interaction from fluid particles to blocks.

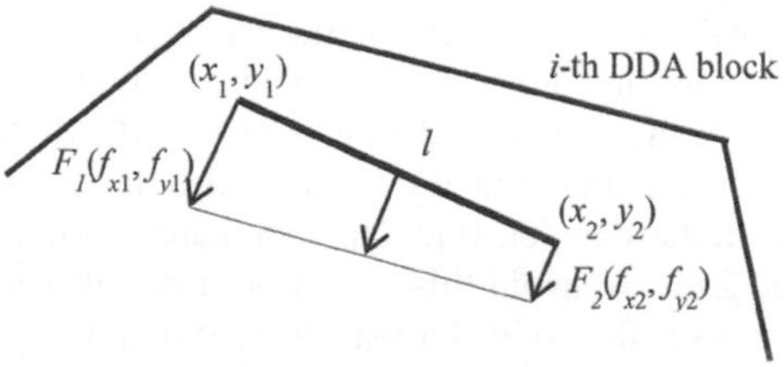

Figure 10. Line load along the point (x_1, y_1) to the point (x_2, y_2).

4.3.4 *Procedure of coupling analysis*

Before starting the coupling analysis, the initial state must be set both in MPS analysis also in NMM-DDA analysis. In the NMM-DDA, an initial block alignment should be obtained under the condition of self-weight and contact forces between blocks and/or elements. In the MPS, an initial velocity and water pressure of the particles should be constructed. An initial setting analysis of MPS and NMM-DDA is individually performed at first. Next, the coupling model is constructed by superposing the results of the initial setting analysis of MPS and NMM-DDA. Water pressure obtained by MPS analysis works to the DDA blocks and NMM elements, and the motion of the blocks and elements is controlled. On the other hand, movement of the blocks changes the flow path of the water particles and it regulates velocity and pressure of water particles. Coupling analysis is completed by iterating this process alternatively.

In order to avoid excess overlap between water particles, time step increment should be chosen carefully. Movements of the block per one step have to be held in the permissible range of particles overlap in MPS. Also, excess overlap between water particle and wall particle inside the DDA block and NMM element should be avoid. Therefore, the time step interval in NMM-DDA and MPS needs to be set small enough. Coupling analysis is done by alternatively performing MPS analysis and NMM-DDA analysis. The flow chart of coupling analysis is shown in Figure 11 schematically (Kuno et.al., 2016).

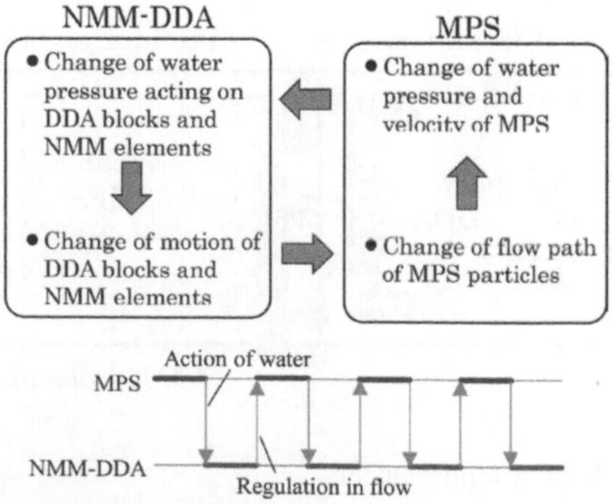

Figure 11. The flow of coupling analysis.

5 EXAMPLES OF THE COUPLING ANALYSIS

5.1 *Falling disk blocks with running water*

The analysis model for disc blocks is shown in Figure12. It is the analysis model to see the effect of the running water on rocks when they fall off on a slope. For MPS model, fluid particles are unlimitedly suppled at the left side top of slope at a certain water level and water (fluid particles) goes down on the slope with a certain velocity. Water particles diminish at the bottom of the slope to maintain a water level on the slope. The slope inclines 20 degree. Table 2 and 3 show physical properties and parameters used in the MPS and DDA analysis.

Before the coupling analysis, the initial state setting analysis was carried to make stable output. 2000 time steps iteration was necessary number in MPS analysis to reach the steady condition. At this stage the water pressure and flow velocity were obtained. In addition, DDA alone analysis was conducted to compare with coupled analysis and no water DDA analysis. Physical properties and parameters in this analysis is same as that of coupling analysis (Sasaki et.al., 2015).

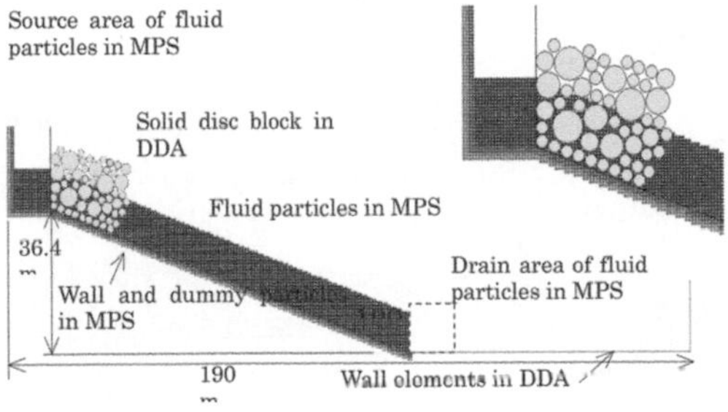

Figure 12. Analysis model for disc block.

Table 2. Physical properties and analytic parameters in MPS.

Initial distance among particles (m)	0.5
Density of fluid (kg/m3)	1000
Coefficient of kinematic viscosity (m2/s)	1.0×10^{-4}
Compressibility (1/Pa)	0.45×10^{-9}
Gravitational acceleration (m/s2)	9.8
Time step (s)	0.001

Table 3. Physical properties and analytic parameters in DDA.

Radius of disc elements (m)		1~2.5
Density of solid element (kg/m3)		2650
Friction angle of surface (degree)		10
Cohesion of surface (N/m2)		0.0
Velocity damping in collision		0.85
Penalty spring constant (N/m)	Normal	1.0×10^{10}
	Shear	1.0×10^{8}
Max. time step (s)		0.001
Number of time step	Initial	0
	Coupled	50000

Coupling analysis and DDA alone analysis results are shown in Figure 13 and 14 for comparison. In the coupling analyses, number of time steps is 50000. The result shows that water particles in MPS flows unevenly and some part of water was dammed up by DDA disc blocks. And some of the water overflows over the disc blocks. This means that interaction between fluid and solid is noticed clearly. Comparing to DDA alone analysis result, the distribution of the disk block is different at a same elapsed time. It shows water pressure influences on the movement of the disk blocks.

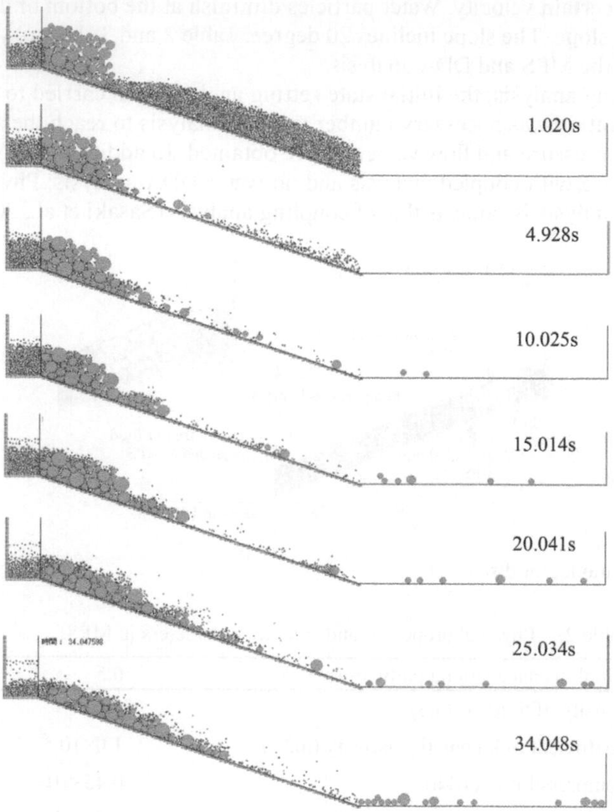

Figure 13. Coupled analysis results of DDA and MPS for debris flow.

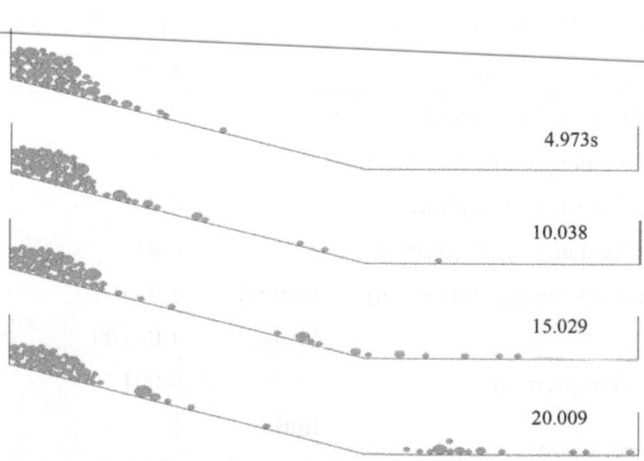

Figure 14. Analysis of DDA alone with no water for debris flow.

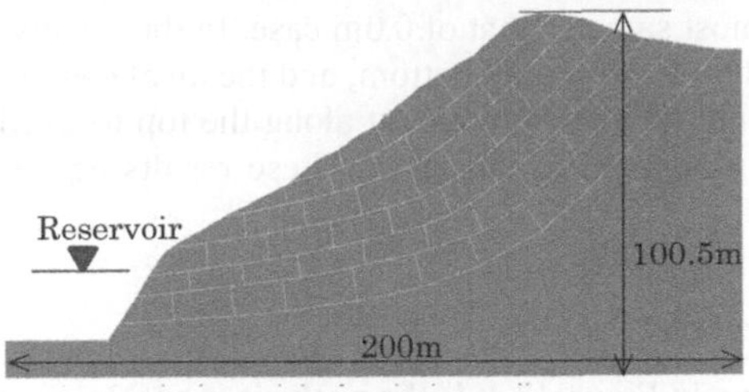

Figure 15. Analysis model of rock slope adjacent to a reservoir.

At the beginning of coupling analysis, it seems that water on the slope flows smoothly to the sink hole at the end of the slope and then water level gradually decreases because supply of water particles is restricted by disc blocks. At the time of 10 seconds after the beginning, the remarkable movement of the disc blocks start, then water and blocks flow down on the slope together. At around 20 seconds, disk blocks and the water particles rapidly move down together to the bottom. This is basically similar to debris flow which is often seen at the mountain slope with a heavy rain. Comparing coupled analysis and DDA alone analysis, it is noticed that there is not so much difference in the falling speed of disk block, but disc blocks in coupled analysis are rolling down further to a long distance and spread widely. Those means debris flow with water (rain) extend the damages.

5.2 *Landslide stability in rising water level*

The analysis model is shown in Figure 15. It is the analysis model to show the effect of the water pressure acting on joints between the DDA blocks. In the model, the DDA blocks are arranged as layered and dip-slope, and the width of the joint between wall particles inside the DDA blocks is approximately 2m, and its width is twice size of representative diameter of water particles in MPS. However, the width of the joint is artificially defined in order to give a water pressure to the DDA blocks. The total number of the DDA blocks is 100. Table 4 show physical properties and parameters used in the MPS and NMM-DDA analysis.

The analysis was made for four case of water level, where the water level is 0.0m, 10.0m, 30.0m and 50.0m. Before the coupling analysis, the initial setting analysis was carried out to build the contact forces between the DDA blocks, and 4000 iterations in NMM-DDA analysis was performed. On the other hand, the initial state setting analysis in MPS analysis was omitted because static water pressure was gradually applied to the DDA blocks after the establishment of contact forces. In the coupling analysis, NMM-DDA and MPS analysis were alternatively repeated for the time steps in 4000 iteration.

The results for 0.0m in water level show the remarkable displacement along the top and second layers. In the 10.0m case, water particles enter inside the lower joints. However, the displacement of

Table 4. Physical properties and analytic parameters in NMM-DDA.

Young's modulus(N/m2)		1.0×109
Poisson's ratio		0.3
Density of solid element (kg/m3)		2500
Friction angle of surface (degree)		43
Cohesion of surface (N/m2)		0.0
Penalty spring constant (N/m)	Normal	1.0×109
	Shear	1.0×108
Max. time step (s)		0.001
Number of time step	Initial	1000
	Combined	4000

top and second layer is almost same as that of 0.0m case. In the results of 30.0m case, water particles enter inside the 4 joint layers from the bottom, and the displacement of third layer is observed. For the 50.0m case, the remarkable displacement along the top to third layers, however the slight displacement is observed along 4th to 6th layer. These results agree with the general landslide movement.

5.3 *Collapse of landslide dam*

The analysis model is shown in Figure17. It is the analysis model to see the effect of the water pressure along the void between the blocks when the landslide dam collapses. As shown in Figure 17, the basement is modeled as NMM elements, and the landslide dam is modeled as DDA blocks. The void, of which width is 2.0m, is built along the basement of landslide dam, and its width is twice size of representative diameter of the water particle. Moreover, the width of the joint between wall particles inside the DDA blocks is set to 1m. The water supply area is also built to maintain the water level behind the landslide dam. The number of DDA block and NMM element are 18 and 1261, respectively. The physical properties and analysis parameters used in the MPS and DDA analysis are shown in Table 2 and 3.

Before the coupling analysis, the initial setting analysis, where total number of iteration time was 500, was carried out to build the contact forces between the DDA blocks and NMM elements. On the other hand, the initial state setting analysis in MPS analysis was omitted because static water

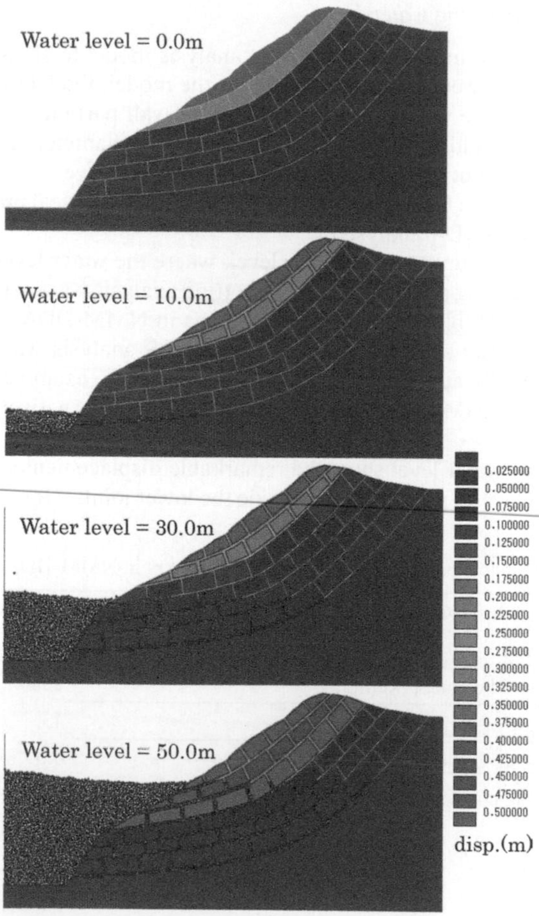

Figure 16. Results by coupled analysis for mountain slope.

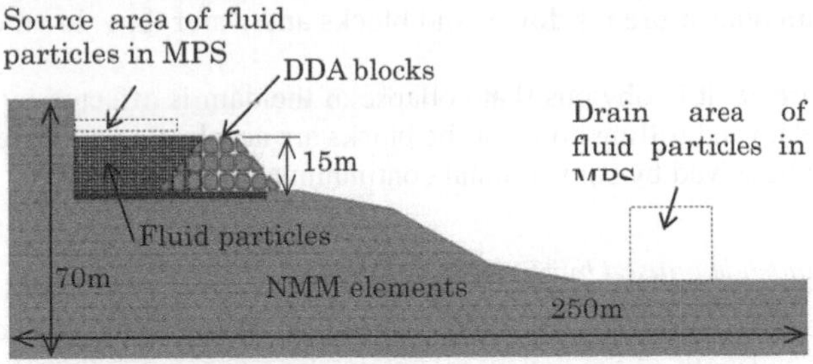

Figure 17. Landslide made dam model of coupling analysis in MPS, NMM and DDA.

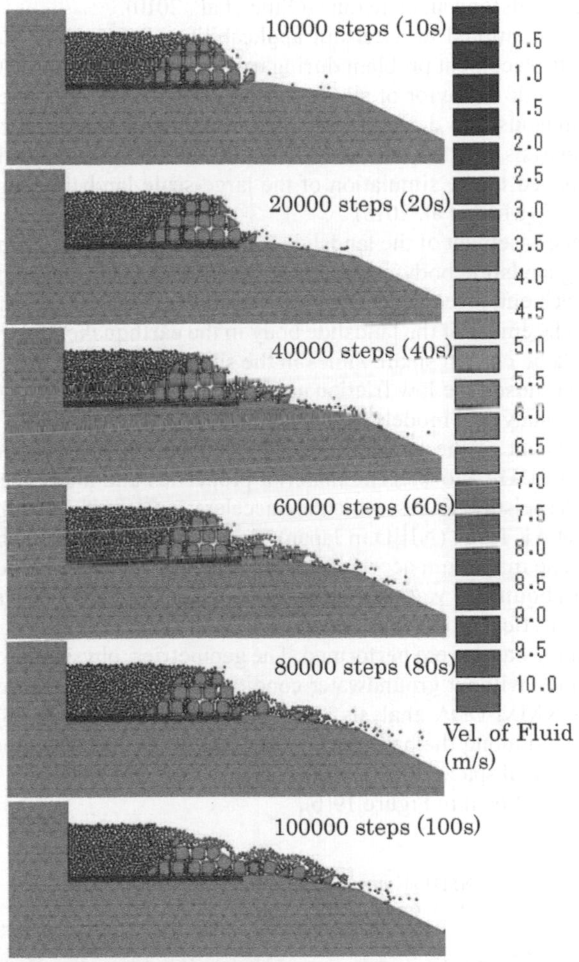

Figure 18. Landslide made dam results of coupling analysis (failure process).

pressure was gradually applied to the DDA blocks and NMM elements after the establishment of contact forces. In the analysis, number of time steps iteration was 200000.

The results of coupling analysis are shown in Figure 18. At the beginning of coupling analysis, opening between the blocks and inflow of the water particles are observed at the right side of the dam as shown in Figure8 at the 10000 steps. It is considered that the water pressure acting along the void beneath the dam advances opening between blocks. After 40000 steps, the right side of landslide dam almost breaks down, and blocks and water flow down on the slope around 100000 steps.

As shown in Figure 18, it is obvious that collapse of the dam is affected by the water pressure. The collapse of the dam and rolling down of the blocks are accelerated by water pressure and flow. This process cannot be solved by conventional continuum analysis method.

5.4 *Large-scale landslide caused by strong earthquake*

The large-scale landslide (Aratozawa landslide) trigged by Iwate-Miyagi inland earthquake (M7.2) occurred in 2008. The length of the moved mountain slope is about 1.3km, the volume of moved landslide body amount to $6.7\times10^{7}m^{3}$, and the maximum displacement of rock masses in the middle and lower part of the landslide is about 300m. The dip angle of main sliding surface is 0-2°. The formation around the landslide, which is Neogene sedimentary rocks, consists of welded tuff, pumice tuff and alternation of sandstone and siltstone (Ohno et al., 2010).

The simulation was attempted to ascertain applicability of the newly coupling method to the large scale and large displacement problem during earthquake (Irie et al., 2009). For the numerical simulations of the dynamic behavior of slopes during earthquakes, it is necessary and preferable to consider both continuous and discontinuous deformations of fractured rock masses appropriately. The coupling analysis of DDA and NMM (Miki et al. 2010) was introduced and this method (NMM-DDA) was applied to the simulation of the large-scale landslide trigged by Iwate-Miyagi inland earthquake, 2008 (Miki et al. 2013).

In the simulation, the basement of the landslide was divided by NMM elements (continuous rock mass) and the moving landslide body consisted of DDA blocks. The seismic forces were given to the basement NMM elements to consider the local variation of seismic forces. However, in order to simulate the large displacement of the landslide body in the earthquake, the friction angles between the DDA blocks should be chosen small values in the simulation. It has considered that the lack of ground water affection caused the low friction angles.

Figure 19 shows the analytical models. The length and height of the model are 1900m and 325m, respectively. The basement of the landslide was divided into NMM elements, and the landslide mobile body consists of DDA blocks. The material properties and analytical conditions are summarized in Table 5. The seismic forces, which are acceleration records during Iwate-Miyagi inland earthquake observed at AKTH04 (NIED in Japan), were given to the basement NMM elements as a dynamic body force. The maximum acceleration of horizontal motion was beyond 1000 gal. In the simulation, the viscous boundary was applied along both right and left sides, and the displacements along bottom of the analytical model were fixed.

In the simulation, three cases were performed. The geometries, physical properties, seismic force and boundary conditions without groundwater conditions were equal in three cases. As shown in Figure 19, Case1 was NMM-DDA analysis alone (no interaction with MPS). In Case 2, the void (line space) was arranged along the basement of landslide body of DDA blocks on the NMM element. However, the artificial spaces between DDA blocks were not arranged. The initial groundwater condition in Case 2 is shown in Figure 19(b).

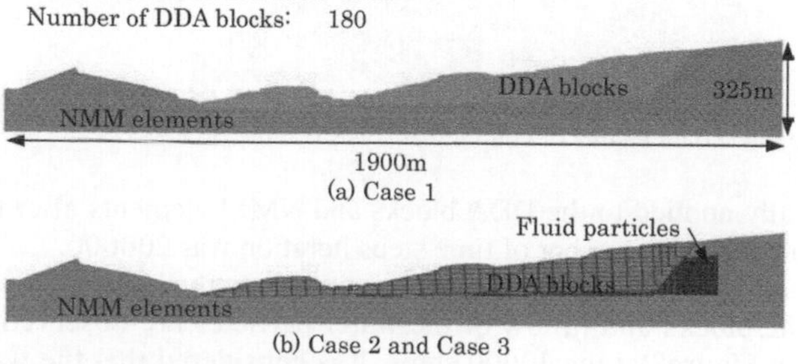

Figure 19. The analytical model of the large-scale landslide by earthquake.

Table 5. Physical properties and analytic parameters in MPS.

Properties and parameters	MPS
Initial distance among particles (m)	5.0
Density of fluid (kg/m^3)	1000
Coefficient of kinematic viscosity (m^2/s)	1.0×10^{-6}
Compressibility (1/Pa)	0.45×10^{-7}
Gravitational acceleration (m/s^2)	−10.0
Time step (s)	0.01

Fluid particles for MPS were distributed in right side of the NMM basement to apply water pressure to DDA blocks. In the analysis, interaction between MPS fluid particles and NMM elements were ignored, and interaction between fluid particles and DDA blocks were taken into account. In Case 3, the space line was arranged along the basement of landslide body of DDA blocks and the artificial line spaces between block was arranged.

The width of artificial line space between DDA blocks, where a water pressure is given to the DDA blocks, was set about twice size of representative diameter of water particles in MPS. The initial groundwater condition in Case 3 was equal to that in Case 2. Table 5 shows physical properties and parameters used in MPS analysis. Before the coupling analysis, the initial setting analysis was carried out to build the contact forces between the DDA blocks, where 1000 iterations in NMM-DDA analysis was performed before applying the seismic force.

Figure 20 shows the simulation results after 40,000 steps. The displacement of DDA blocks in Case 1 and Case 3 was nearly equal. The displacement in Case 2 was different from that in Case 3. Comparing Case 2 with Case 3, the block displacement in upper part of the landslide body in Case 2 was larger than that in Case 3. However, the block displacement along the basement in Case 2 was slightly larger than that in Case 3.

In Case 3, fluid particles moved through the artificial void between blocks, and the groundwater surface after 40,000 steps was almost flat as shown in Figure 20(c). On the other hand, the groundwater surface of Case 2 was not flat because fluid particles could not move between blocks. The water pressure along the basement in Case 2 was larger than that in Case 3 consequently. From this reason, it is considered that the block displacement along the basement in Case 2 showed larger displacement.

Figure 21 shows the simulated block displacement in Case 2 after the earthquake. The large displacement of DDA blocks appeared in the center and upper part of the landslide body, and maximum displacement was beyond 300m. The water pressure acting on the bottom of the landslide body was about 800kPa. In early stage of the simulation after the earthquake, the subsidence appeared at the head part of the landslide, which is the right-hand side in the model. The toe part of the landslide in

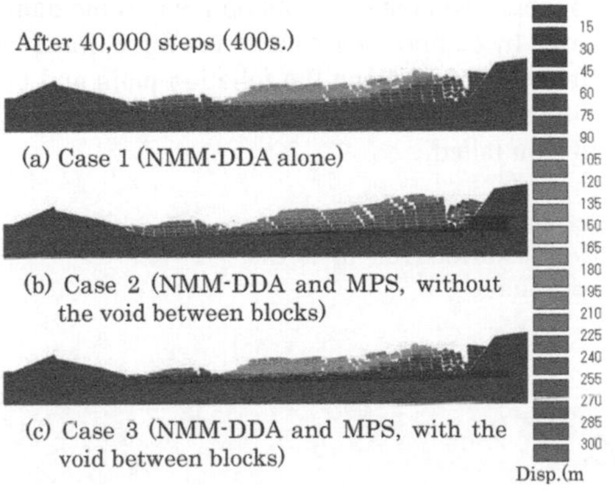

Figure 20. Displacement distribution of DDA blocks after 40,000 steps.

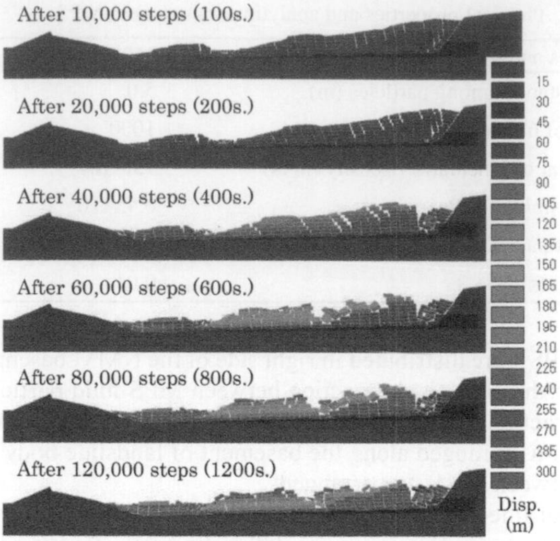

Figure 21. Displacement distribution of DDA blocks in Case 2 at landslide.

the left-hand side of the model was thrust up to the left side slope in late stage of the simulation. A remarkable thrust up in Case 2 did not appear in Case 1 and Case 3. It is clear that a water pressure affected the landslide movement.

Figure 22 shows the simulation results after 120,000 steps in Case 2. The ground surface line by geological survey after the earthquake is also shown in Figure 22. The surface line of the simulation in Case 2 is similar to that of the geological survey. It is considered that the simulation results are consistent with the geological survey. These results show the potential ability to simulate the large-scale landslide movement during earthquake. However, the simulation results also show complex relations among strength of sliding surface, water pressure and permeability of joint.

5.5 *Malpasset dam failure: Preliminary case study*

5.5.1 *Malpasset dam accident*

Complex geology and water in rock masses may be major sources of problems in dam safety. Water seeping in rock masses can affect the safety of dams in essentially two ways: erosion and uplift. Several dam failures have been attributed to excessive uplift.

This year 2019 is the 60th memorial year when the Malpasset dam in France failed in 1959. Failure mechanism of Malpasset dam is estimated as shown in Figure 23. That is at the downstream of the dam site the rock stress increased along the foliation due to the dam water filling. The permeability of these parts decreased by compression of external force (force from dam gravity and filled reservoir water). The uplift was exerted along the foliation plain and pushed up the wedge shape rock mass at the bottom of the dam. Consequently, move of the base rock caused the twist deformation of the dam surface and dam failed.

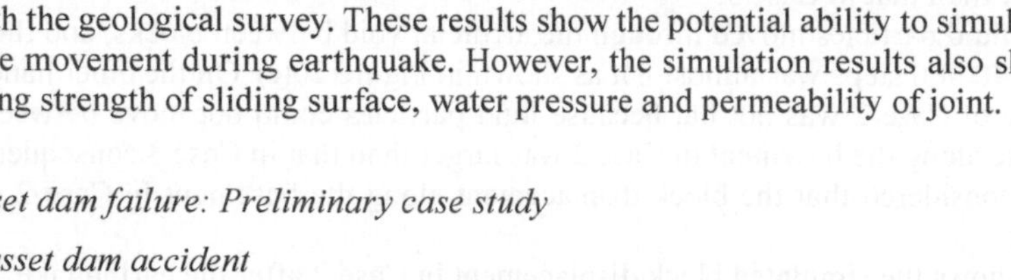

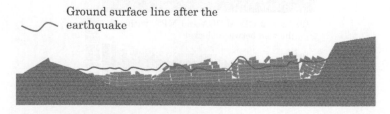

Figure 22. Comparison between simulation results and ground surface line after the earthquake.

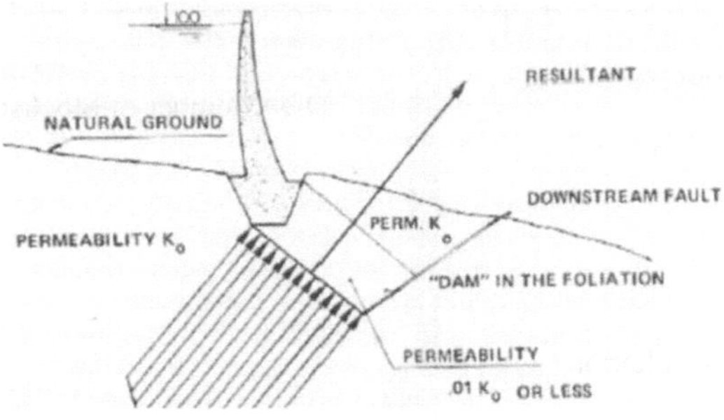

Figure 23. Estimated mechanism of Malpasset dam failure(Londe 1987).

Many investigation and numerical analyses have been done, based on the assumption that rock masses behave like porous continua (e.g. soils) with respect to seepage and ignore the influence that geology could have on uplift. It is suggested that the geology of dam foundations be fully understood prior to selection of an uplift distribution.

The importance of geology in controlling uplift and dam stability has been emphasized by several authors over the past years. Casagrande (1961) emphasized how unfavorable geology could lead to unusual high uplift. He found that problems with excessive uplift pressures could be attributed to isolated, highly pervious geological features such as faults, seams and shear zones which are located in the foundation near the base of the dam. Dam engineers should keep in mind the importance of following two factors.

(1) the effect of rock mass discontinuities, anisotropy and block movement on dam stability, and
(2) the effect of drains and grouting on uplift, under both static and dynamic conditions.

Predicting the stability of a dam on jointed rock must be done using a combination of hydraulic and mechanical models. Indeed, as a jointed rock mass deforms under gravity, reservoir and seismic loads and seepage forces, rock discontinuities open or close and their permeabilities change. This, in turn, creates a new distribution of water pressure in the rock mass and induces new deformations and stresses in the dam-foundation rock system.

In the discrete models, a rock mass is modeled as a network of finite planar hydraulic conduits which are allowed to deform in directions normal and parallel to their planes. The discontinuities have zero tensile strength and therefore can open when subjected to tensile stresses. Also, they are allowed to shear and their shear strength can be mobilized. Opening and sliding of discontinuities change their hydraulic properties and the pressure distribution in the rock mass. The new pressure distribution creates more opening and sliding. For each network, a discrete finite element mesh could be constructed, a seepage analysis carried out and the uplift calculated. Minor geological details could indeed create larger or smaller uplift forces than those calculated with the conventional approach.

We introduced coupling analysis of the discontinuous models using NMM-DDA and particle MPS to solve a complex solid and fluid interaction problems. Here we will perform preliminary study to check the applicability of 2-D DDA-NMM and MPS coupling method whether the dam failure can be simulated by this coupling method.

5.5.2 *DDA-NMM and MPS coupling analysis*

5.5.2.1 DDA block model

Dam structure is assumed to be one block for simplicity. Foundation rock masses consist of convex blocks formed by cross discontinuities, foliation and joints. There is an inclined fault at the downstream side in the rock mass. Number of DDA blocks are 66. Dam width is set to be wider than real size in order to put wall particles. Three types of particle conditions are set for apertures to percolate water. The hydraulic aperture used here is artificial for calculation purpose.

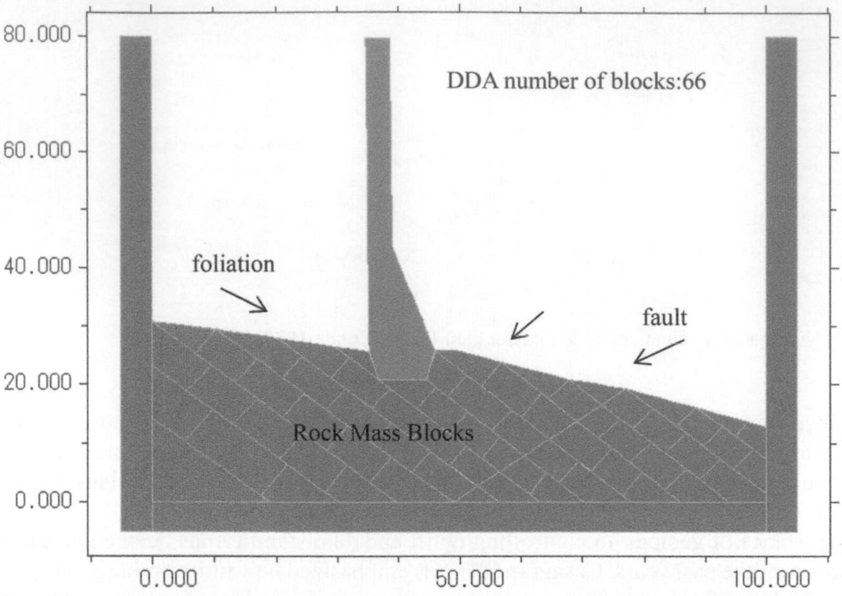

Figure 24. Model for Malmasset dam example analysis (block part).

1. Type 1: Water goes through only along foliation which is directly connected with dam reservoir water. Water pressure will be applied to the perpendicular to the foliation plane.
2. Type 2: Discontinuities at the left side of fault can take water inside, but path is not connected to the ground surface on the left side. Water can go through (along) the fault but cannot across the fault.
3. Type 3: Type 3 condition is almost same as Type 2. Water path is closed at the ground surface. Density of wall particle is sparse, then water particles may leak a little.

5.5.2.2 Result of analyses

3 cases of simulation were performed as shown in Table 6 with different types. Figure 25 shows stress distribution in the rock mass. In Case 1, stress concentration may occur along foliation direction at the area of bottom of dam toward downstream (righthand side in the figure). Since there is no water leakage in Case 1, hydraulic pressure is applied to each rock blocks at the dam foundation along foliation direction. In Case 2, water washed out a few small blocks near surface and water continuously leaks. Therefore, water pressure is released and is not as big as Case 1. In Case 2 and Case 3, stress concentration tends to release and in Case 3 stress distribution is close to original no-water condition. In Case 3 water gradually leaks at the downstream surface. Water pressure accumulated in rock joints is bigger than in Case 3, but smaller than Case 1.

Table 6. Analysis cases for different type.

Condition(MPS model)	type 1	type 2	type 3
Young's Modulus(rock) : 0.7×10^9(N/m^2) Friction angle(fault) : 10° Friction angle(foliation, joint) : 20°	Case-1	Case-2	Case-3

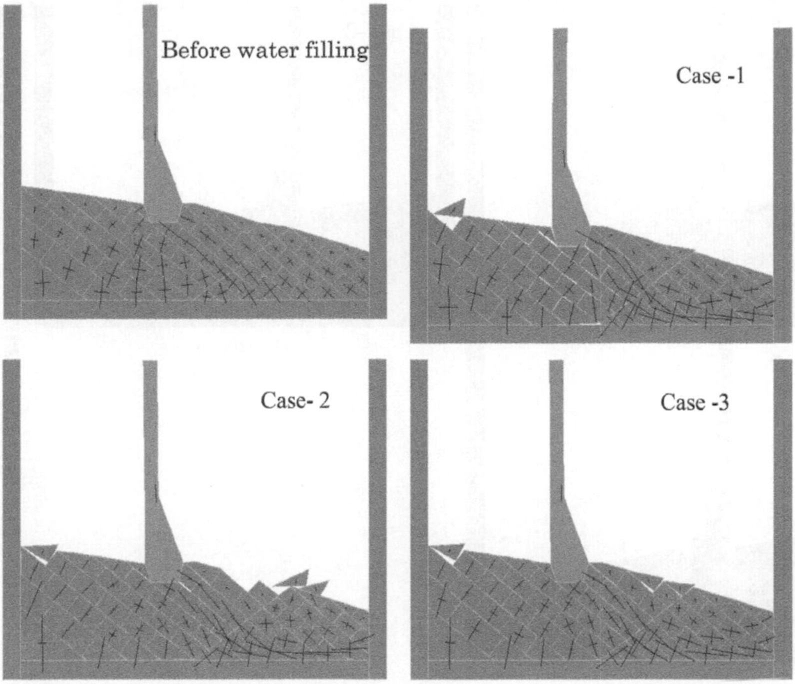

Figure 25. Stress distribution after reservoir filled and water goes through discontinuities (3000 times repetitions).

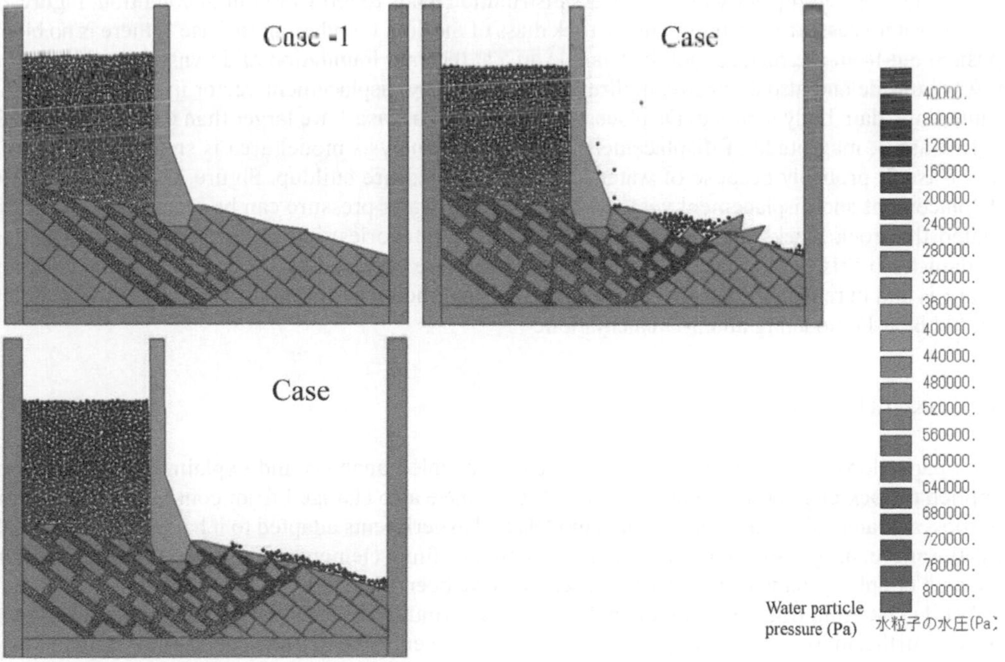

Figure 26. Water pressure distribution in the rock mass (Coupling iteration 3000).

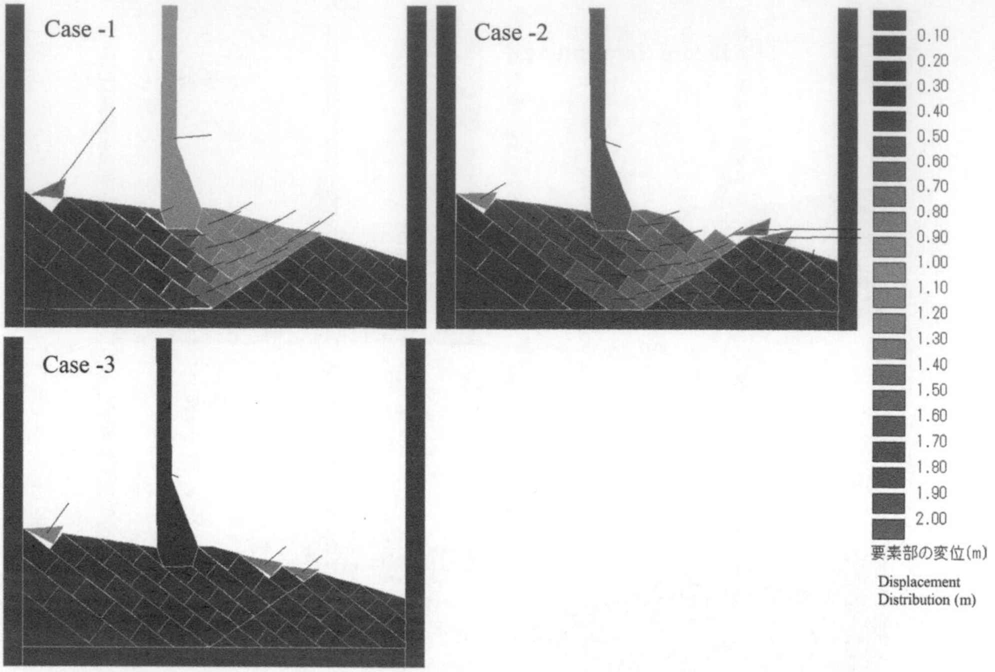

Figure 27. Classified displacement and displacement vector (coupling iteration 3000 times) (displacement vector magnification 10).

Stress distribution after reservoir water filling depends on water flow condition of MPS. If water is not blocked by sealing of exit joints, stress distribution tends to return to initial condition. Figure 26 shows water pressure distribution in the rock mass of the dam foundation. In Case 1, there is no block pushed out from the surface, but the triangle area at the dam foundation of downstream side moved toward upside and also downstream direction as shown by displacement vector in the figure. At the same time, dam body inclined. Displacements of blocks in Case 1 are larger than Case 2 and Case 3.

In Case 3, magnitude of displacement vector in the analysis model area is smaller than Case 1 and Case 2, probably because of water leak and less pressure buildup. Figure 27 shows classified displacement and displacement vector. Moved area by water pressure can be recognized. It is estimated that rock blocks displacement (different color categories) induce unfavorable deformation in dam from this preliminary analysis. The results were obtained by 2-Dimensional dam section analysis and in reality, 3-D distortion mechanism is not shown in the result. The future investigation should be 3-D modeling and 3-D calculation.

6 CONCLUSIONS

This paper looks back on the past and present of coupled analysis and explains how it has been applied to rock engineering. The rock models used have also changed from continuum to discontinuities, and analysis using sophisticated models and experiments adapted to it have been conducted. Furthermore, progress of numerical analysis methods, (finite element method etc.), has a great influence on coupling analysis, and various attempts have been reported.

While changing from continuum analysis to discontinuity analysis, DEM, DDA, etc. was developed. Furthermore, since the particle method has recently been introduced into the fluid and new analysis visualization has begun, this report summarizes some examples introduced in rock engineering.

The coupling phenomenon that we first learn in the Geotechnical Engineering class is consolidation. Consolidation is a phenomenon in which excess pore water is gradually squeezed out from the medium and deformation (generally subsidence) occurs, but the relationship with groundwater flow is not explained at all (with emphasis on changes water pressure, not groundwater flow). In the meantime Terzaghi's one-dimensional consolidation equation has emerged and is extended by Biot to 3D.

Since the medium is assumed to be an elastic body, and hydraulic pressure expressions for such as effective stress and total stress appears on the front, it is hard to understand that this phenomenon is also one end of the groundwater flow. The phenomenon of consolidation is not a special thing, and it is not clearly understood that it is a case of groundwater and soil interaction that occurs everywhere when water gets caught in the soil or rock masses.

However, complex coupling analysis has appeared in response to the needs from the field. With regard to various coupling actions, if you look closely at natural phenomena, you can see the related figures. Soil or Rock (stress: Mechanical)-seepage water (water pressure: Hydraulic) coupling, rock mass-seepage water-heat flow (temperature: Thermal) coupling, rock-seepage water-thermal-chemical (Chemical) coupling are generally considered. However, the more coupling terms, the more complex the solution becomes and difficult.

Over the years, theories and experiments have shown synergy effects, and the development of the finite element method, which is a representative of numerical analysis, has progressed. In the stress-flow coupling problem, the basic equation for the water flow and the basic equation for the stress-deformation analysis are solved by coupling, and it is clarified that the one connecting these two equations is the "effective stress". The Biot 3-dimensional consolidation equation was introduced and the research rapidly went on. As a result, flow and stress deformation has been integrated, and the coupling problem has come to be applied in practice.

In civil engineering, slope stability during rainfall, hydraulic uplift analysis of dam foundations, stability during flooding of tunnels, groundwater flow analysis around underground power plants (underground cavities), practical application went on and with long-term integration among a wide range of different field fusion. These techniques were extended to the knowledge and safety evaluation etc. to understand the behavior of the rock around the cavity for high-level radioactive waste geological disposal.

The development of THMC coupled with the four elements of THMC is largely due to the contribution of DECOVALEX, a collaborative research host on the geological disposal of high-level radioactive wastes, which examines all elements and requires precise and advanced analysis. DECOVALEX continues to be active even now, and many relevant international conferences and symposia are held to disseminate information. In the field of civil engineering, chemical coupling events are occasionally found, and examples include stability problems of slopes and dam foundations taking into consideration of aging and deterioration (for example, weathering phenomenon).

Starting from porous media and continuum models, in coupled analysis for rock mass, DFN models with discontinuities have been introduced and applied in various cases. Many national research institutes involved in high-level radioactive waste geological disposal projects commonly use the most complex THMC full-coupled models.

In addition to the refinement and experiments of THMC model, a new attempt in analysis methods is coupled analysis with particle method, discrete element method (DEM) and discontinuous deformation method (DDA). The particle method expresses a fluid-like continuum as a collection of many particles, analyzes its behavior based on the Navier-Stokes equation, and can express dynamic fluid motion well. Discontinuous analysis methods such as DEM and DDA model the rock mass as a large number of polyhedral or spherical particle aggregates, but by tracking the phenomenon that occurs when the groundwater particles flowing through the space interact with the solid particles, it is possible to understand in detail the mechanism of ground failure by seepage flow.

However, a major issue that has emerged as analysis technology advances rapidly is the issue of the quality of data input in analyses. In numerical analysis, modeling (geometrical features) of the target medium (soil and rock) is performed, and physical property values are input. As models become more complex, input parameters increase, and advanced experiment, testing and proper judgment are required to determine the parameters.

That is, the result of analysis means that it depends on the accuracy of input values, the accuracy of modeling, the analysis technique, etc. If the input parameters are not balanced, the validity of the analysis results will become questioned.

Even if analysis is performed, the results cannot be verified without actual measurement data. Recently the importance of on-site measurement or monitoring has been particularly strong. It is because the demand to grasp the actual situation visually becomes strong in actual measurement. In response to this, various high-performance measuring devices have been developed, and real-time measurement results has become possible through wireless communication and the Internet, so the entity of complex phenomena is becoming visible.

In particular, the miniaturization of inexpensive sensors makes it possible to easily measure at a large number of points, and by being able to display them three-dimensionally, the comparison with the analysis results becomes quick, and the analysis accuracy has been improved greatly.

These large amounts of acquired data are called big data, and it is expected that phenomena that have not been seen until now will be highlighted, leading to a deeper understanding of coupling behavior. It is hoped that developing the analysis and measurement methods in cooperation with each other in the future will create synergetic effects, and further understanding of coupling behavior will be achieved.

ACKNOWLEDGEMENTS

My special appreciation goes to Dr. Shigeru Miki and Dr. Takeshi Sasaki for their long-time cooperative works in developing various models. Also special thanks to Prof. Hideki Yasuhara and Prof. Guowei Ma for providing me valuable suggestions and recent papers. These helps a lot to write this paper.

REFERENCES

Bear J. 1972. *Dynamics of fluids in porous media*, Elsevier, New York.

Bear J. and Corapcioglu, M.Y. 1981. A mathematical model for consolidated in a thermoelastic aquifer due to hot water injection or pumping, *Water Resour. Research*, Vol. 17, No. 3, 723–736.

Biot M.A. 1941. General theory of three-dimensional consolidation, *J. Applied Physics*, 120–155.

Casagrande, A. (1961). Control of seepage through foundations and abutments of dams, *Geotechnique*, vol. XI, pp.161–181.

Cundall P.A. 1971. A computer model for simulation progress, large scale movement in block system. In Proceedings of ISRM Symp., Nancy, France, 11–18.

Dershowitz W., Herbert, A. and Long, J. 1989. Fracture Flow Code Cross- Verification Plan, *STRIPA Project Technical Repo*rt No. 89–02, Swedish Nuclear Fuel and Waste Management Co.

Domenico P.A. and Schwartz F. W. 1990. *Physical and Chemical Hydrogeology*, John Wiley & Sons, Inc.

Elsworth D. and Yasuhara, H. 2006. Short-time scale chemo-mechanical effects and their influence on the transport properties of fractured rock, *Pure Appl. Geophys.*, 163, 2051–2070.

Gringarten A., Witherspoon, P.A., Ohnishi, Y. 1975. Theory of heat extraction from fractured Hot Dry Rock, *Journal of Geophysical Research* Atmospheres 80(2), 1120–1124.

Hudson J.A. and Harrison J. P. 1997. *Rock mechanics interactions and Rock engineering systems (RES)*, In Engineering Rock Mechanics, An introduction to the principles, Pergamon Press.

Irie K., Koyama, T., Nishiyama, S., Yasuda, Y. and Ohnishi, Y. 2012. A numerical study on the effect of shear resistance on the landslide by Discontinuous Deformation Analysis (DDA). *Geomechanics and Geoengineering, an International Journal*, 2012; 7(1): 57–68.

JAEA, AIST (National Institute of Advanced Industrial Science & Tech), NUMO, CRIEPI (Central Research Institute of Electric Power Industry) 2019. Advanced development of safety evaluation technology in coastal area for HWL in Japan, *Annual Summary Report*.

Jing L., Ma, F., Fang Z. 2001. Modeling of fluid flow and solid deformation for fractured rocks with DDA method, *Int J. Rock Mech Min Sci*, 38, 343–355.

Koshizuka S. and Oka. Y. 1996. Moving particle semi-implicit method for fragmentation of incompressible fluid. *Nuclear Science and Engineering*, 123, 421–434.

Koyama T., Tsukahara, T., Jing, L., Kawamura, H. and Ohnishi, Y. 2008. Numerical simulation of laboratory coupled shear-flow tests for rock fractures – comparative study on difference in using Reynolds and Navier-Stokes equations to simulate the fluid flow. *Thermo-hydromechanical and chemical coupling in geomaterials and applications, Proc of the 3rd International Symposium GeoProc 2008*, Lille, France, Wiley, pp. 525–532.

Kuno M., Miki, S., Ohnishi, Y. & Sasaki. T. 2016. Coupling analysis of rock mass and water for debris flow on a rock slope by DDA (Discontinuous Deformation Analysis) and MPS (Moving Particles Semi-implicit) method, *ARMS9*, Indonesia, Bali.

Kurikami K., Kobayashi, A., Tijimatsu, S. and Ohnishi, Y. 2003. Influence of Groundwater to Coupled Heat-Stress-Flow Phenomenon in High Level Nuclear Waste, *Journal of Geotechnical Engineering, JSCE*, No. 736, pp.261–271.

Londe, P. 1987. The Malpasset Dam failure, *Engng Geology*, Vol.24, 259–329.

Long J. C. S., Remer, J. S., Wilson, C. R. and Witherspoon, P. A. 1982. Porous Media Equivalents for Networks for Discontinuous Fractures, *Water Resources Res*., V.18/3, pp. 645–658.

Lucy, L.B., 1977, A numerical approach to the testing of the fission hypothesis, *Astron. J.* 82, pp1013–1024.

Ma G., Li, T., Wang, Y. and Chen, Y. 2019a. A mesh mapping for simulating stress-dependent permeability of three-dimensional discrete fracture networks in rocks, *Computers and Geotechnics*. 19–07.

Ma G., Li, T., Wang, Y. and Chen, Y. 2019b. Numerical simulations of nuclide migration in highly fractured rock masses by the unified pipe-network method, *Computers and Geotechnics*, 03.24.

Miki S., Sasaki, T., Koyama, T., Nishiyama, S. and Ohnishi, Y. 2010. Development of coupled discontinuous deformation analysis and numerical manifold method (DDA-NMM). *International Journal of Computational Methods*, Vol. 7, No. 1: 131–150.

Miki S., Ohnishi, Y. and Sasaki. T.: 2017. Water flow and rock mass coupling analysis of debris flow on a rock slope by DDA and MPS (Moving Particle Simulation) method, In *Proceedings of 51st US Rock Mechanics Geomechanics Symposium, ARMA*17–630.

Min K-B, Jing L., Stephansson O. 2004. Determining the equivalent permeability tensor for fractured rock masses using a stochastic REV approach: method and application to the field data from Sellafield, UK, Hydrogeology J., 12:497–510.

Mustapha H., Dimitrakopoulos, R., Graf, T. and Firoozabadi A. 2011. An efficient method for discretizing 3D fractured media for subsurface flow and transport simulation, *Int J. Numer Mech Fluids*, 67, 671–70

Oda M. 1986. An equivalent continuum model for coupled stress and fluid flow analysis in jointed rock masses. *Water Resour Res*;22:1845–56.

Ohnishi, Y. and S. Miki. 1996. Development of circular and elliptic disc elements for DDA, In *Proceedings of 1st. Int. Forum on Discontinuous Deformation Analysis (DDA) and Simulation of Discontinuous Media, USA*, 44–51.

Ohnishi Y., Sasaki, T., Koyama, T., Hagiwara, I., Miki. S. and Shimauchi, T. 2016. Recent insights into analytical precision and modelling of DDA and NMM for practical problems, *Geomechanics and Geoengineering*, Vol. 9, No. 2, pp.97–112, (Taylor & Francis, Engineering & Technology Readers' Award 2016).

Ohno, R., Yamashina, S., Yamasaki, T., Koyama, T., Esaki, F., & Kasai, S. 2010. Mechanism of a large-scale landslide triggered by the earthquake in 2008 – A study of Artosawa landslide. *Journal of the Japan Landslide Society*, Vol. 47, No. 2: 8–14. (in Japanese).

Priest S.D. and Hudson, J.A. 1981. Estimation of discontinuity spacing and trace length using scanline surveys, *Int. J. Rock Mech. Min. Sci. Geomech. Abstr*., 18.

Priest S. D. and Samaniego, A. 1983. A Model for the Analysis of Discontinuity Characteristics in Two Dimensions, *Proc. 5th Congress Int. Soc. Rock Mech*., Melbourne, Australia.

Priest S.D.1993. *Discontinuity Analysis for Rock Engineering*. Netherlands, Springer.

Ren F., Ma G., Wang Y., Li T., Zhu H. 2017. Unified pipe network method for simulation of water flow in fractured porous rock, *J Hydrology*, 547: 80–96

Sasaki T., Hagiwara, I., Iwata, N., Miki, S., Ohnishi, Y., and Koyama, T. 2015. Parameter studies of practical rockfall problems with fluid flow by 3D-DDA. In *Proceedings of 12th International Conference on Analysis of Discontinuous Deformation*, Wuhan, China 48–56.

Shi G.H., and Goodman, R.E. 1989. Generalization of two-dimensional discontinuous deformation analysis for forward modeling. Int J. Numer. Anal. Meth. Geomech, 13, 359–380.

Shi G.H. 1991. Manifold method of material analysis, *Transactions of the 9th Army Conference on Applied Mathematics and Computing*, In Report No.92-1, U.S. Army Research Office.

Tsang C.F. 1987. *Introduction to coupled processes*, Chapter 1, Coupled Processes associated with Nuclear Waste Repositories, edited by CHIN-FU TSANG, Academic Press.

Wu J.H., Ohnishi, Y., Nishiyama, S. 2004 Simulation of the mechanical behavior of inclined jointed rock masses during tunnel construction using Discontinuous Deformation Analysis (DDA), *International Journal of Rock Mechanics & Mining Sciences*, Vol. 41, No. 5, pp. 731–743.

Yoshida H., Kojima, K., Ohnishi, Y., Tochiyama, O., Nishigaki, M., Tosaka, H., Sugihara, K., Ogata, N. 2014 Examination of realistic conceptual model of near-field process in HLW repository, *Japan Geoscience Union Meeting*

Muller Award

Rock Mechanics for Natural Resources and Infrastructure Development – Fontoura, Rocca & Pavón Mendoza (Eds)
© 2020 Taylor & Francis Group, London, ISBN 978-0-367-42284-4

From common to best practices in underground rock engineering

P.K. Kaiser
Laurentian University and President, GeoK Inc., Sudbury, Ontario, Canada

ABSTRACT: Common practices are not necessarily best practices when judged from an economic or workplace safety perspective. As in other engineering disciplines, it is necessary to systematically improve engineering design practices. This lecture addresses some deficiencies in common practice that may lead to flawed or ineffective rock engineering solutions. More than ever, as we go deeper in underground construction, are rock engineers challenged by a number of opportunities that exist for improvements. In the past, common practices that worked well at shallow depth may need to be replaced as the rock mass behavior has changed and poses new hazards at depth. This lecture focuses specifically on opportunities resulting from better means to assess the vulnerability of excavations, to characterize the rock mass, for ground control, and rockburst damage mitigation. Theoretical considerations and field observations are used to justify the proposed changes and highlight practical implications and benefits. In the spirit of Prof. L. Müller, this lecture aims at pointing the way to future improvements in rock engineering, i.e. 'im Felsbau', and offers guidance on how to move from common to best practices.

It is an honor to present the 8th Müller lecture in memory of Prof. Leopold Müller and his achievements as the enthusiastic founder and first President of the ISRM in 1962. Today, we are remembering a man, whose vision moved rock engineering to a new science that influences geomechanics engineers all over the world. By creating the Society, he aimed at aggregating scientists interested in a new-born branch of science, rock mechanics, with the purpose of bringing together the scattered knowledge obtained by groups working more or less in isolation.

I like to start by acknowledging that many of my achievements, as well as the content of my presentation today, must be attributed to long-term collaborations with academic and industrial colleagues and also to many students whose careers have started in rock mechanics research. My mentors have taught me to question the status quo and to advance the state-of-the-art, and I like to dedicate this lecture to my most influential mentors Drs. N. Morgenstern, E. Hoek, E.T. Brown (6th ISRM President), and also to my wife Kathi, who has always supported my journey of discovery[1].

1 THE STORYLINE

Müller (1963), in the preamble to 'Der Felsbau', stated "Der Felsbau ist auf dem Wege, eine wissenschaftliche Disziplin zu werden. (Construction in rock ... is on the way to become a scientific discipline.)" Over the last 50+ years, the science of rock mechanics and its implementation through rock engineering has evolved into a mature discipline, and new challenges emerge as we tackle larger and deeper excavations in civil construction and mining.

He elaborated "Mit der Größe von Projekten wuchs die Verantwortung des Ingenieurs und des Baugeologen ganz bedeutend und zwang dazu, die bis anhin geübte gefühlsmäßige Behandlung dieser Aufgabe mehr und mehr zu verlassen und eine Theorie des Felsbaues zu erarbeiten." (With the size of projects, the responsibility of the engineer and the construction geologist grew quite significantly and forced us to leave intuitive treatment practices behind to develop a theory of rock engineering). In the spirit of his vision, it follows that rock engineering evolves over time and

1. https://videostream.laurentian.ca/Mediasite/Play/bf2879ebbe874213b42762775d593bc11d

common practices need to be questioned and improved to arrive at best practices that lead to safer and more economic solutions.

Prof. Müller also suggested that it is sometimes necessary to go back to fundamentals to identify deficiencies in order to pave the road for progress.

Enormous progress has been made over the last decades and the new knowledge has found application through engineering standards, 'ISRM suggested methods', or common engineering practices. This is also reflected in the recent change of ISRM's name to 'International Society of Rock Mechanics and Rock Engineering'. Today, as in 1963, new challenges force us to leave our comfort zone of common practices and develop more sophisticated and, at the same time, simple, more efficient and effective rock engineering practices.

"Everything should be made as simple as possible, but not simpler" (attributed to A. Einstein). Common practices are often too simple and have frequently led to failures and costly mistakes, largely because the fundamental understanding of rock behavior and response to construction or mining are not fully reflected in experience-based approaches. Best practices make it simple, reflecting all essential or dominant Engineering Design Parameters (EDPs) – no more, no less. This means that unnecessary complexities have to be eliminated and representative rock engineering approaches adopted.

During the MTS lecture at the 50th US Rock Mechanics Symposium in 2016 on 'Underground rock engineering to match the rock's behavior – Challenges of managing highly stressed ground in civil and mining projects', the author suggested that dichotomies exist and gaps between reality and current practices have to be closed by the application of recent advances in rock mechanics research to arrive at sound rock engineering solutions.

"A robust rock engineering solution in underground mining or construction must respect the complexity and variability of the geology, consider the practicality and efficiency of construction, and provide safe and effective rock support. For this purpose, it is essential to anticipate the rock mass and excavation behavior early in the design process, i.e. at the tender stage before excavation techniques are chosen and designs are locked-in in construction contracts. Whereas it is possible in most engineering disciplines to select the most appropriate material for a given engineering problem, *in rock engineering, a design must be made to fit the rock, not vice versa.*

Lessons learned from excavation failures (Figure 1a) tell us that stressed rock at depth is less forgiving and that advances in rock mechanics demand a sound comprehension of the behavior of stress-damaged rock near excavations. Comprehension in this context means explaining all observations such that fiction can be separated from reality and engineering models and methods become congruent with the actual behavior of a rock mass." (Kaiser 2016b).

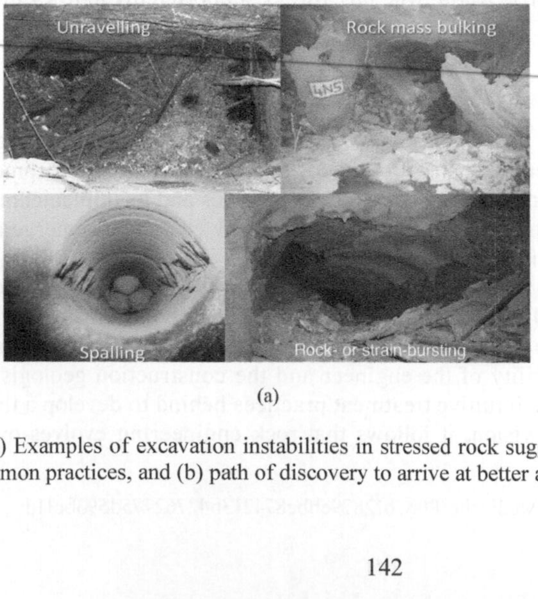

Figure 1. (a) Examples of excavation instabilities in stressed rock suggesting flawed or ineffective solutions based on common practices, and (b) path of discovery to arrive at better and best practices.

Because many engineering approaches and common practices are flawed or ineffective, and may even be obsolete, best practices have to be developed based on hypotheses of the underlying causes for the deficiencies (Figure 1b). Once these hypotheses have been verified by testing and field observations, improved solutions can be found and implemented. Eventually, new knowledge and experiences may be acquired and the innovation cycle may restart as indicated in Figure 1b.

Rather than elaborating on flaws of common practices, this lecture focuses on best practices in three key areas of rock engineering for safe and cost-effective underground mining and construction:

- Excavation vulnerability and fragility assessment;
- Rock mass behavior and characterization for rock mass strength determination; and
- Ground control and support selection.

2 UNDERSTANDING EXCAVATION BEHAVIOR

Understanding excavation and rock mass behavior is a prerequisite for successful rock engineering and therefore for the development of best practice guidelines.

2.1 *Excavation failure mechanisms*

Only if all potential excavation failure mechanisms (Figure 2a) are understood can the vulnerability of an excavation to failure and the potential severity (extent or violence) of damage be anticipated, and the resulting critical load, displacement and energy demands be established (by empirical, analytical, or numerical modeling). Because of the author's recent experience from major mining (caving) operations, this lecture is built on lessons learned from moderate to severe excavation instabilities in mining. The cartoon-like images illustrate the four mechanisms described in the figure caption.

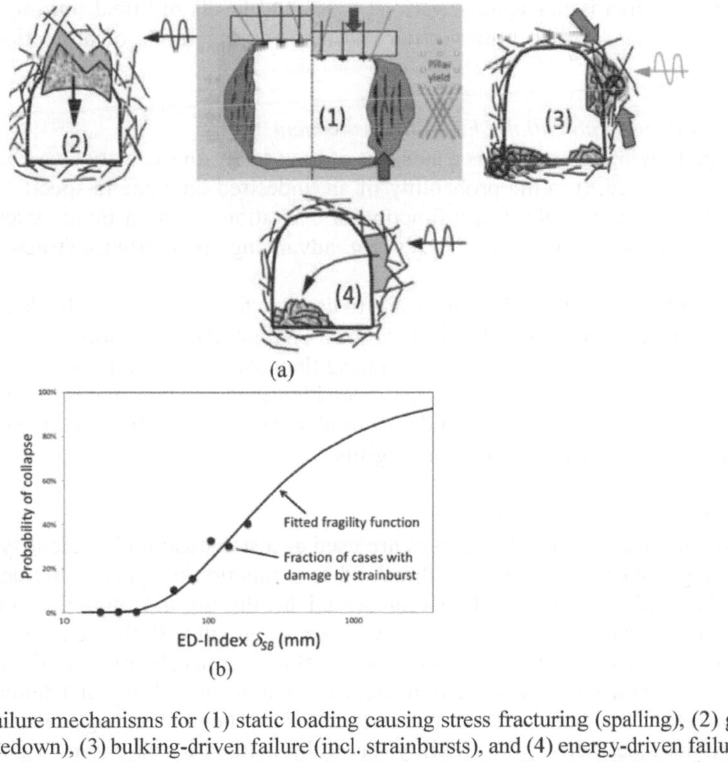

Figure 2. (a) Failure mechanisms for (1) static loading causing stress fracturing (spalling), (2) gravity-driven failure (incl. shakedown), (3) bulking-driven failure (incl. strainbursts), and (4) energy-driven failure (incl. stress wave reflection, etc.); and (b) Generic example of a fragility curve defined by the ED-index δ_{SB} for damage by strainbursts (3).

2.1.1 *Defining safety margin – Vulnerability assessment*
By comparing load, displacement and energy demands to the capacity of rock support systems, the proximity to failure, i.e. the probability of failure or factor of safety, can be assessed and appropriate support systems can be selected. In this manner, the vulnerability of an excavation to failure is described by answering the question 'How close is it to failure?', i.e. 'How full is the glass?' for each of the four failure modes shown in Figure 2a.

In rock engineering, depending on the anticipated excavation behavior, one or more of the following three Factor of Safety (FS) have to be assessed:

$$FS_{Load} = \frac{\text{Support Load Capacity}}{\text{Load Demand}} \quad \text{(e.g. for (2) in Figure 2a)} \tag{1}$$

$$FS_{Disp} = \frac{\textit{Support Displacement Capacity}}{\text{Displacement Demand}\textit{is}} \quad \text{(e.g. for (3) in Figure 2a)} \tag{2}$$

$$FS_{Energy} = \frac{\textit{Support Displacement Capacity}}{\text{Displacement Demand}\textit{is}} \quad \text{(e.g. for (4) in Figure 2a)} \tag{3}$$

Once the vulnerability is established, more questions arise: 'How severe or violent will the failure be?', i.e. 'How much water will spill?', or 'How fragile is the excavation?' The fragility of an excavation describes the likelihood of damage at a defined severity, e.g. in terms of volume or weight of displaced rock, cumulative displacement imposed on the support, or the violence of energy release.

In other words, critical Engineering Demand (or Design) Parameters (EDP) have to be selected to establish the vulnerability and fragility of an excavation. The development of best practices therefore starts with the identification of dominant EDPs (Kaiser & Cai, 2019).

The vulnerability of an excavation describes the state of exposure to physical damage. Strictly speaking, it is the probability of damage to occur at a given site without consideration of the severity of the resulting damage. The vulnerability assessment aims at identifying potential damage locations, evaluating design issues against various types and levels of threat (mining-induced stress, ground motions in burst-prone mines, etc.), and determining levels of protection by mitigation measures.

2.1.2 *Defining damage potential – Fragility assessment*
The *fragility* (antonym: robustness) is a measure of how easily an excavation can be broken or how much damage is caused. It is the probability of an undesired outcome (a specified damage level, e.g. R1 to 5 (Potvin et al. (2009)) as a function of excitation, i.e. a particular excavation demand parameter (EDP) or a combination of EDPs (e.g. advancing stress front, extraction ratio, ground motion, etc.).

An excavation may or may not be vulnerable to damage for a given load, displacement or energy demand, but it may be more or less fragile (or robust). Vulnerable excavations are likely to be damaged but robust excavations will suffer less damage than fragile excavations.

Once a deficiency in a common practice has been identified, the first step is to establish a hypothesis of how to overcome it (Figure 1b) and how to identify critical EDPs that are needed to characterize and assess the vulnerability and the fragility.

2.1.3 *Identification of critical EDPs*
In earthquake engineering, fragility curves are used as a statistical tool to identify the probability of exceeding a given damage state or a threshold as a function of specific engineering demands. For earthquakes, this demand is often represented by the ground motion (preferably spectral displacement at a given frequency). A fragility curve is a graph with the demand (horizontal axis) defined by a representative EDP (e.g., peak ground velocity or acceleration (*PGV* or *PGA*), mining-induced stress, dynamic stress wave increments, etc.), and the probability of a defined damage level (cracked shotcrete, failed bolts, etc.) on the vertical axis. When the damage severity is affected or dominated by multiple factors, the fragility has to be defined based on an Engineering demand index (ED-index) that considers the relative contribution from all relevant demands.

The same concept can be applied to static demands (loads or displacements). Some examples of EDPs or ED-indices are listed below:

- EDPs for FS_{Load}: Geological structure geometry for wedge volume or weight estimation, or an ED-index linking the stress level *SL* for depth of failure d_f estimations to the weight of stress-fractured rock;
- EDPs for FS_{Disp}: an ED-index reflecting the peak and post-peak strength of the rock mass and its dilation or rock mass bulking behavior;

  and for rockbursting ground:
- EDPs for FS_{Energy}: ground motions (*PGV* or *PGA*) for damage dominated by strong remote seismic events (e.g. shakedown from seismic events or earthquakes) and, for strainbursts, an ED-index combining stress level *SL*, the depth of strainbursting d_{SB}, the bulking factor, and stored strain energy.

For example, when assessing the fragility to spalling (subscript S) or self-initiated strainbursting (subscript SB), the severity of damage by excessive bulking of stress-fractured rock can be described by an ED-index that reflects the displacement demand:

$$\delta_S \, or \, \delta_{SB} = d_f \bullet BF = (bSL - c) \bullet a \bullet BF \tag{4}$$

where, d_f = depth of failure, static $SL = (3\sigma_1 - \sigma_3)/UCS = \sigma_{max}/UCS$, a = the equivalent tunnel radius, and b and c are constants ($b = 1.37$ and $c = 0.42$ for static loading; for dynamic loading see Kaiser (2006))[2]. For dynamically loaded strainbursts, the stress level is temporarily increased to $SL_{SB} = SL + \Delta SL_d$, where ΔSL_d reflects the change in SL due to dynamic loading of the excavation (Kaiser et al. 1996).

A generic example of a fragility curve for the strainburst severity is presented in Figure 2b (note: log-scale for EDP = strainburst displacement). In this case, there is a 40% probability of collapse by a strainburst if the displacement induced by the strainburst reaches 200 mm. The vulnerability to strainbursting is largely controlled by the rock mass's stress to strength ratio (or *SL*) and the fragility is dominated by the depth of failure (static plus dynamic) and the bulking factor (Equation 4).

For relatively small seismic events, the fragility is not or only vaguely related to the ground motion (Kaiser & Cai, 2019). Hence, fragility curves for strainbursts have to be developed in terms of load, displacement, and energy release demands (excluding *PGV* or *PGA* as they only trigger the failure process and do not contribute to the severity of damage).

For falls of ground or seismically triggered shakedown failures, respectively, gravitational and dynamic accelerations (ground motions measured by representative *PGA*) define the demand and the fragility is controlled by the brittleness of the support system. An ED-index reflecting the ground motion and ductility of the support is therefore needed to assess the fragility of an excavation to shakedown failure.

2.1.3.1 *From common to best practice*

Identification of EDPs is a prerequisite to determine whether common practices are applicable or flawed. For excavation design and support selection, EDPs or ED-indices characterizing vulnerability and fragility must be clearly identified and then used to recognize potential failure modes to obtain relevant rock mass parameters.

---

2. smax = 3s1 – s3, with s1 and s3 representing the in-situ, or mining-induced major and minor principal stresses (s1m and s3m), in the vicinity of the tunnel in a plane perpendicular to the tunnel axis. smax represents the excavation-induced stress at the location with the highest tangential stress near the wall of an equivalent circular excavation in elastic rock. The stress level SL as a EDP is therefore an index rather than a measure of the actual stress to strength ratio. When mining changes the stress field near a tunnel the stress level SL changes. The combined effect is called 'mining-induced stress' and is used on the right scale of Figure 3a. This approach does not account for the intermedia principal stress that may affect the failure mode, e.g., at or near the tunnel face or near intersections.

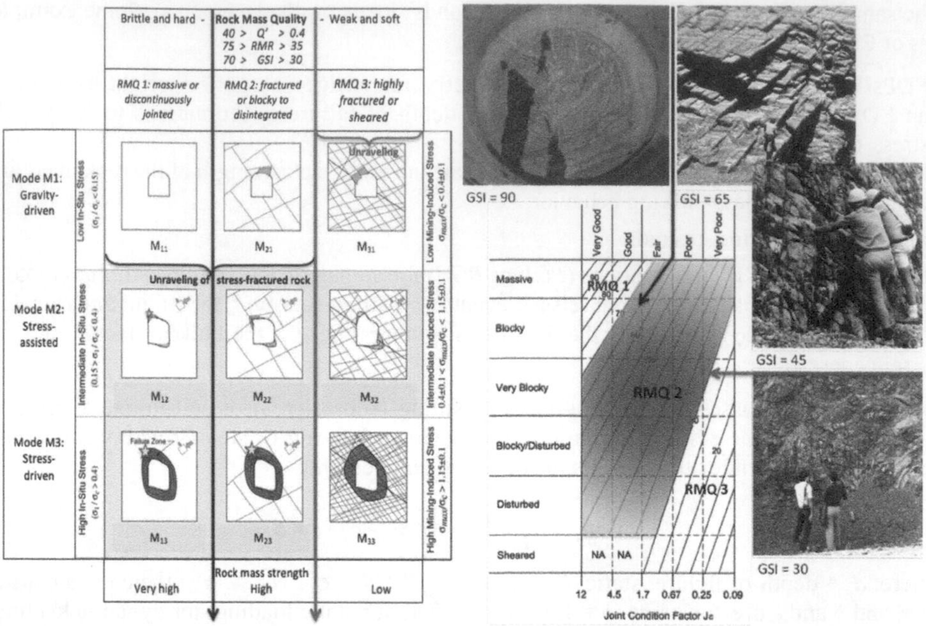

(a) (b)

Figure 3. (a) Excavation behavior matrix showing expected rock mass failure modes M_{11} to M_{33} as a function of rock mass quality (RMQ) or strength, and in-situ stress (left scale) or the excavation- and mining-induced stress (*SL* scale on the right) (modified from Kaiser et al. (2000)). The red and black arrows link to (b) the corresponding rock mass quality in the Geological Strength Index (GSI) chart (Kaiser (2017) and Kaiser & Cai (2019) with images courtesy E. Hoek).

2.2 *Static failure modes of unsupported excavations*

Under static loading, the excavation behavior can be characterized by two dominant EDPs as illustrated by the 3 × 3 excavation behavior matrix presented in Figure 3 (modified from Kaiser et al. (2000) and presented by Kaiser (2017) and Kaiser & Cai (2019). The horizontal axis represents the EDP = Rock Mass Quality (RMQ) or rock mass strength and the vertical axis the EDP = stress level.

The rock mass quality (RMQ) is grouped: RMQ1 for massive to discontinuously jointed rock, RMQ2 for fractured and blocky to disintegrated ground, and RMQ3 for weak and soft, highly fractured or sheared rock. The boundaries have been slightly adjusted from Kaiser et al. (2000) based on experiences with excavation failures at depth to:

- 40 > Q' < 40; with Q' = modified rock tunneling quality index Q for $J_w/SRF = 1$;
- 75 > RMR (Rock Mass Rating) < 75; or
- 70 > GSI (Geological Strength Index) > 70).

The stress intensity is defined on the left side if Figure 4a by the in-situ stress to strength ratio σ_1/σ_c and on the right by the stress level $SL = \sigma_{max}/UCS$. Using the stress level as an EDP is beneficial as the impact of the in-situ stress ratio $k = \sigma_1/\sigma_3$ or mining-induced ratio $k_m = \sigma_{1m}/\sigma_{3m}$ is accounted for. This distinction is particularly relevant when mining changes the mining-induced stress field, and failure modes may change, e.g. from M_2 to M_1 due to relaxation.

Excavations in massive to discontinuously jointed rock masses are prone to stress fracturing near the excavation: (M_{11}: Elastic response; M_{12}: Localized brittle failure of intact rock adjacent to excavation boundary; M_{13}: deep brittle failure of intact rock; potentially surrounding the entire excavation.)

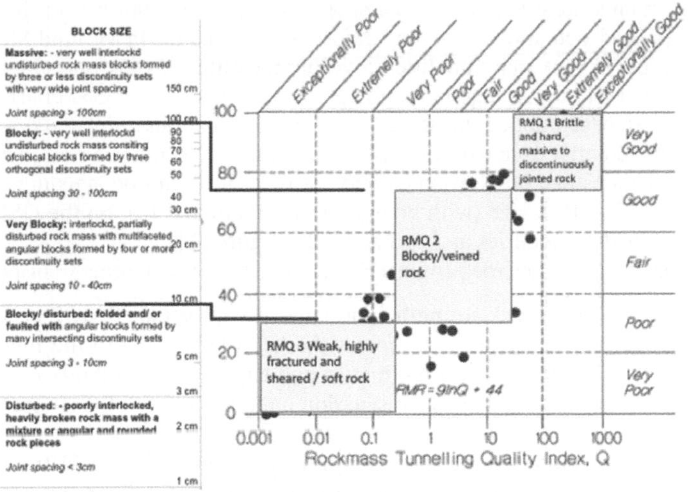

Figure 4. Rock mass quality grouping for excavation failure and EDP selection. Left: GSI block size description. Right: simplified grouping of rock mass quality RMQ1 to 3 superimposed pm chart from Hutchinson & Diederichs (1996) with linear trend after Bieniawski (1979).

Failure modes with localized spalling or stress-fracturing are observed once the stress level $SL = \sigma_{max}/UCS$ exceeds 0.3 to 0.5 (M_{12} and M_{13}) or 0.25 to 0.35 in defected or veined rock (Bewick et al. 2019). At intermediate mining-induced stress levels ($SL \leq 1$), the fracture zones typically remain localized (i.e. notches form). The stress-fractured zone becomes continuous in rock masses with low tensile strength at very high stress levels and when k is approaching unity.

Strainbursting may occur at the excavation wall or at some distance from the wall as indicated by the stars in Figure 3a. They occur at the interface between the excavation damage zone and the more competent elastic rock mass surrounding it. Furthermore, they may occur in supported rock and in rock of quality RMQ2 if joints are oriented such that shear slip is prevented along these joints. Consequently, as the depth of stress-fractured ground increases, excavations become increasingly more vulnerable to falls of stress-fractured ground and to strainbursting. According to Barton & Grimstad (1994), minor bursting (rock 'spitting') is to be expected during a tunnel advance at excavation-induced stress levels $SL > 0.65$ and severe rockbursting is to be anticipated, largely due to larger burst volumes, at $SL > 1.0$.

Excavations in fractured or blocky to disintegrated rock masses are prone to failure with some structural controls: (M_{21}: Falling or sliding of rock blocks and wedges; M_{22}: Localized brittle failure of (weak) intact rock blocks bound by open discontinuities or joints that facilitate movement of rock blocks; M_{23}: Brittle failure of intact rock around an entire excavation combined with movement of rock blocks bound by open joints.)

At low stress, this rock mass requires support. At elevated stress (M_{22} and M_{23}), blocks formed by open joint sets tend to fracture near excavations due to extension straining caused by stress heterogeneities. As a consequence, the natural block size is reduced and the rock mass becomes prone to unraveling. The extent of rock mass disintegration increases with depth and the resulting rock mass damage zone is prone to failure involving interactions with natural joints. The EDPs are typically defined by the depth of failure. In hard brittle rock, geometric bulking (due to a geometric non-fit of rock blocks or fragments) may impose large deformations on the support. The safety margin is best assessed in terms of an EDP describing the displacement demand.

Excavations in highly fractured or sheared ground are prone to falls of ground, unraveling and excessive plastic deformation: (M_{31}: Unraveling of blocks from the excavation surface; M_{32}: Localized failure of rock, weak blocks bound by open joints, and unraveling along discontinuities; M_{33}: Squeezing in an elastic/plastic continuum with or without swelling potential.)

At elevated stress levels (modes M_{32} and M_{33}), the behavior is further characterized by stress-driven plastic yield (Martin et al. 2003). This rock mass class (RMQ3) always requires support with a robust retention system (e.g. mesh and shotcrete) to prevent unraveling between bolts. The dominant EDPs are load-demands in M_{31} and deformation demands in M_{32} and M_{33}.

Four typical images of rock mass qualities covering the range of rock mass quality from RMQ1 to RMQ3 are shown in Figure 3b together with applicable ranges in the GSI chart.

As indicated earlier, best practices must be simple but reflect all essential and dominant EDPs. For the identification of potential excavation failure modes and for the determination of relevant EDPs, unnecessary complexities can be eliminated and the rock mass can be, as suggested by Terzaghi (1946), characterized by block size (with boundaries at 10 cm and 1 m on the GSI scale; Figure 4) and joint condition (with boundaries at $J_c = 3$ and 0.25; Figure 3b).

For the selection of appropriate rock engineering approaches, it is, in order of priority, necessary to:

- identify whether the intact rock strength dominates: RMQ1 versus RMQ2 or 3.
- According to Figure 4, in 'good' rock masses the key EDPs relate to the intact rock and rock block strength. The block size does not matter and engineering methods applicable to blocky rock mass models are not suitable (see discussion on applicability of GSI-strength equations in Section 5.1.1).
- identify whether inter-block characteristics dominate: distinguish RMQ2 from RMQ3.
- According to Figure 4, definitions of what constitutes poor ground vary widely.
  - In RMQ2, interlock contributes to the rock mass strength and the key EDP describes the block size as it dominates the excavation behavior (e.g. unraveling depends on excavation size).
  - In RMQ3, the characteristics of infilling dominate and the joint condition becomes the key EDP. Recognizing this sensitivity led to the development of the GSI with particular focus on weak and soft rock (Hoek et al. 1995).

From common to best practice – failure under static loading

Excessive effort is often expended in common practice to collect and rate rock mass details that have little impact on the excavation behavior and consequently on a design. Meanwhile, dominant factors, e.g. persistence, veining, and rock alteration, are either ignored or underrepresented.

Best practices must ensure that rock mass quality classes RMQ1 to 3 and the stress level *SL* as well as its evolution over the duration of a project are properly defined. In mining, it is particularly important to reflect on the stress path causing changes in the *SL* that may lead to changes in failure modes. For example, relaxation may move the failure mode from $M_{22 \text{ or } 23}$ to $M_{21 \text{ or } 31}$; whereas an increase in extraction ratio may move it from $M_{22 \text{ or } 32}$ to $M_{23 \text{ or } 33}$. Both may drastically change the support requirements and render common practices in support selection as inapplicable.

2.3 *Dynamic failure modes of unsupported excavations*

By comparing Figure 2a to Figure 3a, it follows that dynamic disturbances from earthquakes or mining-induced seismic events can lead to (a) strainbursts (in $M_{12 \text{ to } 13}$ and M_{23}) and (b) to seismically induced falls of ground or shakedowns (in $M_{31 \text{ to } 33}$ and $M_{22 \text{ to } 23}$). This is illustrated by Figure 5. The listed EDPs depend on the respective failure mode (see Section 2.3.3).

2.3.1 *Shakedown damage*

For shakedown, there are two phases to consider: (1) the fall is initiated or triggered, and (2) the support dissipates the kinetic energy of the falling mass until a new equilibrium is established. If unsupported, the trigger limit defines the failure point. If the support is effective, it survives the load and kinetic energy of the falling rock (thus called survival limit). For more details on shakedown failure analysis, the reader is referred to Chapter 8 in Kaiser et al. (1996).

2.3.2 *Strainburst damage*

Strainbursts are sudden, violent failures of the rock mass in a 'burst volume' near an excavation. They occur in highly stressed rock, often at stress raisers as illustrated by Figure 6a (in red), i.e., at the transition from damaged to relaxed ground (in blue) or at locations of elevated stress due to geological structures. The burst process, simulated by Gao et al. (2019a, b), involves the creation of

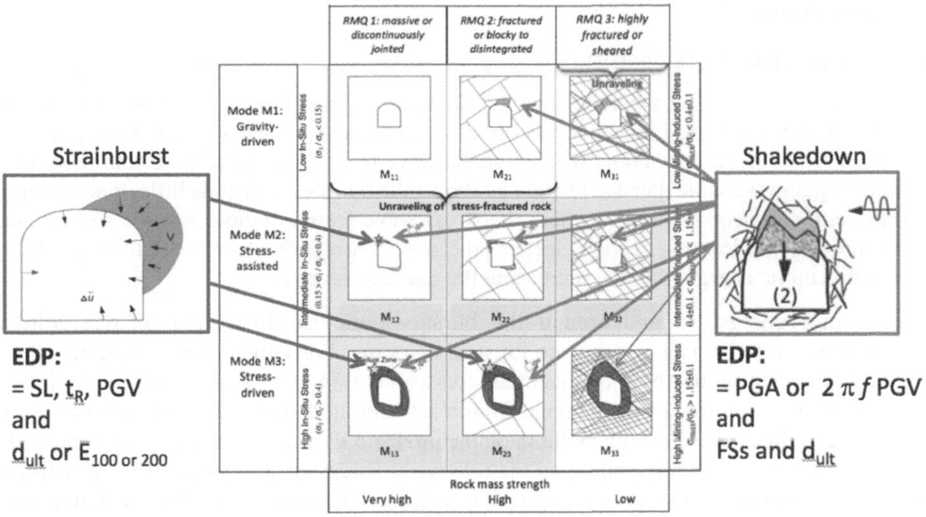

(a) (b)

Figure 5. Excavation behavior matrix with two predominant dynamic failure modes: (a) strainburst, and (b) shakedown.

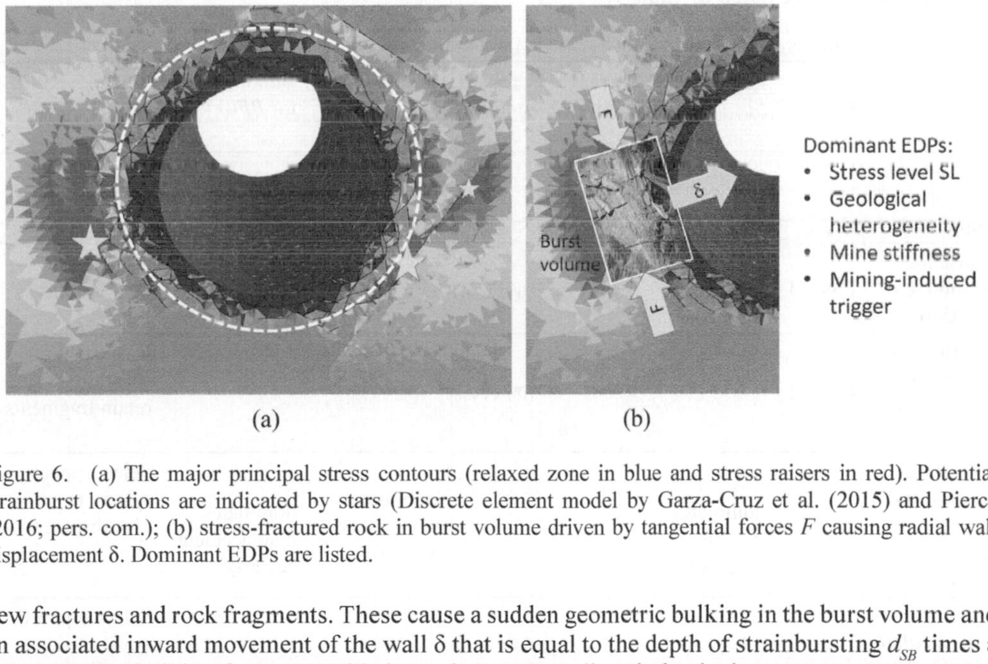

(a) (b)

Figure 6. (a) The major principal stress contours (relaxed zone in blue and stress raisers in red). Potential strainburst locations are indicated by stars (Discrete element model by Garza-Cruz et al. (2015) and Pierce (2016; pers. com.); (b) stress-fractured rock in burst volume driven by tangential forces F causing radial wall displacement δ. Dominant EDPs are listed.

new fractures and rock fragments. These cause a sudden geometric bulking in the burst volume and an associated inward movement of the wall δ that is equal to the depth of strainbursting d_{SB} times a representative bulking factor *BF*. This inward movement directly loads the support components at the plate and causes indirect loading or straining of tendons that penetrate the burst volume and the undamaged rock beyond.

The violent wall displacement may cause rock ejection if the burst volume and the 'burden' of relaxed rock (shown in blue in Figure 6b) is not effectively supported. The bulking velocity v_B at the inner edge of the burst volume depends on the time of rupture t_R (the time it takes to fail the burst volume). The initial velocity can be approximated by $v_B = \delta/t_R$. It eventually reduces to zero if the rock is effectively supported. If not, some rock and support components may be ejected at velocities v_{ej} less than v_B.

Types of strainbursts

There are three types of strainbursts:

- *Self-initiated strainburst* where excavation damage is caused by the release of stored strain energy when the mining-induced stress (load) exceeds the supported rock mass capacity;
- *Seismically triggered strainburst* where a stress wave from a remote seismic source initiates the failure (triggers the same damage process as above; the remote event adds little or no energy); and
- *Dynamically loaded strainburst* where the stress wave from a remote seismic source causes a substantial dynamic stress change[3] deepens the zone of stress-fractured rock to $d^{d}_{SB} = d_{SB} + \Delta d_{SB}$, and adds kinetic energy, thereby increasing the damage severity.

Stored strain energy, in and around the burst volume (intrinsic and relative brittleness; Tarasov & Potvin(2012)), is a prerequisite for strainbursting. For vulnerable excavations, the burst volume is brought to failure by the tangential stresses near an excavation reaching the local rock mass strength. The severity of the damage from a self-initiated strainburst is related to the strain energy stored in the bust volume (depending on the intrinsic stiffness) and the energy released from the surrounding rock mass (depending on the relative stiffness or the loading system stiffness). In other words, the severity of strainbursting is dominated by the brittleness of the rock mass and the loading system stiffness. The severity is only related to the energy release from large remote seismic events if the magnitude $M_L > 2$ (dynamically loaded strainbursts; Kaiser & Cai, 2019).

2.3.3 *EDP for dynamic failure processes*

The most relevant EDPs for the assessment of the vulnerability and the fragility of excavations to dynamic failures are listed in Table 1.

Table 1. EDPs for dynamic failure modes (2) to (4) in Figure 2a.

	Static *FSs*	Stress level *SL*	Support ductility d_{ult}	Ground motion PGV	Strainburst rupture time t_R (ms)	Bulking factor $BF(p_s)$	Note
Shakedown Unsupported		√		√			Always check load capacity
Supported	√		√				
Loose rock or shotcrete **ejection**	FSs = 1			$(n*PGV)^2$			Use mesh to hold fragments
Spalling by tangential stress increment	FSs = 1	√		n*PGV			Use mesh or shotcrete to retain fragments
Triggered strainburst (energy and displacement)	√	*FSs* = 1 dominates d_{SB} (burst volume)	√		√ dominates severity or violence	√ dominates displacement severity	Select support system to survive displacement and energy
Dynamically loaded strainburst	√	√	√	√ Adds D*SL* = *f*(*PGV*)	√	√	Select support system to survive displacement and energy
Momentum transfer	√	√	√	√		√	Design for large events $M_L > 2$

3. Ortlepp (2005) suggested that strainbursts with M_L = -0.2 to 0 are usually undetected but that buckling bursts (also caused by stored strain energy release) generate seismic events of M_L = 0 to 1.5. For mining, Kaiser & Cai (2019), based on data by Morissette et al. (2012), show that seismic events of magnitudes M_L < 1.5 to 2 at a distance R > 10 m add insufficient energy to affect the severity of strainbursts.

Depending on the dominant failure process, different EDPs need to be considered. With reference to Figure 2a:

- For failure mode (2), i.e. shakedown, the weight (gravitational force) is enhanced by dynamic forces caused by dynamic acceleration. Ground motions from a remote seismic event (*PGA* or *PGV*) are controlling EDPs for failure initiation, i.e. for vulnerability assessment of unsupported excavations. If supported, the survival limit (Kaiser et al. 1996) depends on the static factor of safety FS_s before the shakedown event, and the ductility of the support, i.e. the remnant displacement capacity of the support at the time of dynamic loading.
- For failure mode (3), i.e. dynamic bulking failure during a strainburst, the wall displacement δ^d_{SB}, which may depend on the ground motions *PGV* (if $M_L > 2$) and the bulking velocity $v_B = \delta/t_R$, are the controlling EDPs.
- For failure mode (4), i.e. dynamic stress reflection or energy-enhanced or -driven failures by strong or close seismic (fault slip) events, causing ejection or shakedown of loose rock or poorly bonded shotcrete (shotcrete rain), the ground motions *PGV* is the controlling EDP (not covered here).

From common to best practice – dynamic failure

Common ground-motion-centric or pure energy-based support selection with *PGV* as a single EDP ignores or underestimates the often dominant rock mass bulking process near the excavation as well as the associated displacements. These displacements cause direct and indirect impact loading of support components. Best practices have to consider the impact of pre-burst support deformation and bulking displacements during the burst (see Section 6).

When dynamic failure processes are triggered, energy can be released from the burst volume, i.e. "when the glass is full, the severity of damage depends on the size of the reservoir". Other EDPs, such as rupture time and bulking factor as well as the static factor of safety and support ductility, have to be considered (Kaiser et al. 1996).

3 REPRESENTATIVE ROCK AND ROCK MASS BEHAVIOR FOR DESIGN

Once the excavation behavior and failure processes are understood and relevant EDPs are identified, the designer has to obtain representative rock mass strength envelopes and then capture the rock and rock mass behaviors using representative yield or failure criteria.

3.1 *Rock strength or failure envelopes*

If the rock or a rock mass behaves like a cohesive and frictional material (soil), the linear Mohr–Coulomb failure criterion, specifying that the shear strength is composed of two simultaneously mobilized strength components, cohesion and frictional strength, is applicable:

$$\tau = c + \sigma' \tan\phi \tag{6}$$

As Schofield (1998) pointed out, strain affects the two strength components differently even in soil tests, rendering Equation 6 inapplicable for materials with interlock and with strain-dependent cohesion. Consequently, Equation 6 is not valid for intact rock or interlocked rock masses. It is only applicable for failure mode M_{33} in RMQ3.

The Mohr–Coulomb error

Schofield (1998), by reference to Taylor's work and to Terzaghi's use of the Mohr–Coulomb (MC) criterion, pointed out that it is a serious error to assume that the cohesion and frictional strength are simultaneously mobilized. He correctly concluded "In adopting the Mohr–Coulomb equation Terzaghi made the error of regarding apparent cohesion as a soil property independent of strain". Based on work by Martin (1997), it is now well-understood that the mobilization of the cohesive and frictional strength components of rock and rock masses is strain-dependent. Accordingly, Coulomb's strength equation needs to be rewritten as shown below with cohesion, 'effective' stress internal to the rock, and the dilation angle depending on the plastic strain ε_p:

$$\tau = c(\varepsilon_p) + \sigma'(\varepsilon_p)\tan(\phi + i(\varepsilon_p)) \tag{7}$$

As a consequence, rock mass failure forecasting becomes highly sensitive to the assumed stress–strain characteristics. For rock engineering applications, Schofield's statement can be paraphrased as 'adopting the Mohr–Coulomb equation and regarding apparent cohesion as a rock property independent of strain is a serious error'. Overcoming this deficiency is an essential step to move from common to best practice. Unfortunately, common practices of using elastic-perfectly brittle models ignore this and often lead to predictions of excessive depth of yield. Best practice strength models have to properly represent the gradual transition from peak toward the residual strength. The word 'toward' is intentionally used here to highlight that the residual strength is rarely reached in underground construction (see Section 3.1.6 on post-peak strength).

3.1.1 *Peak strength envelopes*

Recognizing the nonlinearity of triaxial test data (in the $\sigma_1 - \sigma_3$ space) and the need for a tension cut-off led to the development of the now commonly adopted Hoek–Brown strength envelope for rock and rock masses. The resulting peak strength equations are also independent of strain (as in Equation 6) for MC criterion) and thus provide best-fit envelopes to peak strength data.

Based on a careful examination of the low confinement range of most brittle rocks, Kaiser & Kim (2014) introduced an s-shaped peak strength envelope to better represent the cohesion loss at low confinement due to extensional cracking, and the high strength gain near the spalling limit due to high dilation. The resulting equations for the s-shaped peak strength are again independent of strain but provide more representative peak strength envelopes for brittle rock (see Section 3.1.2 concerning elevated HB parameters).

Unconfined compressive strength (UCS)

While the UCS test is one of the most common index tests for site characterization, producing meaningful results remains a challenge. The indiscriminate use of UCS data without careful screening of failure modes may mislead when used in rock mass classifications, semi-empirical depth of failure estimations, TBM penetration rate charts, or when determining rock strength envelopes.

Bewick et al. (2015) reviewed critical factors impacting the UCS[4] and provide guidance for processing and the interpretation of UCS test data from homogeneous to heterogeneous rock. Common practice of UCS testing typically leads to (excessively) high variability in test data: coefficients of variation (CoV) generally exceed 25% for homogeneous rock and 35% for heterogeneous rock. This variability can be attributed to variations in specimen failure modes in heterogeneous rock. To improve consistency, testing must be carried out by a qualified laboratory using appropriate QA/QC procedures, and the ISRM suggested or ASTM methods. Best practices as outlined by Bewick et al. (2015) need to be adopted:

- The UCS for homogeneous rock should be based on 5 to 10 or more intact specimens; filtering of test data is required to get the intact strength of the homogeneous component of rock blocks.
- For heterogeneous rock a minimum of 10 to 40 specimens are required.
  - If the CoV exceeds 30 to 35%, the data are likely bi- or even multi-modal due to unique characteristics. Respective mean strength values should be defined for sub-datasets (e.g. Figure 7).
  - The complete dataset is representative of the strength and variability of a heterogeneous rock block.
- Depending on the empirical design approach, either the homogenous or the heterogeneous UCS or both may be relevant; e.g. for the stress level to establish the depth of spalling, the mean UCS for the homogeneous specimens should be used (Martin 2019).

An example is presented in Figure 7. The mean UCS = 120 MPa with an excessive CoV = 59% suggests that the sample population consisted of at least two populations. After identifying specimens that mostly failed with intact rock breaks, the bi-modal distribution shown in green was derived with a mean UCS = 220 MPa for the homogeneous part and 96 MPa for the heterogeneous part. The practical implications of ignoring the bi-modal nature can be serious. For example, the disk consumption may be unexpectedly high and the TBM penetration rate may be severely compromised if estimates are based on mean UCS-values for the entire sample population. The extreme values defining the homogeneous rock strength will dominate the disk performance.

Best practices in UCS data testing and reporting must include filtering of UCS data by failure mode separate homogeneous and heterogeneous data.

4. Note: UCS ≠ σci, the Hoek-Brown unconfined strength parameter (Kaiser & Kim (2014).

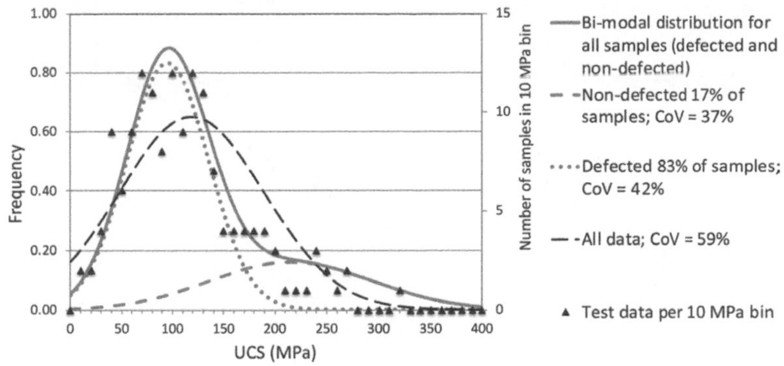

Figure 7. Frequency distributions for a heterogeneous, brittle Quartzite.

3.1.2 *S-shaped failure envelope for brittle rock*

As indicated above, Kaiser & Kim (2008, 2014) introduced the s-shaped or tri-linear peak strength envelope concept to better represent the rock strength in the low confinement range (Figure 8a). For the example shown in Figure 8b, the mean UCS (or UCS_I) depicts the average expected UCS in the low confinement range ($\sigma_3 < UCS/10$). The apparent AUCS (or UCS_{II}) provides the y-intercept of the shear strength envelope at higher confinements. The transition occurs at the spalling limit $k_s = \sigma_1/\sigma_3$ ($k_s = 15$ in this case).

When numerical modeling tools do not facilitate s-shaped or tri-linear failure envelopes, it is necessary to approximate the s-shaped peak strength envelope by selecting appropriate Hoek–Brown (or Mohr Coulomb) parameters. As illustrated by Figure 8c (Kaiser & Kim, 2014), uncommon high m_i-values emerge ($m_i = 72$ in this case). Similarly, if MC parameters were fitted, abnormal high slopes ($\phi + i$) would appear in the low confinement range.

Best practice for such materials is to ignore common Hoek–Brown parameter tables (with σ_{ci} and $m_i = 24$ for Quartzite; Hoek et al. (1995)) and to adopt a representative mean UCS-value (e.g. 95 MPa) and a high, best fit m_i-value (e.g. $m_i = 72$).

3.1.3 *Post-peak strength (PPS)*

In soil mechanics, the term 'residual strength' was introduced to represent the shear strength at very large strains, e.g. for slope stability analyses. As indicated above (Equation 7), the rock strength is strain-dependent, and cohesive and frictional strength components are mobilized at different rates (Martin 1997). Near underground structures, however, the actual plastic strains are rarely sufficient to reach the residual strength. This is illustrated on a triaxial test data set with a 1% axial post-peak strain limit (Figure 9a; the equivalent s-shaped peak strength with AUCS = 250 MPa is shown in Figure 9 c). The Hoek-Brown peak strength is defined by UCS = 130 MPa and $m_i = 18$.

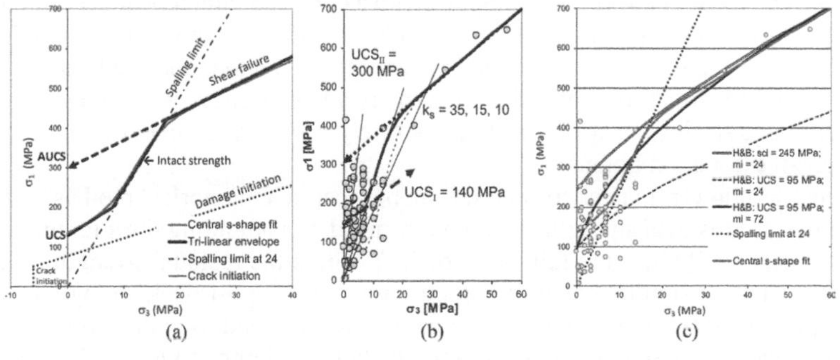

Figure 8. (a) Schematic tri-linear failure criteria; (b) fitted s-curve (dotted for lower limit) with respective linearized approximations; and (c) Hoek–Brown envelopes obtained by various fitting approaches. The resulting parameters are shown in the legend, i.e., UCS and m_i (modified from Kaiser & Kim, 2008).

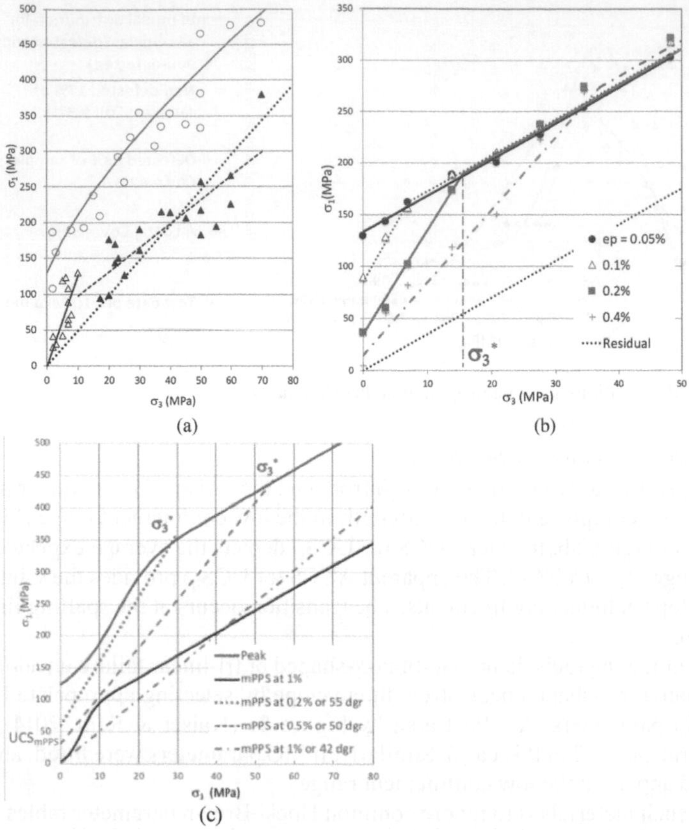

Figure 9. (a) Triaxial test data from tests on Quartzite (o for peak) with 1% axial strain limit (triangles); (b) bi-linear post-peak strength (PPS) envelopes for a Marble with brittle-ductile transition at $\sigma_3{}^* = 15$ MPa for a plastic strain $\varepsilon_p = 0.2\%$ (data from Wawersik & Fairhurst (1970)); the peak strength and three mobilized PPS envelopes are shown; residual at $\varphi = 34°$); and (c) s-shaped peak and mobilized post-peak strength (mPPS) at 1% strain for (a). Bi-linear mPPS envelopes with transitions at $\sigma_3{}^* = 22$ and 60 MPa for 0.2 and 0.5%, respectively, are also shown.

At 1% strain and $\sigma_3 <$ UCS/10 (Figure 9a), the cohesion is lost and the PPS is defined by a high 'friction plus dilation' angle due to interlock ($k = 15$ or $\varphi + i = 55°$). At higher confinements, the apparent cohesion AUCS(1%) = 65 MPa or $k = 3.3$ for $\phi = 33°$.

For a Marble (Figure 9 b modified from Kaiser (2016b)), the PPS is highly strain-dependent, leading to bi-linear mobilized PPS (mPPS) envelopes with a cohesion intercept and a change in the slope at a strain-dependent brittle ductile transition (at $\sigma_3{}^*$). With increasing strain, the mPPS decreases and eventually reaches the residual strength. The cohesion intercept and the location of the transition point of the mPPS are strain-dependent and the residual strength is, for most rock types, only reached at very large strains, i.e. at strains that are rarely encountered in underground engineering, except close to an excavation boundary.

Whereas the transition from peak to residual can be steered in numerical models with appropriate constitutive laws, the residual strength should not be used as a bounding limit for underground rock engineering; except for RMQ3 and failure mode M_{33}, where the residual strength (Figure 9b) may be reached. Rock failure simulations utilizing purely frictional post-peak strength envelopes, with the PPS set equal to the residual strength with zero cohesion, tend to underestimate the mobilized PPS and consequently lead to predictions of excessively large depth of yield. When elasto-perfectly brittle plastic transition models are used (e.g. with RS^2) strain-limited mPPS have to be adopted to prevent excessive loss of strength and yield zone propagation. It is recommended that the mPPS envelope honors three points as illustrated by Figure 9c: the tensile strength (not shown), UCS_{mPPS} and the brittle-ductile transition point at σ_3*. For the Quartzite, the mPPS at 1% strain is represented by the s-shaped mPPS curve in Figure 9c. It can be approximated by a purely frictional mPPS with a slope of $(\phi + i) = 42°$.

From common to best practice

Ample opportunities and economic as well as safety benefits can be derived by moving from common to best practices in rock strength determination. Common practices of fitting peak and residual strength data are clearly not best practice.

- Much care has to be taken to obtain representative UCS-values by identifying failure modes and separating homogeneous from heterogeneous specimen (Bewick et al. 2015).
- Failure envelopes are often s-shaped or tri-linear when cohesion and frictional strength components depend on plastic straining. This is valid for the peak and post-peak strength.
- Unless numerical models are calibrated to capture post-peak strength degradation, mPPS limits should be defined for anticipated plastic strain limits.

Figure 9c summarizes best practices in peak and mPPS definition for the test data shown in Figure 9a. For this case, the peak strength can be described by the conventional Hoek–Brown envelope (Figure 9a) or by a s-shaped envelope (Figure 9c). The mPPS at small plastic strains exhibits a remnant cohesion (or $AUCS_{PPS}$ = 60 and 30 MPa) and a high slope $(\varphi + i)$ ranging from 55 to 50° at 0.2 and 0.5% plastic strain, respectively. At 1% plastic strain, the mPPS is bi-linear with $AUCS_{PPS}$ = 65 MPa and a slope angle of $\phi = 33°$ at $\sigma_3 > 10$ MPa.

3.2 *Rock mass behavior near excavations*

3.2.1 *Failure criterion for brittle rock masses*[5]

As explained by Kaiser (2016a & b) based on work by Diederichs (2007), tensile stresses induced during deviatoric loading of heterogeneous rock lead to Griffith-type extension fracturing with the consequence of a depressed failure envelope (Figure 10) in the low confinement zone where fracture propagation causing spalling is not suppressed by the available confining pressure. The resulting failure envelope for a rock mass is s-shaped and the stress space can be divided into two behavioral zones: spalling dominated stress fracturing at low confinement (to the left of the spalling limit), and shear rupture dominated behavior at high confinement (to the right). This divides the rock mass surrounding an excavation into two zones, an 'inner' and an 'outer' shell (Figure 10).

3.2.2 *'Inner' versus 'outer' shell behavior*

The threshold between the inner and outer shell is defined by the spalling limit, approximately near $\sigma_3 = UCS/(10 \text{ to } 15)$ as shown in Figure 10. From a practical perspective, there is a need to differentiate between engineering problems dominated by stress-fracturing (in the inner shell) and shear rupture (in the outer shell).

Inner shell engineering problems are those dominated by the behavior of the rock mass in the zone immediately surrounding an excavation where the confinement is low, i.e., in the zone where stress-fracturing can occur and blocks or fragment can rotate. Engineering challenges of support design, strainbursting, etc. fall into the class of inner shell problems.

On the other hand, engineering problems related to pillar instability, including pillar bursting, fall in the outer shell class where shear rupture dominates because spalling is partially or fully suppressed by sufficient confinement.

5 Strength envelopes define the rock strength (e.g. s-shaped envelopes) as obtained by testing, and failure criteria (e.g. bi-modal criteria by Diederichs (2007)) represent the rock or rock mass behavior for numerical modeling purposes.

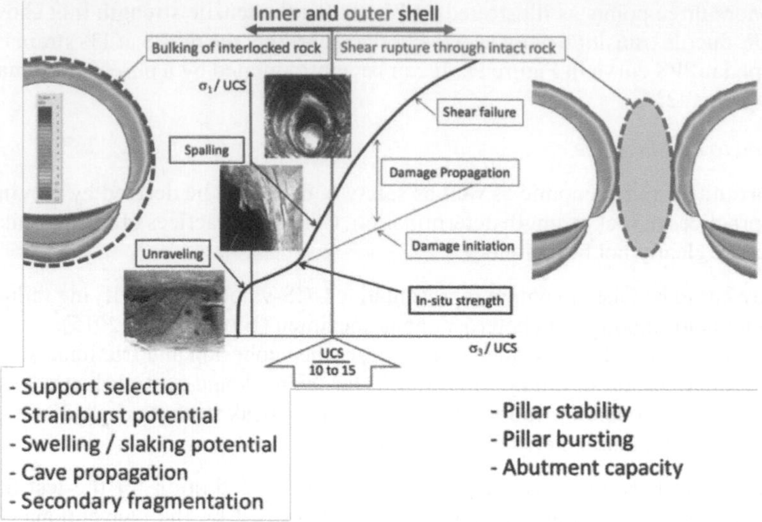

Figure 10. S-shaped failure criterion for brittle rock masses (center) and zoning of stress space for inner (left) and outer (right) shell behaviors in underground rock engineering (red contour $\sigma_3 = 12$ MPa).

3.2.3 *Depth of yield, failure and strainbursting*

Depth of yield d_y

Continuum models typically show indicators of yield and thus can be used to establish the depth of yield around an excavation (x in Figure 11; also shown are confining pressure contours for $\sigma_3 = 0$ to 10 MPa).

The three colored points (red, green, blue) in Figure 11 indicate that the rock at these locations still has a substantial cohesive strength and thus, while yielding, will not fall apart and will not fail under gravity loading alone, i.e. will not unravel. Only near the excavation, at locations with tensile failure (o), is the cohesion fully lost and unsupported rock would unravel. The depth of yield is therefore not the same as the depth of failure.

3.2.3.1 *Depth of failure* d_f

The depth of failure defines the depth to which a rock mass fails and unravels if unsupported. The normalized 'extreme' depth of failure d_f^e/a, i.e. the maximum depth of notch formation by spalling recorded in a tunnel domain with otherwise equal properties, increases linearly as a function of the stress level (up to $SL = 1$ or $\sigma_\theta = \sigma_{max}$ = mean UCS (Martin 2019)).

The initial linear trends of the normalized extreme depth of failure d_f^e/a for $SL \leq 1$ is defined by:

$$\frac{d_f^e}{a} = 1.25.SL - 0.51 \pm 0.1 \tag{8}$$

The extreme depth of failure increases at a lesser slope of 0.75 to a maximum at $SL = 1.5$.

The range for the mean depth of failure d_f^m, including locations with no failure, was established by calibrated numerical modeling by Perras & Diederichs (2016) (see Kaiser (2016c)). It typically ranges between 20 to 30% of d_f^e:

$$d_f^m \approx \frac{d_f^e}{3.5 \text{ to } 4.5} \tag{9}$$

Based on these semi-empirical relations, it is possible to anticipate the mean and extreme depths of failure in brittle failing rock.

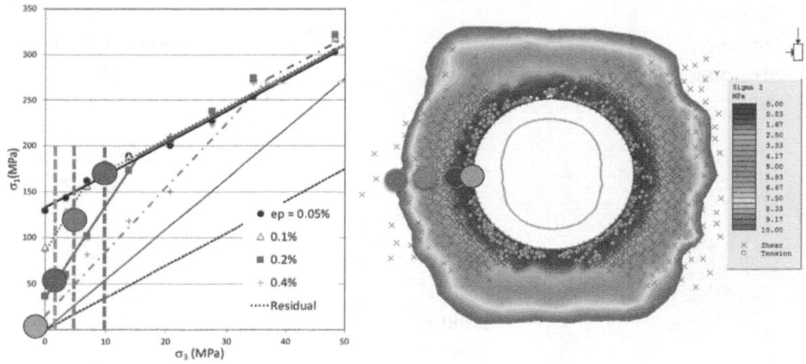

(a) (b)

Figure 11. (a) Illustration of four states at locations indicated by circles in principal stress space for Marble (Figure 9); and (b) continuum model of tunnel showing yield locations (x for shear and o for extension) for k_o = 0.5 together with confining stress σ_3 contours.

3.2.3.2 *Depth of strainbursting d_{SB}*

The depth of strainbursting is difficult to predict because instability not only depends on stress and strength but on stress gradients and geometric factors, too. It can be as little as the depth of an individual spall or as deep as the extent of highly stressed rock. Strainbursts occur at locations of stress raisers as illustrated by the principal stress contours in Figure 6. These stress raisers occur at the outer limit of the inner shell and can easily involve 25 to 35% of the tunnel radius (e.g. 0.75 to 1 m for a tunnel with a radius a = 3 m).

By reference to Equations 8 and 9, it can be argued that the depth of strainbursting falls somewhere between d_f^m and d_f^e; i.e. for $d_f^m = 1/4\, d_f^m$, d_{SB} = 0 to 3/4 d_f^e (e.g. between 0 and ≤ 1.65 m for a = 3 m and $SL \leq 1$). In gradually spalling ground, d_{SB} is near zero and for typical tunnel sizes in massive to moderately (unfavorably) jointed ground d_{SB}-values of 1 to < 2 m must be expected. This is consistent with observations from strainburst-prone mines. The total depth of failure after a strainburst, may be larger as some burden may be ejected and blocky ground may unravel after the burst.

3.2.4 *Geometric bulking of stress-fractured rock*

Stress-fractured and bursting ground bulks when deformed past the peak strength of the rock mass. This leads to unidirectional bulking deformations that are controlled by the excavation geometry and the imposed tangential strain (Figure 12 a). This directional bulking process is not captured by dilation models that relate the strength to the volumetric strain (bottom of Figure 12a).

Estimation of geometric bulking displacement at wall

The geometric bulking deformation can be estimated following the semi-empirical approach outlined by Kaiser (2016c) whereby the estimated depth of failure d_f (not d_y) is multiplied by a confining pressure dependent bulking factor $BF(\sigma_3)$ (Figure 12b). In bursting ground, pre-burst fracturing to d_f imposes a wall displacement $d_f * BF_{static}$ on the support and the strainburst adds further displacements $d_{SB} * BF_{SB}$ (note: BF_{static} and BF_{SB} are not necessarily equal). For example, at an average confining pressure σ_3 = 0.1 MPa, excavation-induced bulking is estimated at BF = 3%. For a = 3 m and SL = 0.8, a depth of failure of d_f = 1.47 m causes d_{wall} = 44 mm. This displacement consumes some of the displacement capacity of the installed support before a potential burst.

The mining-induced bulking is larger because of the increased geometric non-fit at larger deformations. For a mining-induced or strainburst associated strain of 1 to 2% the chart indicates a bulking factor BF = 8%. Hence, a strainburst of d_{SB} = 1 m would add 80 mm. If the support does not have a remnant displacement capacity of 80 mm, it will fail. The total displacement after the burst will be 124 mm.

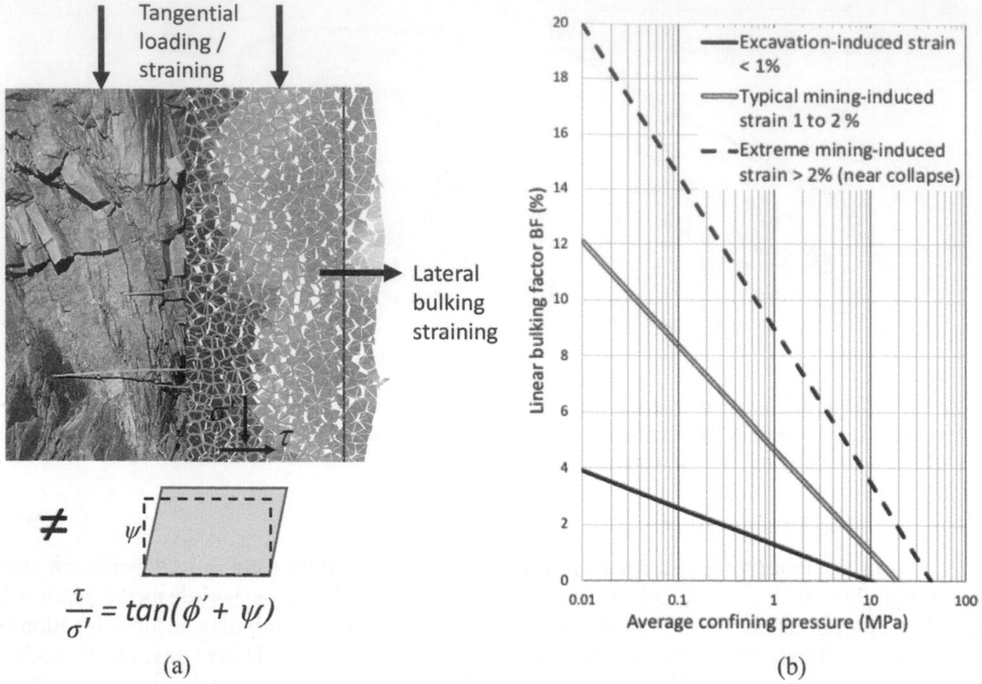

Figure 12. (a) Unidirectional rock masses bulking due to stress-fracturing (photo) reflected in cartoon-like Voronoi model ≠ non-representative dilation model; and (b) semi-empirical bulking factor charts (modified from Kaiser 2016c) for excavation and mining-induced bulking.

From common to best practice

Again, there are ample opportunities and economic, as well as safety benefits, that can be derived from differentiating between near wall (inner shell) and outer shell behaviors. This is particularly valid for support designs where bulking deformations from statically and dynamically deformed stress-fractured rock dominate the rock mass behavior.

Different EDPs dominate the behavior in the inner and outer shell. Separate EDP-value sets need to be defined to assess the vulnerability to damage and the fragility in order to select appropriate rock support systems.

4 ROCK MASS CHARACTERIZATION

"Den Werkstoff Fels in seinem Zustande und seinem Verhalten zu beschreiben, ist die erste Aufgabe dessen, der seine Felsbauwerke sicher anlegen, zweckmäßig konstruieren, schön gestalten und wirtschaftlich ausführen möchte." or "Describing the construction material, the rock mass, in its condition and its behavior is the first task … to safely lay out, sensibly construct, …, and economically execute works in rock." (Müller 1963)

For this purpose, the goals of rock mass characterization have to be tailored to the technical objectives of an investigation. For example, for cave engineering, the workflow includes the assessment of five engineering aspects (Brown, 2007): caveability; fragmentation; cave performance; extraction level stability; and mine construction (rock support). Each has to address specific questions and each design component is influenced by a different rock mass behavior and thus is dominated by different rock mass properties. For example, the vulnerability to stress fracturing (spalling), the bulking characteristics for flow control and support selections, and the confining stress impacts on pillars or support design are dominated by different EDPs. As discussed previously, the development of best practices in rock mass characterization therefore starts with the identification of critical EDPs including related rock mass characteristics and properties.

In underground construction, the rock mass has to be described long before access to observe its behavior is possible. This demands a systematic approach of rock mass quality quantification moving from 'inferred' to 'proven' rock mass quality (Section 4.1), a practice that is often ignored or rarely systematically executed.

With increasing depth, the changing rock mass behavior creates additional challenges for the mining engineer. At depth, the rock is highly stressed, closer to failure, and often more confined leading to a higher interlocking and elevated strength. This issue is covered in Section 5 on rock mass strength estimation.

An effective rock mass characterization program, including logging, mapping and laboratory testing, needs to collect and interpret features that are relevant for clearly defined purposes. How to overcome challenges of rock mass characterization for underground construction in deep mining is covered by Kaiser et al. (2015). The recommended methodology is not repeated here.

In the following, two aspects are addressed to assist in the application of best practices in rock mass characterization: (1) stages of characterization to move from 'inferred' to 'proven' quality descriptors, and (2) the suitability of classifications for rock mass characterization.

4.1 *Stages of rock mass characterization – from assumption to fact*

Rock mass characterization involves many model-building components: geological, structural, rock mass, and hydro-geological models. These models are being developed in an incremental and iterative manner with data initially collected during scoping studies from boreholes. The data is then gradually refined during follow-up studies and eventually during the construction phases by mapping and back-analyses of monitoring results. What is often ignored in common practice is to clearly define and register assumptions such that the characterization program can focus on eliminating one assumption after another and on replacing them with factual, reliable information.

As for mineral resource definition increased knowledge and confidence needs to be built from 'inferred' to 'indicated' to 'measured' or from 'inferred' to 'probable' to 'proven' (on left of Figure 13).

Assumptions have to be made in the early stages of a characterization program, assumptions that are inferred from comparable rock formations and existing empirical rules from similar rock masses. Next, uncertainty has to be incrementally removed by reducing the variability and move toward 'probable' data. For this purpose, targeted data collection focused on specific assumptions is to be used and documented to ensure the credibility of information. Finally, assumptions can be removed and replaced by 'measured' data or 'proven' quality ratings when sufficient factual data

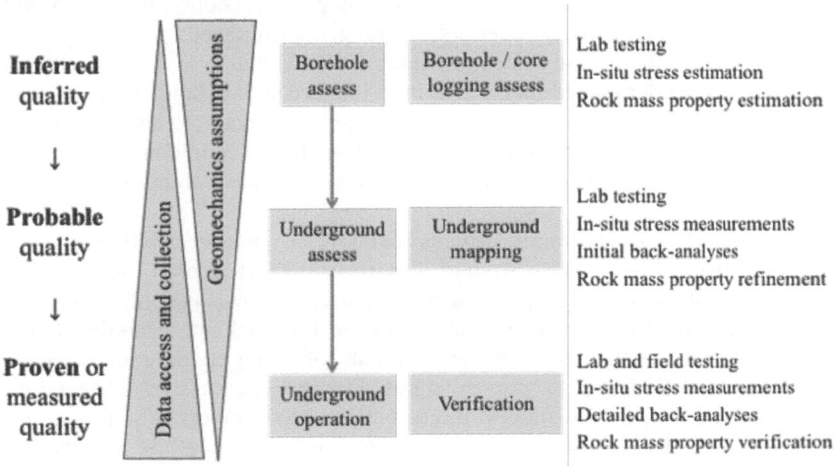

Figure 13. Evolution of data collection to eliminate or confirm geomechanics assumptions at different stages of a rock mass characterization program.

is available (likely after underground observations are available). Some assumptions may never reach the status of 'measured', e.g. in-situ stress, and back-analyses of other monitoring data may be required to reach the 'proven' class. This approach is illustrated by the wedges in Figure 13.

The use of qualified professionals to make necessary assumptions is a very important principle of geotechnical engineering as it is better to extrapolate from experiences and to plan based on comparable information. However, boundaries between assumed and measured (or proven) information are often blurred and it is essential to clearly define and then replace assumptions by facts as the site characterization proceeds. Eventually, the validity of all assumptions has to be proven.

Best practice in rock mass characterization includes a registry of assumed EDPs (incl. their variability) and tracks the path from assumption to fact. While EDP-values have to be inferred during the preliminary design phase, good engineering practice demands that they are raised to probable and then to a proven status with minimal remnant uncertainty.

4.2 *Characterization versus classification*

Ideally, a rock mass is characterized in a comprehensive manner such that all engineering questions can be answered. However, because different engineering tasks require different EDPs, most rock mass classification systems focus on one or two applications (e.g. support selection). Strictly speaking only part of the classification deals with the characterization of the rock mass (block size and interlock, condition and persistence of joints, etc.), the remainder deals with the application for a defined purpose (support selection, rock mass strength determination, etc.). For this reason, and because of the fact that each classification system considers different parameters and assigns different weights, specific classification indices (Q, RMR, MRMR, GSI, etc.) should not be correlated. They should be used independently for the purposes defined by the developers and for the ground conditions that form the foundation for the classification system.

Even the GSI (Geological Strength Index), while not providing guidance for support selection, was developed with a bias toward the characterization of a rock mass in the low confinement range near excavations (slopes or inner shell of tunnels). Furthermore, the GSI was introduced for collecting field information for rock mass strength determination and "to address two principal factors considered to have important influences on the mechanical properties of a rock mass, i.e., the structure (or blockiness) and the condition of the joints" (Hoek & Brown, 2019). Strictly speaking, the GSI-system was developed only for rock mass strength estimation and for conditions of blocky ground (see Section 5).

While each classification system stands on its own merit, the indiscriminate use of rock mass classification without consideration of the limit of applicability obtained from the implicit data sources and without respect for the types of engineering projects (slopes, tunnels, mining (pillars and caves)) has led to serious consequences and is strongly discouraged.

For example, conventional rating systems such as RMR, Q and GSI were developed and calibrated for conditions not dominated by large mining-induced stress changes and stress-fracturing of strong rock blocks. Hence, they are often not applicable, for example, for defected rock and in large strain environments. If the GSI is indiscriminately applied to conditions other than those used to develop the GSI-based strength equations, the resulting rock mass strength tends to be underestimated (see Section 5).

Other deficiencies of rock mass rating systems are addressed in more detail in the ISRM on-line lecture on "Challenges of Rock Mass Strength Determination" (Kaiser 2016a) and the impact of rock mass heterogeneity on in-situ stress variability is discussed by Kaiser (2016b).

Furthermore, because of the common practice of collecting rock mass classification data without having identified the controlling EDPs and without following an 'assume, revise and verify' approach to move from 'inferred' to 'proven' status (Figure 13), excessively detailed information is often collected and then not distilled to the essentials. More importantly, properties that may dominate the behavior of the rock mass are ignored (e.g. tensile strength to anticipate spalling or rock fragmentation).

Following the rationale outlined in this article, it is of primary importance to identify whether a rock mass belongs to one of three classes (RMQ1 to 3; refer to Figures 4 and 5). This is particular significant when estimating the rock mass strength (Section 5).

4.2.1 *From common to best practice*

Again, opportunities and economic as well as safety benefits can be derived by moving from common to best practices in rock mass characterization. Too many parameters are frequently collected that, in the end, do not matter, and essential parameters are ignored or only collected late in a project when unexpected rock mass qualities (changed conditions) are encountered.

Targeted rock mass characterization should be guided by identified EDPs and by the eventual intent (engineering tasks). For example, the Q-system was originally intended for support selection and then expanded for TBM and other applications, whereas the GSI-system was developed for rock mass strength determination. Both should be used with discretion for the intended purposes. Most importantly, they should be used to identify RMQ-classes that matter for the performance of underground excavations (RMQ1 to 3). The influence of stress (and water) should be treated separately at the characterization stage and only considered when used for engineering designs. Limits of applicability need to be respected (see Section 5 for GSI).

Best practice in rock mass characterization has to follow a systematic process to move from inferred to proven rock mass quality designations. Many 'unexpected' problems could be anticipated if the full spectrum of possible ground conditions was properly described by clarifying inferred from probable and proven conditions. In this manner, claims for 'changed conditions' can be prevented because construction techniques suitable for uncertain conditions can be selected.

Similarly, the development of in-situ testing and monitoring programs should focus on critical components of the path from assumption to fact.

5 ROCK MASS STRENGTH ESTIMATION

Some of the challenges in rock mass strength estimation for the design of deep underground excavations are covered by Kaiser (2016a) building on the work by Bewick et al. (2015) and Kaiser et al. (2015). More recently, the GSI-approach has been 'updated' not because fundamental changes were required, but to "discuss many of the issues of its utilization and to present case histories to demonstrate practical applications ..." (Hoek & Brown, 2019). In a companion paper Bewick et al. (2019) provide guidance for rock mass strength estimation when the limitations of the GSI-strength equations are reached.

"Why do we tend to underestimate the rock mass strength for underground construction at depth?" was the lead question raised during the ISRM lecture. The basic answer is that common practices fail to capture the strengthening effect of interlock in non-persistently jointed rock where failure through intact rock adds strength, and geometric bulking leads to a more rapid strength gain at elevated confining stress. As outlined in detail in the on-line lecture, best practice in rock mass strength determination certainly does not always follow common practice.

For situations where blocks exist (blocky ground with three persistent joint sets) and block rotation is possible (even if interlocked to some extent), Hoek & Brown (1997 and 2019) present GSI-based strength equations for isotropic rock masses containing block forming joints and blocks without defects. They indicate that the underlying GSI-experience stems from excavations in rock masses where block rotation contributes to the failure process. The following GSI-strength equations are therefore valid for RMQ2 and 3:

$$\sigma_1' = \sigma'_3 + \sigma'_{ci}\left(m_b \frac{\sigma_3'}{\sigma_{ci}} + s\right)^a \tag{10}$$

where, $m_b = m_i exp\left(\dfrac{GSI-100}{24}\right)$; $s = exp\left(\dfrac{\text{GSI}-100}{24}\right)$, and $a = \dfrac{1}{2} + \dfrac{1}{6}\left(e^{\frac{-GSI}{15}} - e^{\frac{-20}{3}}\right)$.

However, if rock blocks are strongly interlocked and block rotation is restricted, the rock mass will be able to mobilize extra strength, particularly when it is highly confined ($\sigma_3 > UCS/10$). As illustrated by Figure 14a, the rock mass strength, extrapolated using Equation 10, from low to high confinements can be as low as 50% of the anticipated confined rock strength with interlocked and shear rupture behavior (compare black to red arrow at $\sigma_3 = 20$ MPa). The shaded area in Figure 14a illustrates the difference in strength obtained by extrapolation from inner shell values.

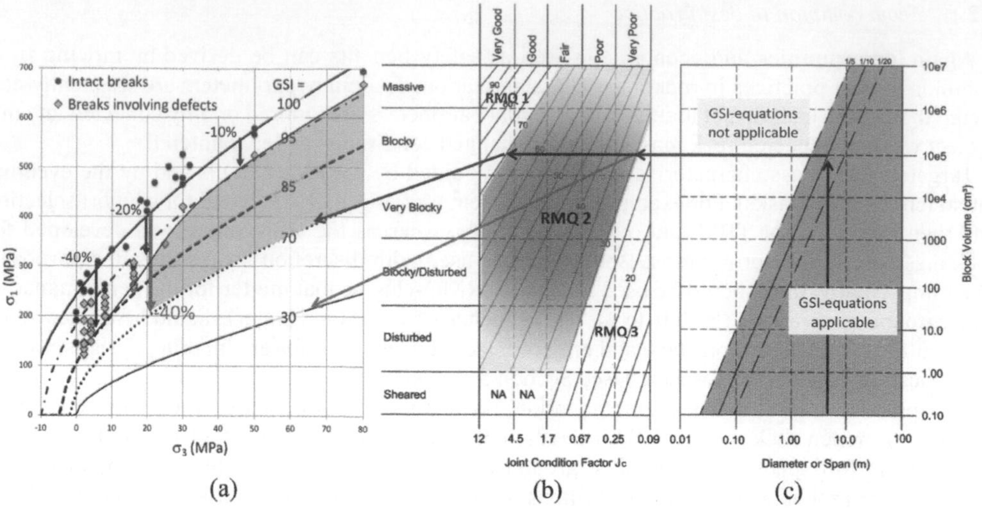

Figure 14. (a) Example of strength degradation and underestimation by GSI-strength equations; (b) applicability range for RMQ2 boundaries; and (c) scale applicability limits for GSI-strength equations.

When Equation (10) is adopted, it is implicitly assumed that the degree of interlock is sufficiently small such that the rock blocks formed by open joints can rotate during the failure process. In massive to moderately jointed rock this is not the case and the strength is controlled by stress-fracturing of rock blocks, rock bridges and asperities, and by the dilation of highly interlocked rock fragments. Hence, the strength degradation from the intact rock strength is much less at elevated confining pressures than conventionally assumed (Bahrani et al. 2013 and 2016). Consequently, the rock mass is much stronger than anticipated by the standard models as elaborated by Kaiser (2016a) and illustrated by Figure 14a.

5.1 *Applicability limits of GSI strength equations*

At depth, the rock mass is often massive to moderately jointed with non-persistent joints and joints are highly compressed leading to further strengthening effects due to 'over-closure' (Barton et al. 1985). Consequently, the conventional GSI-strength equations are rarely applicable for RMQ1, not even for RMQ2 at high J_c-values (Figure 14c) when the rock mass is highly confined as encountered at depth. Similar deficiencies have to be overcome when rock mass classifications are used for the design of excavations at depth.

There is another basic assumption of the GSI approach that needs to be respected, i.e. it is only applicable for rock masses that can be assumed to be isotropic and simplified to a continuum, i.e., if the block volume is small relative to the excavation size. For common engineering problem scales, the GSI-strength equations are applicable when a rock mass is made up of blocks with edge lengths smaller than 1/10 (range 1/5 to 1/20) of the problem dimension (i.e., tunnel diameter, excavation span and pillar height). Based on this criterion, the applicability of the GSI-strength equations can be assessed using Figure 14c.

They are applicable for:

- shafts or raises of 1 to 4 m diameter for very blocky to sheared ground with block size < 10^4 to 10^5 cm^3;
- standard tunnel sizes and pillar heights (5 to 10 m) for blocky ground with block size < 1 m^3; and
- large caverns and caves (span > 20 m) as long as the persistence of joints in very strong rock is sufficient to create non-interlocked blocks in the 10 to 1000 m^3 range.

For example for a 5 m wide tunnel, GSI-equations can be used for very blocky or worse ground (as indicated by arrow in Figure 14c).

As mentioned previously, Bewick et al. (2019) provide guidance for rock mass strength estimation when the limitations of the GSI-strength equations are reached. It is not that the GSI is not applicable, as erroneously implied by the wording in previous publications by the author (e.g., by mentioning the GSI applicability limits in the ISRM on-line lecture), but the applicability of the GSI-strength equations is limited.

From common to best practice
The common practice of indiscriminate use of the GSI-strength equations (Equation 10) for rock masses of RMQ1 and part of RMQ2 tends to underestimate the strength. They are applicable for 'soil like' behavior in RMQ3 for failure mode M_{32} and M_{33}, and for RMQ2 of very blocky ground with low J_c-values. For massive to moderately jointed rock masses with GSI > 65, the systematic methodology for estimating equivalent rock mass strength parameters, outlined by Bewick et al. (2019), should be adopted. This methodology compliments the HB-GSI approach for rock mass strength estimation of a massive to moderately jointed rock mass.

The best practice to obtain reliable rock mass strength include the elements explained by Kaiser et al. (2015). Because best practice approaches must overcome the deficiency of common practices that tend to underestimate confined strength of highly stressed rock, it is advisable to adopt an observational approach, starting with experience-based, inferred assumptions, replacing them increasingly with verified, proven facts.

6 GROUND CONTROL – SUPPORT SELECTION

6.1 *Introduction*

Various rock mass classification systems (e.g., Q, RMR, etc.) have found wide application for support selection. These approaches are suitable if conditions at a project match those that form the underlying classification databases, i.e. civil tunneling data. They may however not be suitable when conditions change as they do in mining applications where stresses and failure modes change, or when the rock mass behavior at great depth differs. Unfortunately, rock mass classifications are frequently applied beyond their range of applicability. For example, they may be applicable for the advance of mining tunnels during mine development but are inapplicable when mining-induced stress changes dominate the excavation and support behavior.

Whether a support is selected based on risk or on a more conventional factor of safety-based design approach, load, displacement, and energy demands are compared with respective support system capacities. For static conditions, the support design is commonly dominated by load equilibrium (wedge stability) or displacement compatibility (squeezing ground) considerations (Equations 1 & 2). For dynamic earthquake or rockburst loading, it may be necessary to also consider energy equilibrium conditions (Equation 3). It is therefore mandatory to estimate three demands (load, displacement and energy) and the capacity of the integrated support system.

One important step in rock support design is to identify potential failure modes as explained in Sections 2. Only when the anticipated failure modes are correctly identified, can the most appropriate methods for demand estimation be applied.

In each design domain, the load, displacement and energy demands on the rock support are calculated individually by considering all possible excavation damage mechanisms. This is achieved by evaluating the damage severity in terms of depth of failure, rock mass bulking, and the anticipated impact velocity (if applicable). It is often difficult to know in advance which type of damage mechanism will eventually dominate. Hence, it is advisable to analyze all reasonably possible damage mechanisms and then identify the critical support demands, i.e., the possible worst-case scenario.

Once the demands on the support are identified, it can be evaluated whether a rock support system can be designed to control the failure process. If excessive demands are identified, other means of ground control management such as destressing, hydro-fracturing, etc., may have to be deployed to help reduce the support demand.

6.1.1 *Capacity estimation*
The load, displacement, and energy capacities of individual support components (mesh and shotcrete, or bolts and cable bolts) are conventionally obtained from pull-out (direct loading) or split tube (indirect loading) tests in the laboratory or in the field. Respective component capacity data are available from the literature (e.g. Cai & Kaiser, 2018) or site-specific values are obtained from field tests.

Unfortunately, these test results are highly dependent on the adopted test method and the ability of the support to resist internal loading is often unknown. Furthermore, the ultimate capacities reported in the literature may in practice not be reached (e.g. due to operational constraints such as allowable convergence) and, most importantly, the respective ultimate capacities are not simultaneously reached. A methodology to establish the capacity of integrated support systems utilizing all individual support components is discussed in Section 6.2).

6.1.2 *Demand estimation*

Load demands are obtained by estimating the volume of anticipated unstable ground (wedges, volumes or depth of stress-fractured ground) and by assessing the remnant load capacity after accounting for the simultaneously available bolt and surface support capacities. For dynamic conditions, the shakedown potential is established by adding a dynamic acceleration to the gravitational acceleration.

Displacement demands are commonly obtained by the use of analytical models (ground-reaction curves) or continuum numerical models with non-linear constitutive models. Both models are deficient when geometric bulking of stress-fractured ground in the inner shell of the excavation dominates the displacement demand. As a consequence, commonly adopted capacity models to represent individual support components are often limited in that the interaction between rock support and rock, particularly stress-fractured rock, is not fully captured; i.e. the displacement demands imposed by the fractured rock are underestimated.

Energy considerations are adopted when kinetic energy demands from earthquakes or rockbursts are anticipated. Unfortunately, energy cannot be measured rendering the energy-based designs unverifiable. However, because the product of displacements times bolt forces defines the energy consumption of a support component and the integrated support system, displacements provide an indirect measure of the energy consumed by the support system. It is for this reason that a deformation-based support design approach was introduced by Kaiser (2014) (discussed further in Sections 6.4). Displacement-based designs have a major advantage in that both displacement demands and capacities can be measured and compared with model outputs and field measurements of displacements (e.g. from laser surveys).

6.2 *Support system capacity (SSC) estimation*

An integrated support system is made-up of compatible support components (bolts, cable bolts, mesh, shotcrete, etc.) with load–displacement characteristics of individual support components obtained from static or dynamic laboratory or field tests. These components have to work together (in parallel) to provide rock retention, reinforcement, and holding functions (Cai & Kaiser, 2018). The capacity of each component is first mobilized and then consumed until it fails. The impact of the installation sequence on the differential loading or straining of support components has to be considered and all possible weak links in the rock support system have to be eliminated.

Much effort has been expended around the world to establish direct and indirect load, displacement, and energy dissipation capacities of support components. However, there is no systematic engineering approach to estimate the capacity of the integrated support system consisting of a bolting and surface support system.

A means to estimate the direct and indirect load capacities is introduced for direct loading by Cai & Kaiser (2018 in Chapter 4) and expanded for direct and indirect loading by Kaiser & Cai, (2019 in Chapter 3). For this purpose, it is assumed that all bolts work in parallel, meaning that they are simultaneously loaded at the plate through a surface support system (direct loading) and strained along the bolt by the relative movement of fractured rock blocks (indirect loading). For the prototype model, the load–displacement characteristics of individual support components are approximated by an equivalent perfectly plastic model (Figure 15a) to generate the cumulative support system load profile (Figure 15b). The corresponding displacement and energy dissipation profile of the support system is presented in Figure 16.

The support system's displacement capacity can be expressed as $\delta_{system} = \delta_{bolt\ (direct\ and\ indirect)} + \delta_{surface}$, where δ_{bolt} represents the bolt's displacement capacity (with the most deformable bolt controlling the maximum capacity) and $\delta_{surface}$ the surface support's displacement capacity. The static or

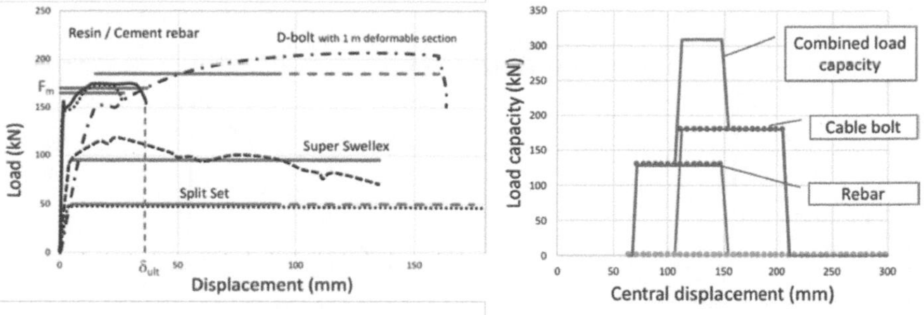

(a) (b)

Figure 15 (b) Approximation approach for various directly loaded bolt types, and (b) Load capacity of a support system consisting of two plastic support components (180 kN rebar and 250 kN cable at 1.2 m spacing) with mesh over mesh-reinforced shotcrete.

dynamic energy capacity of the support system $E_{system} = \Sigma E_{bolt\ (direct\ and\ indirect)}) + E_{surface}$, with ΣE_{bolt} represents the cumulative energy capacity of all installed bolts at a given mining-induced displacement $\delta_{bolt\ plate}$, and $E_{surface}$ represents the surface support's energy capacity at a given bolt plate displacement. This approach greatly differs from commonly adopted approaches in that the energy sharing between support components is accounted for and the deformation compatibility for the entire rock support system is respected. The self-supporting capacity of the reinforced stress-fractured rock mass is ignored in the prototype model described in the following sections.

6.2.1 *Load capacity of support systems*

The load capacity of a support system, obtained by superposing individual support component capacity-displacement characteristics, is illustrated by Figure 15b for a support system consisting of two support components (rebar and cable bolts) with an assumed elasto-plastic load–displacement characteristic. Figure 15a shows various approximations of actual pull test data for direct loading by an equivalent plastic load–displacement characteristic, i.e., a mean load (force) capacity F_m and an ultimate displacement capacity of individual bolts $\delta_{ult(B)}$.

Note: the dashed sections in Figure 15a indicate that the ultimate displacement capacity obtained by pull-out tests may not be available for Split Set bolts that are resisting over an anchor length measuring less than the test length. Furthermore, if test data are derived from indirect or split tube tests much less displacement capacity will be available for directly loading as indicated for the D-bolt.

6.2.2 *Energy and displacement capacity of support systems*

The energy capacity of a support system depends on the imposed displacement. This is illustrated for the same two-component support system of sequentially installed bolts loaded via mesh-reinforced shotcrete by Figure 16a. The bolts in this support system are activated after 65 mm of central deflection and the support system fails at 200 mm when the cable bolt reaches its ultimate displacement capacity.

Figure 16b also shows the decreasing remnant support capacity that is available as the support is deformed. These figures illustrate a most relevant interdependence of energy and displacement capacities.

6.2.3 *Support system capacity consumption (SSCC)*

The effectiveness of support systems can be compromised by quality deterioration (e.g., corrosion; not covered here) and by the consumption of a support system's displacement and energy capacity. Mining not only causes stress changes but also produces associated deformations and tunnel convergence which deform and strain the support. As these displacements increase, part of a support's displacement and energy dissipation capacity gets consumed. This is called support capacity consumption and is reflected in the support energy consumption plot (shown in red in Figure 16b). The support system capacity for this illustrative example is gradually lost until all of its capacity is consumed at 200 mm imposed central displacement. At 150 mm central displacement only 8 kJ/m^2 or 40% energy capacity remains.

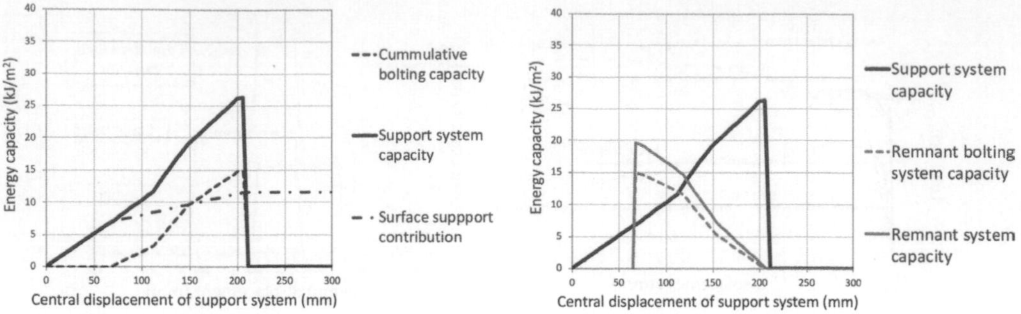

(a) (b)

Figure 16. (a) Energy–displacement characteristics of the same support system as in Figure 15, loaded via a mesh-reinforced shotcrete surface support, and (b) remnant capacity (red) as a function of applied displacement (central deflection between bolts); ‘Central displacement’ refers to the maximum displacement at the mid-span between bolts).

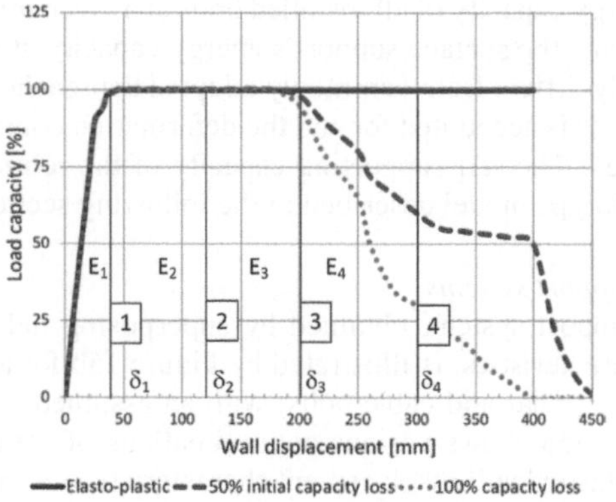

Figure 17. Schematic support system characteristics illustrating four stages of support capacity consumption [1] to [4]. Energy E_1 is the energy used to deform the support from 0 to d_1, E_2 from δ_1 to δ_2, etc. The load capacity is defined as a percentage of the peak load capacity (Cai & Kaiser, 2018).

As schematically illustrated by Figure 17, if an installed support system is deformed to a wall deformation δ_1, the support system has reached its yield load capacity and its elastic energy capacity E_1 has been consumed. The remnant displacement capacity to the first point of the support system degradation at δ_3 is $(\delta_3 - \delta_1)$ or the remnant energy capacity is reduced to $(E_2 + E_3)$. Then, if during mining the support is further deformed to δ_2, the remnant displacement capacity drops to $(\delta_3 - \delta_2)$, and the corresponding remnant energy capacity is reduced to E_3. In this manner, the support capacity is consumed as it is deformed.

At the displacement δ_3, the support system starts to lose its load capacity. Two degradation scenarios are shown in Figure 17 by the dashed and dotted support degradation curves. The system still has some load and displacement capacities but the remnant energy capacity, i.e. the area under these curves, rapidly drops. For this reason and for design purposes, the displacement capacity δ_3 is defined as the ‘allowable’ displacement capacity of the support system.

This example illustrates how the support system capacity is being consumed by mining-induced displacements, i.e. by displacements imposed in a static or gradual manner, or by co-seismic deformations in seismically active mines (Kaiser 2017). How to recognize support consumption in the field is discussed in Section 6.2.5.

The practical implication of the support system capacity consumption is that a support system rarely exhibits its full capacity that was available at the time of support installation.

Consumption of retention system capacity

Similarly, the energy capacity of retention systems can be consumed as they are deformed. According to Figure 18, chain-link mesh could have a capacity of 10 to 13 kJ/m² but only if >300 mm central deflection is acceptable. Because bagging between bolts must be limited for operational reasons and to minimize bending of bolt heads, it is advisable to design for allowable rather than ultimate capacities. For example, at 100 to 200 mm central deflection E_{100} to E_{200} represents a meaningful, allowable energy design capacity (red boxes in Figure 18).

By analogue, if the capacity of a support system can be consumed, it must be possible to restore its capacity by installing additional support after displacements have been imposed on the support. This is called 'proactive or preventive support maintenance' (PSM). This introduces an important support design scenario for situations where the support capacity is lost with 'time'.

6.2.4 *Support system capacity restoration by proactive support maintenance (PSM)*

Once some support capacity has been consumed, the support capacity can be restored by adding bolts offering an extra displacement capacity. By reference to Figure 17, the ultimate displacement capacity may be increased from the ultimate displacement capacity of the support system ($\delta_{ult(SS)} = \delta_3$) to δ_4 or more. For example, if a PSM was conducted at δ_2 to increase $\delta_{ult(SS)}$ to δ_4, the remnant energy capacity would increase from E_3 to ($E_3 + E_4$) with E_4 represented by the area under the full load capacity. The remnant energy capacity would double.

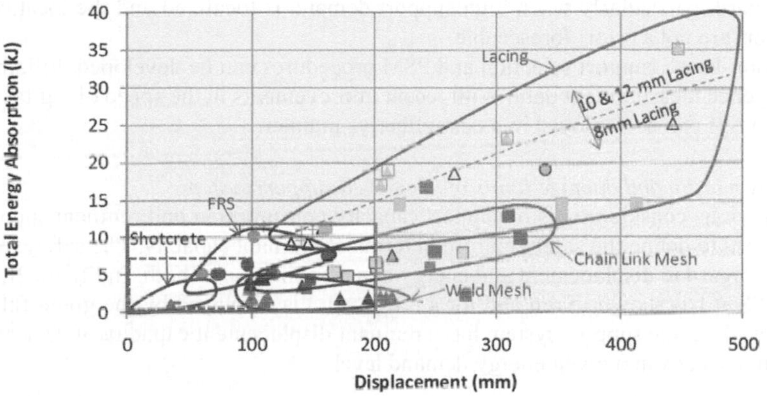

Figure 18. Total energy absorption capacities of retention systems showing ranges of acceptable retention system performance at 100 to 200 mm allowable central deflection (modified after Potvin et al. 2010).

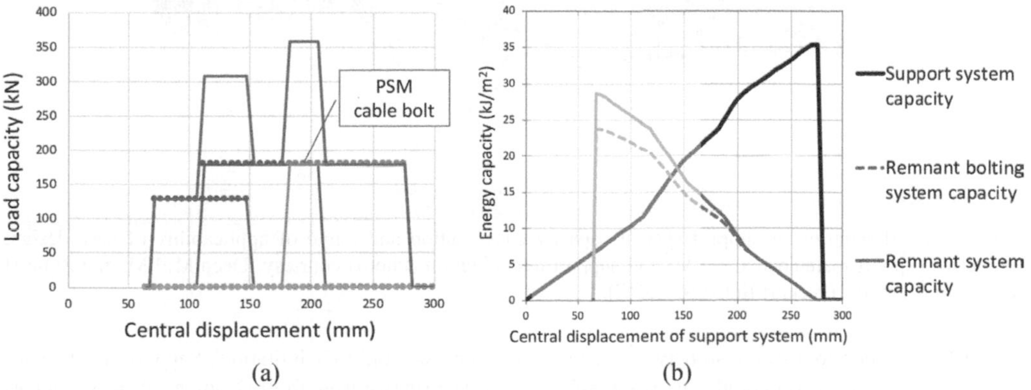

Figure 19. (a) Load capacity after PSM with 250 kN cablebolts at 70 mm bolt head or 170 mm central deflection, and (b) combined capacity and remnant capacity after PSM for comparison with Figure 16b.

If the two-bolt system (rebar and cable bolts) presented in Figure 15b is enhanced by adding cablebolts at a spacing of $s = 1.2$ m after a bolt head displacement of 70 mm or a central displacement of 170 mm, the load capacity line extends to 280 mm (Figure 19a), and the energy capacity increases from 26 to 35 kN/m^2 (compare Figures 16a and 19b). The remnant capacity line is shifted upward by 9 kJ/m^2. Hence, both the remnant displacement and the energy capacities of the support system at 70 mm have increased by a factor of more than 3, i.e. if extra cable bolts were installed at a central displacement of 170 mm.

6.2.5 *Visual recognition of SSCC for implementation of PSM*

Figure 20 demonstrates the concept of the support capacity consumption with photos illustrating increasing support damage (and support capacity consumption) toward the location where the support failed and the excavation collapsed (at the back end of the drift). The displacement scale in the load–displacement graph depends on the composition of the integrated support system.

Displacement ranges are superimposed by arrows to indicate where the original or the baseline design is valid (e.g. to the displacement limit at 100 mm) and when proactive support maintenance is needed or is most effective (e.g. between 80 and 180 mm). If the opportunity is missed to proactively enhance the support, failure may occur and rehabilitation[6] will be required to prevent collapse. The photos illustrate how the support system consumption is reflected by an increasing support damage.

Proactive support maintenance is a practical and often an economical means to increase workplace safety and reduce the potential severity of excavation damage. PSM is particularly beneficial when unexpected large convergences are encountered (measured) as it allows for focused PSM. This may offer an economic alternative to installing burst-resistant or yielding support systems across the board, particularly when high support demand is localized and the locations requiring such a support are not a priori foreseeable.

Deformation-based support selection and PSM procedures can be developed utilizing mine-specific convergence measurement data. With recent improvements in the speed of digital convergence monitoring, PSM can be deployed in a cost-effective manner.

6.2.6 *Safety margin and damage limits of integrated support systems*

Based on previous considerations of support capacity consumption and remnant support capacity, it is meaningful to define the safety margin in terms of remnant capacity. The safety margin can be defined with regard to displacement and energy or a combination of both. In Figure 21, the remnant capacity at FS = 1 is shown in red and for a $FS_{Disp} = 1.3$ is displayed by the green full line. When this line is reached, the support system has a remnant displacement capacity of 23% of the original displacement capacity at a given energy demand level.

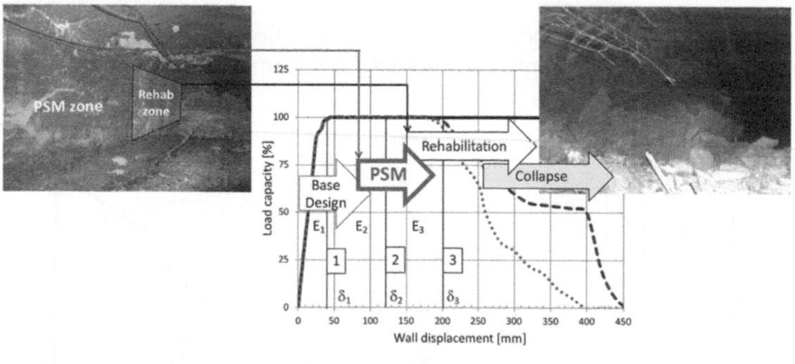

Figure 20. Illustration of support system capacity consumption and range of applicability of base design, proactive support maintenance (PSM) and support rehabilitation (Photos courtesy: Deep Mill Mining Zone at Grasberg Mine, PT Freeport Indonesia 2017).

6. It is important to distinguish between support 'maintenance' and 'rehabilitation'. Support maintenance means that the support is upgraded to 'maintain' a sufficient capacity during future loading episodes. Support rehabilitation implies that the support was damaged to the point where it has to be replaced to 'restore' the original support capacity.

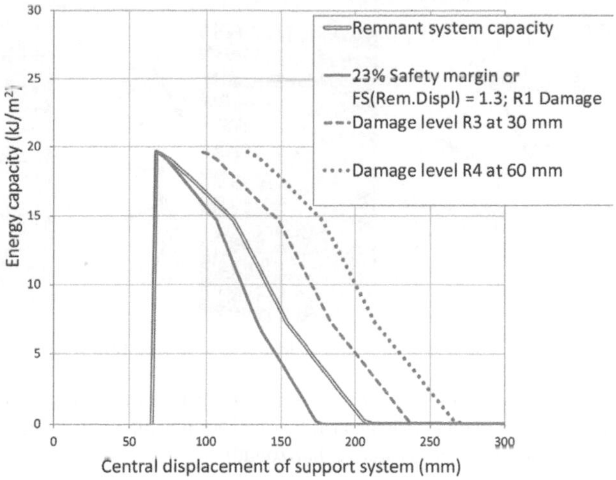

Figure 21. Remnant capacity (red) of the support system used to produce Figure 16b together with three thresholds (green): a displacement-based safety margin (full green line for $FS_{Disp} = 1.3$), and two potential damage limits R3 and R4 at 30 and 60 mm exceedance (dashed and dotted green lines).

In practice, damage levels (e.g. R1 to R5 (Cai & Kaiser, 2018)) are frequently used to assess the status of a damaged support system. When displacements reach the remnant capacity, increasing signs of support deterioration are to be expected (e.g. R2 damage at red remnant capacity line) and then, at increasing thresholds of exceedance, more severe damage is to be anticipated. For example, at thresholds of 30 and 60 mm exceedance in displacement capacity, shown respectively as green dashed and dotted lines, damage levels R3 and R4 are to be expected.

In other words, the remnant capacity curves can be used to assess the safety margin as well as the anticipated damage level in a rational manner. In burst-prone mines, the displacement and energy criteria must be assessed simultaneously (see Section 6.4.2).

From common to best practice

While single pass support systems are cost-efficient, relying on the capacity of the original installed support system can lead to uneconomic and potentially unsafe conditions because mining-induced deformations consume the support capacity. As mining proceeds, both the remnant displacement and energy dissipation capacity, and with it the FS, decrease and the vulnerability of an excavation increases.

When designing a support system best practices consider the potential loss of support capacity over the life of a support system, i.e., the evolution of the demand on the support needs to be defined at the design stage. In otherwise identical ground conditions, support for infrastructures remote from mining, or support in areas affected by extraction, will differ. A proper design focuses on the remnant capacity at the time when loading conditions are most critical, e.g. at the time of dynamic loading by an anticipated seismic event or strainburst.

If support capacity consumption is identified as a critical design criteria the best or most economic practice may involve proactive support maintenance (PSM) based on reliable support performance monitoring (e.g. digital displacement records).

6.3 *Mobilizing the self-supporting capacity of a support system*

The most fundamental principle of rock support is to make the reinforced ground self-supporting by creating a stable rock/support arch. This is to be achieved for static or dynamic loading.

For potentially unstable blocks of representative size B, Lang (1961) proposed a bolt spacing $s \leq 4B$; e.g. for blocky ground, stress-fractured, or veined rock $s \leq 0.8$ to 2 m. He also suggested a bolt length $L \geq 2s$, i.e. $L = 1$ to 3 m for blocky or fractured rock (Figure 22a).

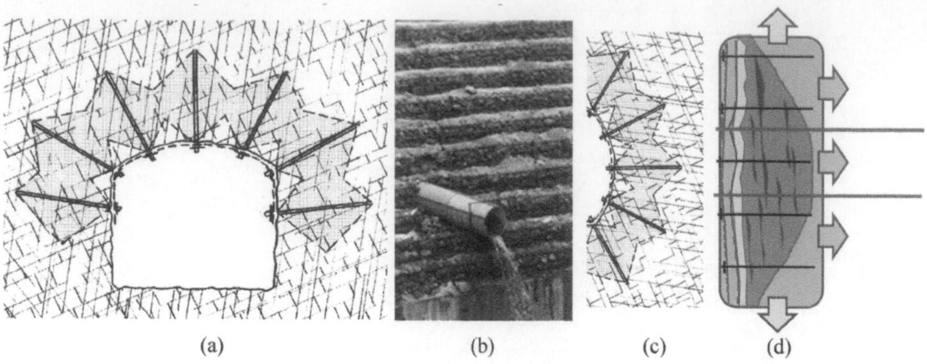

Figure 22. (a) Supported rock arch principle (Hoek et al. 1995); Gabion concept: (b) support of slope; (c) representation of self-supporting wall rock arch, and (d) flat wall equivalent showing tangential resistance forces (yellow, vertical) and radial confining forces (orange, horizontal).

In the inner shell (Section 3.2.2), rock fragments may be smaller than the representative block size and a robust retention system has to be selected to prevent unraveling.

6.3.1 *Gabion concept*

For wall support in cave mining, the equivalent model to Lang's self-supporting arch model is to create 'gabions'. Just like for slope stability (Figure 22b), gabions have to provide:

- immediate retention of broken or fractured rock;
- reinforcement of broken rock in the inner shell;
- surface pressure increasing the self-supporting capacity of broken rock; and
- bulking restraint by reinforcement.

In this manner, a support system is created that behaves like a gabion or an arch (Figure 22c) and

- provides a radial resistance (orange arrows in Figure 22d) to the rock mass (e.g. a pillar) behind the gabion; and
- provides tangential resistance (orange arrows in Figure 22d) to resist the tangential strain driver (HW/FW convergence) causing rock mass bulking.

6.3.2 *Why and how gabions work*

The gabion concept works as an integrated support system consisting of a robust surface support system to ensure full utilization of the capacity of the rock mass reinforcement. A gabion ensures that the surface support is not the weakest link. Furthermore, if combined with relatively dense bolting (typically with $s \leq 1$ m), the gabion also facilitates load splitting by direct and indirect loading of the bolts.

The performance of the gabion support, however, depends on whether the loading is

- by tangential straining (e.g. between HW and FW convergence); or
- by radial loading (e.g. by the sudden bulking inside or behind the gabion).

If loaded inside or via straining of the reinforced annulus, all bolts and cables get activated and yield together in direct and indirect loading. However, if a strainburst occurs behind the gabion (typically at > 2 m in mining), the gabion acts as a 'super plate' or a surface support beam, and dissipates energy as it moves into the excavation. In this case, the support elements inside the burst volume (behind the gabion) dissipate energy by indirect loading and the mass of the gabion consumes energy as it is deformed and laterally translated.

As indicated above, effective gabions provide both tangential and radial resistances and when deformed dissipate energy by tangential straining as well as by internal bulking with frictional strength mobilization of the reinforced broken rock. This is illustrated next by using discrete element numerical models (UDEC or 3DEC) to simulate the behavior of a gabion-supported pillar wall.

6.3.2.1 *Tangential load and energy capacity of gabion-supported wall*

Figure 23 presents tangential stress versus tangential strain graphs for two models, one with strong and the other with weak rock blocks (in black: UCS = 60 MPa; m_i =18; σ_t = 7 MPa, and in red: UCS = 40 MPa; m_i = 15; σ_t = 7 MPa). The discrete element model with Voronoi blocks represents the left wall of a pillar with a zero lateral displacement boundary at the right side of the model. This boundary condition was chosen to approximate conditions of pillars with an elastic core. As a consequence, stress arching occurs, as shown by the stress contours in the insert on the left, and displacements of the stress fractured rock are horizontal (radial) toward the excavation. This leads to the rather ductile response of the pillar wall shown in the graph despite the brittle nature of the simulated fractured rock mass.

The tangential energy dissipation capacity of the reinforced pillar wall is represented by the area under these curves. For a 2.2 m deep gabion of 4.5 m in height, the gabion supported wall in weak rock dissipates 8 kJ/m and in strong rock 14 kJ/m (i.e. per meter along the wall; or 2 to 3 kJ/m^2).

The load (and energy) capacity of the pillar wall supported by a 'weak rock' gabion is roughly half of that with a 'strong rock' gabion. Both exhibit a ductile behavior with < 20% strength drop beyond peak. In this model, the support only adds between 10 and 25% to the tangential load bearing capacity of the gabion. Much of the stabilizing force (yellow up and down arrows) comes from the internal resistance of the confined fractured rock inside the gabion and stress arching behind the gabion.

6.3.2.2 *Radial support provided by gabions*

For Case (3) in Figure 23 with radial support provided by dense bolting, Figure 24 presents the radial stresses at three locations and for four tangential strain levels of 0.5 to 2%:

- Inside the gabion: the average inter-block pressure increases from 6 to 10 MPa and this provides frictional strength to the gabion as it is strained;
- Immediately behind the gabion (at 2.5 m): the average inter-block pressure first increases to 5 MPa and then drops to about 2.5 MPa at 2% tangential strain providing effective confinement to the pillar; and
- 1 m behind the gabion at 3.5 m from the wall, i.e. at the transition from the inner and outer shell of the pillar, the average radial pressure increases rapidly to 20 MPa due to stress arching.

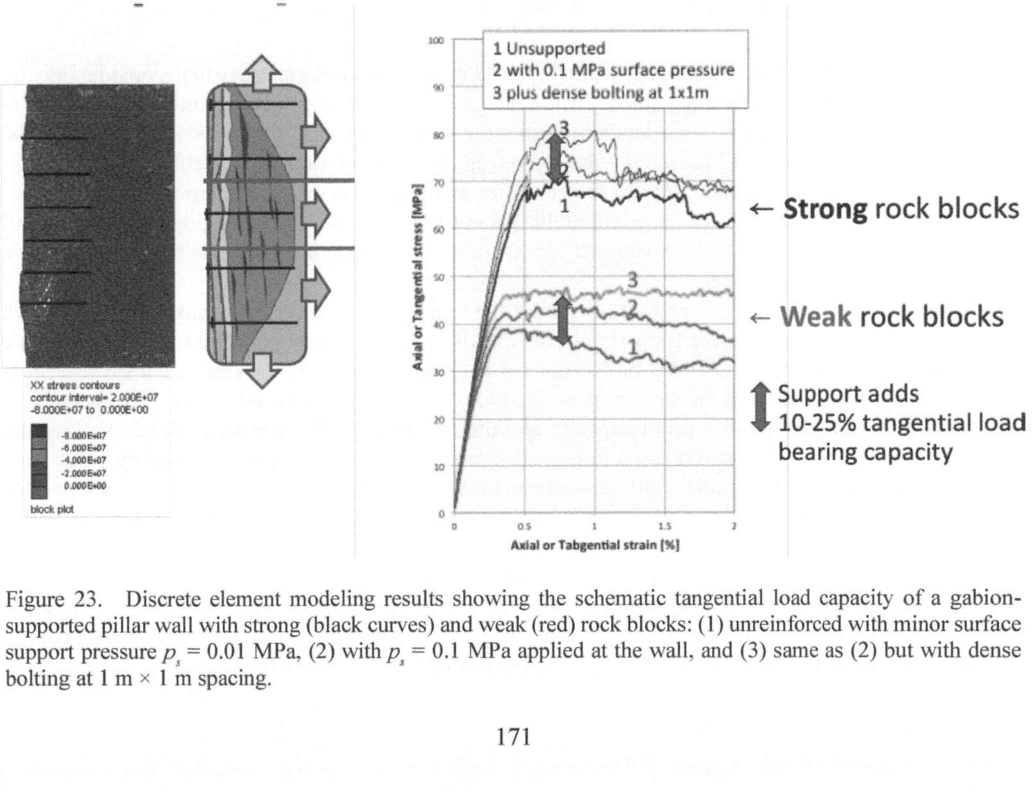

Figure 23. Discrete element modeling results showing the schematic tangential load capacity of a gabion-supported pillar wall with strong (black curves) and weak (red) rock blocks: (1) unreinforced with minor surface support pressure p_s = 0.01 MPa, (2) with p_s = 0.1 MPa applied at the wall, and (3) same as (2) but with dense bolting at 1 m × 1 m spacing.

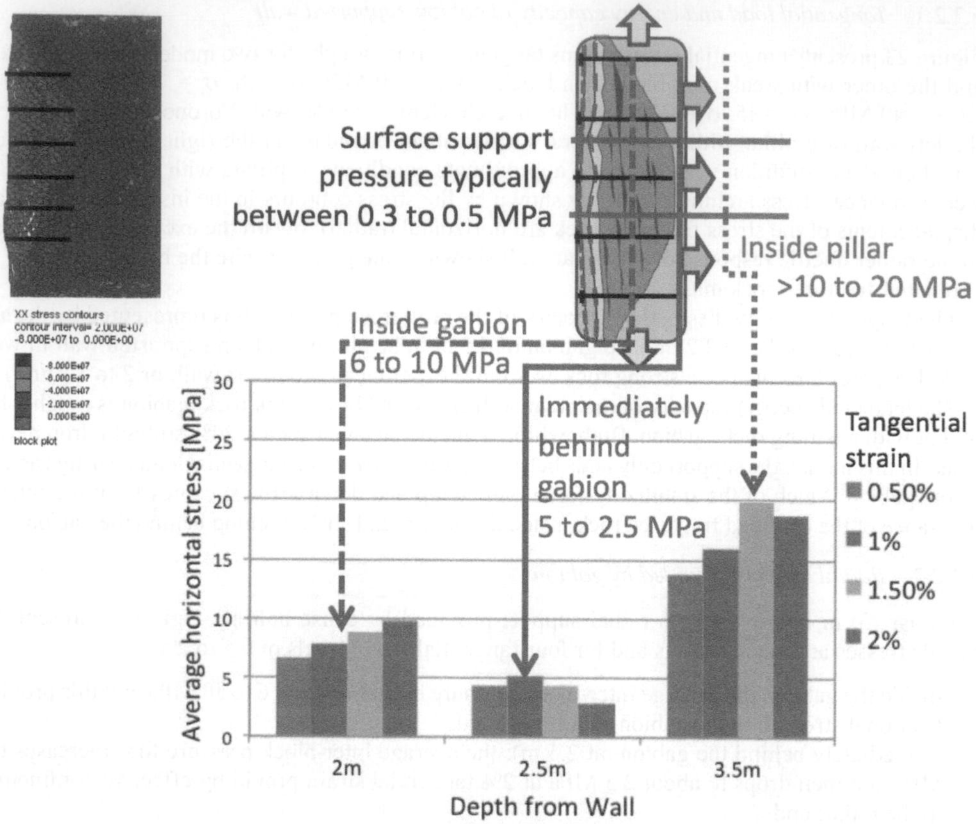

Figure 24. Discrete element modeling results showing the schematic radial pressure capacity of a gabion at increasing tangential strain levels (0.5 to 2%): inside the gabion at 2 m from the wall, immediately behind the gabion at 2.5 m, and 1 m into the pillar at 3.5 m from the wall.

At first sight, these pressures may seem high. They can be understood when considering the effect of dense bolting within the gabion. Inside the gabion, geometric bulking and block rotation is resisted by the reinforcement. This leads to high inter-block stresses and friction that results in the average pressures shown in Figure 24. The pressures are lower immediately behind the gabion in the unreinforced rock, and they drop to lower values as the gabion is forced to move into the excavation. At 2% tangential strain, approximately 50% of these inter-block pressures are lost behind the gabion but the average 'gabion support' pressure is still superior to the pressure generated by the support components alone.

Beyond the gabion, the horizontal stresses increase rapidly because the rock mass is still cohesive due to the limited damage at the transition to the elastic pillar core. The pressure transmitted by the gabion to the rock mass behind the gabion is at least 10-times higher than the equivalent surface pressure (0.3 MPa resulting from the surface pressure plus distributed bolt loads). The gabion acts like a super-plate and transmits the support capacity and the resistance of the confined broken rock to the surrounding rock mass. The gabion therefore greatly enhances the self-supporting capacity of the rock mass in the outer shell. This radial confinement increases the pillar strength. In caving operations, this helps to improve the reliability of a footprint and consequently the reliability of production.

6.3.3 *Application of gabion concept from bursting to squeezing ground*

The gabion concept with varying support system components finds applications in a wide spectrum of rock mass behavior; i.e. from bursting to blocky to yielding and even squeezing ground. Figure 25 presents two examples showing a support system consisting of:

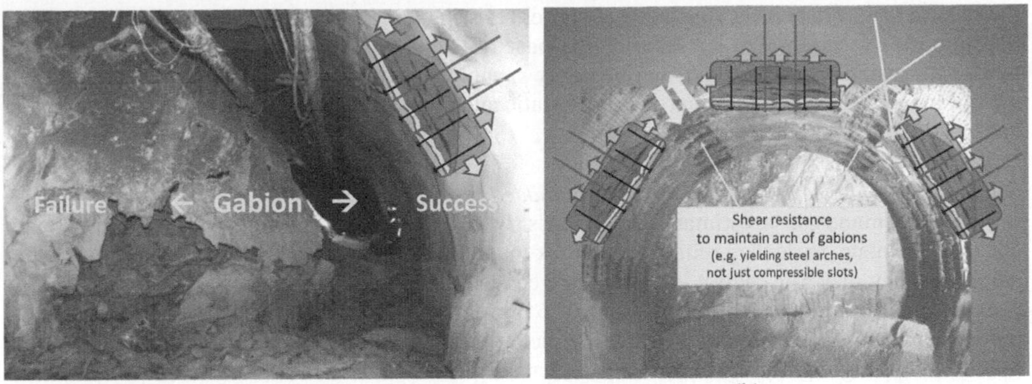

(a) (b)

Figure 25. Gabion concept application for the entire spectrum from (a) bursting to blocky to deformable to (b) squeezing ground.

(a) rockbolts with cable bolts and mesh-reinforced shotcrete: the gabion support successfully prevented damage on the right whereas it collapsed on the left due to excessive bulking during a rockburst; and
(b) rockbolts and shotcrete panels with cables and compression slots between the panels: yielding steel sets or shear resistance cable patterns are needed to ensure the overall stability of the schematically illustrated gabion panel system (Figure 25b).

The gabion concept also finds application in intermediate conditions with blocky and laminated ground.

From common to best practice

Common practice of support design without giving due consideration to the self-stabilizing action of well-constrained broken rock is flawed. The integrated support capacity of a gabion is far superior to the capacity of the support components alone. It provides a most effective support system to resist static and dynamic load demands by creating self-supporting rock arches.

6.4 *Deformation-based support selection for stress-fractured rock*

In stress-fractured ground, two mechanisms affect the excavation performance during construction: (1) raveling of broken rock resulting in short stand-up times, and (b) large deformations caused by geometric bulking imposing large radial deformations on the support. The first is met by ensuring robust rock retention and the second by providing a deformable bolting system.

The challenge of controlling highly stressed brittle rock in civil and mining projects with deformation compatible support was addressed by the author in the written version of the Sir Allan Muir Wood lecture entitled "Ground Support for Constructability of Deep Underground Excavations" (Kaiser 2016). For detailed discussions of the support selection process, the reader is also referred to Kaiser (2014).

A deformation-based support design to manage bulking aims at two fundamental support design axioms:

- Control of the cause for bulking by minimizing the tangential straining of the rock in the immediate vicinity of an excavation (resist HW/FW closure as illustrated by the yellow arrows in Figure 22d and 23); and
- Control of the geometric bulking of stress-fractured ground by rock reinforcement inside the gabion and the application of confining pressure by the gabion to the surrounding ground (Figure 22d).

Because the product of the displacement times resisting force represents the energy dissipated by the supported rock ($E \sim F*\delta$), a deformation-based design implicitly deals with energy dissipation by the integrated support system, and therefore is applicable to conditions where the rock fails in a violent manner during a rockburst. For both static and dynamic loading, the displacement demand has to be estimated (Section 6.4.2).

6.4.1 *How do bolts work – how do they get loaded?*

Before the demand can be estimated, it is necessary to understand 'How bolts work?' Do they carry a load at the plate, or do they reinforce the rock mass to create a self-supporting rock arch. In other words, do they dissipate energy in direct, indirect, or combined loading?

For static mining-induced loading, Figure 26 presents the load distribution along un-plated cable bolts (bar graphs provide bolt load) and the yield (red) and failure locations (black). In two of the four locations highlighted by the red ellipses, tensile yield led to failure (lower two ellipses) whereas in the upper two ellipses, shear contributed to the failure. It follows that both internal axial and shear demands need to be defined. Most importantly, this simulation illustrates that bolts get indirectly loaded and eventually fail internal to the deforming rock mass.

For support in burstprone ground, it is frequently assumed that bolts get loaded via the plate and that the capacity obtained from pullout tests is representative. This is rarely the case because the rock mass fractures and bulks inside the burst volume. This is illustrated by the horizontal displacement profile for three simulated time steps in Figure 27a after a strainburst was initiated at about 2 m from the wall (Gao et al. 2019). The horizontal displacement at the wall is about half of the displacement inside the rock mass.

The corresponding relative displacement or strain over a yield length of 0.2 m is shown in Figure 27b. Extension strain peaks (positive) are first observed at 2.3 m (at 4 ms) and then at 3.6 m (at 6 ms) from the wall. At the same time, a compressive strain peak is encountered at 1.1m depth. A bolt crossing this strain profile will be strained in compression at 1.1 m due to compression of the

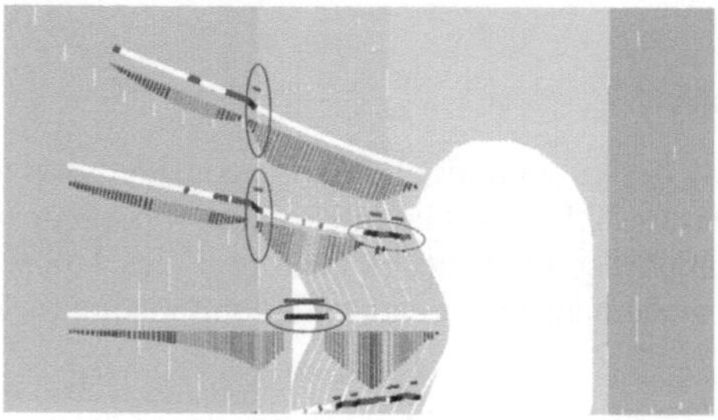

Figure 26. Simulation of cable bolt loading during vertical/tangential straining of laminated rock (Itasca 3DEC simulation; Pierce 2017; pers. com.). Color scheme: Bolts – red = yield, black = failed in tension or shear; bar charts: bolt load from cold to warm. Displacement scale exaggerated.

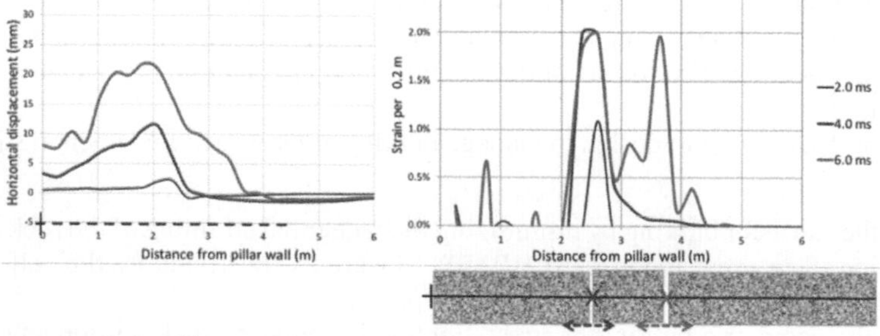

(a) (b)

Figure 27. (a) Simulated horizontal displacement and (b) relative displacement (strain) induced by strainbursts (modified after Gao et al. 2019).

previously fractured (spalled) ground. This reverses static extension straining in the first 1 to 2 m from the wall and unloads the bolt. A bolt will therefore fail at deeper seated locations with elevated extension strain and not necessarily at or near the plate. Interestingly, while wall displacements do occur, the steel near the plate experiences much less strain at this stage in the loading process. A radially installed bolt therefore gets indirectly loaded and dissipates energy inside the rock mass.

Clearly, the displacement demand and its distribution along the bolts need to be understood and quantified to select appropriate support components for an integrated support system.

6.4.2 *Static and dynamic support system demand estimation*

For support design purposes, it is necessary to estimate the displacement and energy demands such that they can be compared with the remnant support system capacity (Section 6.2.6). The displacement demand is obtained from the bulking of the spalling or strainbursting ground (Section 3.2.4). The energy demand for strainbursts is composed of two energy sources, the energy transmitted by a stress wave from a remote seismic source and the energy release at the strainburst location.

6.4.2.1 *Displacement demand*

The displacement demand for continuum behavior can be estimated from analytical or numerical models. However, for brittle rock with geometric bulking most models are deficient and tend to underestimate bulking displacements. For brittle rock, the semi-empirical approach outlined in Section 3.2.4 (as well as by Kaiser 2016c; Kaiser and Cai (2019)) can be adopted whereby the depth of failure (Equations 8 and 9) or the depth of strainbursting is combined with representative bulking factors (Figure 12) to quantify the internal strain and the cumulative displacement over the length of a bolt.

6.4.2.2 *Energy demand*

Energy from triggered strainbursts

As discussed in Section 2.3.2, the bulking velocity v_B at the inner edge of the burst volume depends on the time it takes to fail the burst volume (t_R). This rupture time varies widely depending on the brittleness of the rock. It typically ranges from 1/10 to 1/20 of a second (or less in extreme conditions) and controls the bulking velocity. For example, for 1 m of strainbursting ground with BF = 5% the burden ahead of the burst volume is suddenly displaced by 50 mm, resulting in a bulking velocity v_B between 0.5 and 1 m/s (2.5 m/s for t_R = 1/50 s). If 1 m^3 of spalled rock (burden) is moved at this velocity, the kinetic energy $E_B = ½\, m\, v_B^{\,2}$ ranges from 0.3 to 1.3 kJ/m^2 (8.1 kJ/m^2 for t_R = 1/50 s). This simple example demonstrates that the available kinetic energy is highly sensitive to the rupture time and that high velocities can be attributed to self-initiated or triggered strainbursts.

Energy from dynamically-loaded strainbursts

If combined with a large remote seismic event, the ground motion PGV at the strainburst location can be obtained from applicable ground motion prediction equations (GMPE). Then, the dynamic stresses can be calculated to estimate the incremental deepening of the failure zone and the associated incremental bulking displacement (Kaiser et al. (1996), Kaiser and Cai (2019). Furthermore, the ground motion from the remote seismic event increases the velocity of the burden and possibly also ejects part of the burst volume. The kinetic energy demand therefore is obtained from the combined velocity.

The results of a displacement and energy demand analysis is graphically presented in the displacement versus energy graphs of Figure 28 together with an example of a remnant support system capacity (red line). For a range of variable parameters and for a seismic event of magnitude M_L = 0 to 4 at a distance of R = 30 m, the red path (full line) represents the demand for a scenario where the burden is 'ejected'. The black dashed path is for 'ejection' of the burden and the entire strainburst volume.

For this example, it is assumed that the central deflection before the burst was 62 mm. A pure strainburst adds 62 mm for a total displacement demand of 124 mm. The energy demand ranges from 0.5 to 1.0 kJ/m^2 for the two mass assumptions (Points A and A'). With the remote seismic event of magnitude M_L = 3, the displacement is increased by 14 mmm to 138 mm due to the deepening of the fracture zone and the energy is augmented by the ground motions to 3.7 and 9.1 kJ/m^2 (Points B & B').

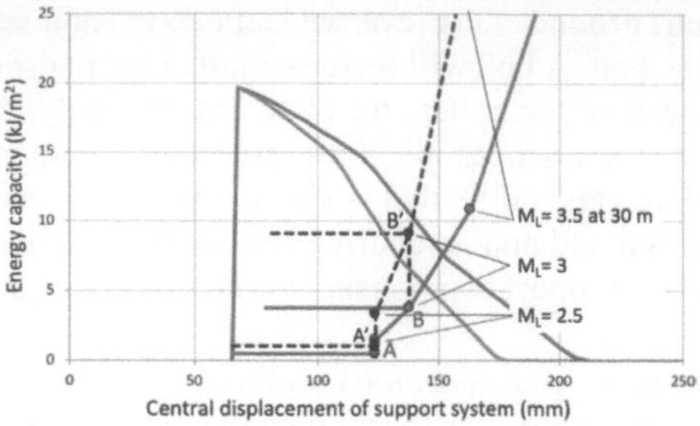

Figure 28. Comparison of displacement and energy demand estimates with simulated support system capacity (red; FS = 1.3 in green): for strainburst at SL = 0.8 and when combined with a remote seismic event for M_L = 0 to 4 at R = 30 m.

The capacity for FS = 1.3 is again shown in green. In this example, the demands do not reach the remnant support system capacity at M_L = 3. If only the burden is 'ejected' (Points B), the FS >1.5 and if the burden and the strainburst volume is 'ejected FS < 1.3 (Point B'). The minimum factor of safety at Point B' in terms of displacements FS_{Disp} = 143/138 = 1.04 and in terms of energy FS_{Energy} = 10.6/9.1 = 1.16. This is far from the commonly but incorrectly estimated FS_{Energy} = remnant capacity before the burst (30 kJ/m^2)/energy demand (9.1 kJ/m^2) = 3.3 for this case.

The path in Figure 28 shows that a strainburst primarily shifts the displacement demand to the right consuming much displacement capacity, and the remote seismic event moves the energy demand up although only for M_L > 2. This is consistent with findings of Morissette et al. (2012), who found that remote seismic events tend to aggravate the damage when the event magnitude exceed 1.5 to 2 (Cai & Kaiser 2018).

This example further amplifies the need for a deformation-based support design, particularly when dynamic perturbations impose an energy demand.

From common to best practice

Again, ample opportunities and economic as well as safety benefits can be derived by moving from common support selection to deformation-based support system design practices. Common practices of support design without giving due consideration to the displacement demand are flawed and common practice of pure energy-based design for burstprone ground is highly flawed. Both can lead to serious safety hazards.

Deformation-based support selection constitutes best practice for ground control in highly stressed, brittle failing ground. In burstprone ground, the displacement and energy demand from strainbursts and from remote seismic events must be simultaneously considered. It is essential to compare the demands to the remnant capacity and not to the capacity of the originally installed support system.

7 CONCLUSION – MOVE TO BEST PRACTICES !

Common practices are often not best practices when judged from an economic or workplace safety perspective, and common practices that worked well at shallow depth may need to be replaced because the rock mass behavior has changed and poses new hazards at depth.

This lecture focused specifically on opportunities resulting from better means to assess the vulnerability of excavations, to characterize the rock mass, for ground control, and rockburst damage mitigation. It is demonstrated that ample opportunities exist to derive benefits from moving from common to best practices. In summary, opportunities are identified in the following areas:

- Identification of engineering design parameters EDPs that characterize the vulnerability and fragility of underground excavations.
- Rock mass characterization that follows a systematic process of moving from inferred to proven rock mass quality designations.

- Grouping of rock mass qualities into three classes (RMQ1 to 3) that reflect three characteristic rock mass and excavation behavior modes.
- Methods to obtain appropriate rock and rock mass strength envelopes for peak, post-peak and residual strength.
- Practices that respect the limitations of classification and characterization systems, in particular, in the use of GSI-strength equations for rock mass strength determination for good rock.
- Differentiation between near wall (inner and outer shell) behaviors for support design and pillar sizing.
- Deformation-based support selection for ground control in highly stressed, brittle failing ground.
- Utilization of the self-stabilizing capacity of well-constrained broken rock by adopting the gabion concept to provide effective support to resist static and dynamic load demands.
- Quantification of support capacity consumption as a critical design criteria.
- Use of proactive support maintenance (PSM) based on support deformation monitoring to restore consumed support capacity.
- Consideration of the impact of pre-burst support deformation and bulking displacements resulting from strainbursts.
- Replacement of energy-based by displacement-based support designs for burst-prone ground.

Best practices take, at the design stage, the evolution of deformation demands over the life of an excavation into account. Proper designs focuses on the remnant support capacity at the time when loading conditions are most critical.

We cannot stagnate and accept the status quo. We have to implement the advanced state-of-the-art by adopting best practices that respect the actual behavior of the rock, excavations and support. Those that 'hide behind' standard or common practice might be liable when others have adopted better or best practices.

ACKNOWLEDGEMENTS

The authors wish to acknowledge the many contributions of industrial sponsors, Rio Tinto through the Rio Tinto Centre for Underground Mine Construction, Freeport McMoran, LKAB, Vale, Glencore, as well as the financial contributions of NSERC (Natural Sciences and Engineering Research Council of Canada).

Much of the presented material is a result of long-term collaborations and the efforts of graduate students. The list of individuals who have contributed in some form to the research reflected in this article is too long, but the author wishes to collectively recognize that much value was derived from their discussions and contributions. The author thanks them all in this anonymous fashion – you know who you are.

Nevertheless, I like to specifically acknowledge a few individuals who have certainly influenced the findings, lessons learned and conclusions presented in this lecture. In alphabetic order, they are Drs. F. Amann, N. Bahrani, R. Bewick, M. Cai, M. Diederichs, E. Eberhardt, H-B. Kim, D. Martin, D. McCreath and M. Pierce.

REFERENCES

Books

Cai, M. & Kaiser, P.K. 2018. Rockburst phenomena and support characteristics, Volume I in Rockburst Support Reference Book. MIRARCO Laurentian University, preliminary ISBN: 978-0-88667-096-2, 191 p. (released May 2018).

Kaiser, P.K. & Cai, M. 2019. Rock support to mitigate rockburst damage caused by dynamic excavation failure, Volume II in Rockburst Support Reference Book. MIRARCO Laurentian University, preliminary ISBN: 978-0-88667-097-9 (to be released late 2019).

Kaiser, P.K. 2017. Excavation vulnerability and selection of effective rock support to mitigate rockburst damage. In X.-T. Feng (Ed.), Rockburst: mechanisms, monitoring, warning, and mitigation. New York: Elsevier. ISBN: 978-0-12-805054-5, 473–518.

Kaiser, P.K., McCreath, D.R. & Tannant, D.D. 1996. Rockburst Support. In Canadian Rockburst Research Program 1990–95. Vol.2, 324 p. (Published by CAMIRO, Sudbury).

Hoek, E., Kaiser, P.K. & Bawden, W.F. 1995. Rock Support for Underground Excavations in Hard Rock. A.A. Balkema, Rotterdam, 215 p.

Hutchinson, D.J. & Diederichs, M.S. 1996. Cablebolting in Underground Hard Rock Mines, Bitech Publishers Ltd., Richmond, BC, Canada, 406 p.

Müller, L. 1963. Der Felsbau. F. Enke Verlag, Stuttgart, Vol.1, 624 ..

Articles

Bahrani, N. & Kaiser, P.K. 2016. Strength degradation approach (SDA) for estimation of confined strength of micro-defected Rocks. ARMA 16–890

Bahrani N., Kaiser, P.K. 2013. Strength degradation of non-persistently jointed rock mass. Technical note: *International Journal of Rock Mechanics & Mining Sciences*, 62:28–33.

Barton, N. & Grimstad, E. 1994. The Q-system following twenty years of application in NMT support selection. 43rd Geomechanics Colloquy. Felsbau, 6/94: 428–436.

Barton, N., Bandis, S. & Bakhtar, K. 1985. Strength, deformation and conductivity coupling of rock joints. *Int. J. of Rock Mechanics & Mining Sciences*, 22(3): 121–140.

Bewick, R.P., Kaiser, P.K. & Amann, F. 2019. Strength of massive to moderately jointed hard rock masses. Rock Mechanics and Geotechnical Engineering, 11(3): 562-575 https://doi.org/10.1016/j.jrmge.2018.10.003

Bewick, R. P., Amann, F., Kaiser, P.K. & Martin, C.D. 2015. Interpretation of UCS test results for engineering design. 13th ISRM International Congress of Rock Mechanics, Montreal, Canada, 14 p.

Bieniawski, Z. T. 1979. Tunnel design by rock mass classifications. U.S. Army Waterways Experiment Stgation, Technical Report No. GL-79-19, 131 p..

Bieniawski, Z. T. 1976. Rock mass classification in rock engineering. In: Proc. Symp. on Exploration for Rock Engineering, Z.T. Bieniawski (editors), Cape Town. 1: 97–106.

Brown, E.T. 2007. Block caving geomechanics. University of Queensland, Second edition, 696 p.

Garza-Cruz, T.V., Pierce, M. & Kaiser, P.K. 2015. Use of 3DEC to study spalling and deformation associated with tunneling at depth. DeepMining'14, ACG (eds. Hudyma and Potvin), 13 p.

Gao, F., Kaiser, P.K., Stead, D., Eberhardt, E. & Elmoe, D. 2019. Strainburst phenomena and numerical simulation of self-initiated brittle rock failure. International Journal of Rock Mechanics and Mining Sciences, 116: 52–63.

Gao, F., Kaiser, P.K., Stead, D., Eberhardt, E. & Elmo, D. 2019. Numerical simulation of strainbursts using a novel initiation method. Computers and Geotechnics, 106: 117–127.

Diederichs, M. S. 2007. Mechanistic interpretation and practical application of damage and spalling prediction criteria for deep tunneling (2003 Canadian geotechnical Colloquium), Canadian Geotechnical Journal, 44(9): 1082–1116. DOI: 10.1139/T07–033.

Hoek, E. & Brown, E.T. 2019. The Hoek-Brown failure criterion and GSI -218 edition. Journal of Rock Mechanics and Geotechnical Engineering, 11(3): 445–463.

Hoek, E., Kaiser, P.K. & Bawden, W.F. 1995. Rock Support for Underground Excavations in Hard Rock. A.A. Balkema, Rotterdam, 215 p.

Kaiser, P.K. 2017. Ground control in strainbursting ground – A critical review and path forward on design principles. 9TH Int. Symp. on Rockbursts and Seismicity in Mines, Santiago, Chile, 146–158.

Kaiser, P.K. 2016a. Challenges in rock mass strength determination for design of deep underground excavations. ISRM on-line lecture (45 min) https://www.isrm.net/gca/?id=1227

Kaiser, P.K. 2016b. Underground rock engineering to match the rock's behaviour – a fresh look at old problems. MTS lecture at 50th US Rock Mechanics Symp., 6., abridged summary in ISRM news 2017.

Kaiser, P.K. 2016c. Ground Support for Constructability of Deep Underground Excavations – Challenge of managing highly stressed brittle rock in civil and mining projects. ITA Sir Muir Wood lecture of International Tunneling Association at World Tunneling Congress, San Francisco, 33 p. www.ita-aites.org or http://www.mirarco.org/grc/#ert_pane1-4

Kaiser, P.K. 2014. Deformation-based support selection for tunnels in strain-burstprone ground. DeepMining'14, ACG (eds. Hudyma and Potvin), 227–240.

Kaiser P.K. 2006. Rock mechanics considerations for construction of deep tunnels in brittle rock. Asia Rock Mechanics Symposium, Singapore, 12 p.

Kaiser P. K., F. Amann & BewickR.P. 2015. Overcoming Challenges of Rock mass Characterization for Underground Construction in Deep Mines. 13th ISRM International Congress of Rock Mechanics, 10–13 May, Montreal, Canada, 14 p.

Kaiser, P.K. & Cai, M. 2013. Critical review of design principles for rock support in burst-prone ground – time to rethink! Ground Support 2013, Potvin, Y. and Brady, B. (eds), Perth, Australia, 3–38.

Kaiser, P.K., Diederichs, P.K., Martin, C.D., Sharp, J. & Steiner W. 2000. Underground works in hard rock tunneling and mining. GeoEng 2000, Melbourne, Australia, Technomic Publ. Co., 1: 841–926.

Kaiser, P.K. & Kim, B-H. 2014. Characterization of strength of intact brittle rock considering confinement dependent failure processes. Rock Mech Rock Eng.; DOI 10.1007/s00603-014-0545–5.

Kaiser, P.K. & Kim, B.H. 2008. Rock mechanics advances of underground construction and mining. Korea Rock Mechanics Symposium, Seoul, 1–16.

Kaiser, P.K., Maloney, S. & Yong, S. 2016. Role of large scale heterogeneities on in-situ stress and induced stress fields. ARMA 16–571.

Lang, T.A. 1961. Theory and practice of rockbolting. Trans. Amer. Inst. Min. Engrs, 220: 333–348.

Martin, C.D. 2019. Stress-induced fracturing (Spalling) around underground excavations: Laboratory and in-situ observations. ISRM on-line lecture, https://www.isrm.net/gca/?id=1359

Martin, C.D. 1997. The effect of cohesion loss and stress path on brittle rock strength. Canadian Geotechnical Journal, **34**(5): 698–725.

Martin, C.D., Kaiser, P.K. & Christiansson, R. 2003. Stress, instability and design of underground excavations. International Journal of Rock Mechanics and Mining Sciences, 40(7-8): 1027–1047.

Morissette, P., Hadjigeorgiou, J., Thibodeau, D. & Potvin, Y. 2012. Validating a support performance database based on passive monitoring data. 6th Int. Seminar on Deep and High Stress Mining, 41–55

Ortlepp, W.D. 2005. A review of the contribution to the understanding and control of mine rockbursts 6th Int. Symp. on Rockbursts and Seismicity in Mines, ACG, Y. Potvin, M. Hudyma (Eds.), Perth, Australia, 3–20.

Perras, M.A. & Diederichs, M.S. 2016. Predicting excavation damage zone depths in brittle rocks. J. of Rock Mechanics and Geotechnical Engng, 8(1): 60–74.

Potvin, Y. 2009. Strategies and tactics to control seismic risks in mines. J. South Afr. Inst. Min. Metall., 109: 177–186.

Schofield, A.N. 1998 (and 2001) The "Mohr-Coulomb" Error. CUED/D-SOILS/TR305; presented in Paris at jubilee of Pierre Habib; published in "Mechanique et Geotechnique", Balkema, 6 p.

Tarasov, B.G. & Potvin, Y. 2012. Absolute, relative and intrinsic rock brittleness at compression. Mining Technology 121(4): 218–225.

Terzaghi, K. 1946. Rock defects and loads on tunnel supports. In Rock tunneling with steel supports, (eds R. V. Proctor and T. L. White) 1, Youngstown, Commercial Shearing and Stamping Company, 17–99

Wawersik, W.R. & Fairhurst, C. 1970. A study of brittle rock fracture in laboratory compression experiments. Int. J. of Rock Mechanics and Mining Sciences, 7(5): 561–575.

Rocha Medal

Rock Mechanics for Natural Resources and Infrastructure Development –
Fontoura, Rocca & Pavón Mendoza (Eds)
© 2020 ISRM, ISBN 978-0-367-42284-4

Characterisation and modelling of natural fracture networks: Geometry, geomechanics and fluid flow

Q. Lei
Department of Earth Science and Engineering, Imperial College London, London, UK
Now at Department of Earth Sciences, ETH Zürich, Zürich, Switzerland

ABSTRACT: Natural fractures are ubiquitous in crustal rocks and often dominate the bulk properties of geological media. The understanding of their geometrical, geomechanical and hydrological properties is a challenging issue relevant to many rock engineering applications. In this paper, I first present a study of the statistics and tectonism of a multiscale fracture system in limestone and propose a conceptual interpretation to the underlying mechanism driving fracture network formation. A critical review is then presented on the state-of-the-art discrete fracture network modelling of coupled hydromechanical processes in fractured rocks. To model the geomechanical behaviour of natural fractures in rock, a joint constitutive model is implemented into the numerical framework of a hybrid finite-discrete element method. The combined formulation can compute the stress/strain evolution in intact rocks, capture the mechanical interaction between matrix blocks, characterise the non-linear deformation of rough fractures and mimic the propagation of new cracks. This geomechanical model is applied to calculate the aperture distribution of various metre-scale fracture networks under in-situ stress conditions, based on which stress-dependent fluid flows are analysed. A novel upscaling approach employing discrete-time random walks is developed to extrapolate fracture network geometries together with their variable apertures into larger scales for permeability prediction. The numerical model is further used to simulate the damage evolution around an underground excavation in a crystalline rock embedded with pre-existing discontinuities. The scope of my research covers the scenarios of both two-dimensional (2D) and three-dimensional (3D) fracture networks. The research findings demonstrate the importance of realistic fracture network representation and systematic geomechanical simulation for modelling the hydromechanical behaviour of fractured rocks.

1 INTRODUCTION

Fractures such as faults, joints and veins are ubiquitous in crustal rocks. These naturally occurring discontinuities often form complex networks and create highly heterogeneous geological conditions. The widespread presence of natural fractures raises a fundamental question about the underlying mechanisms that drive such complicated evolutionary and collective phenomena. Fractures that nucleate from initial flaws (Pollard & Aydin 1988) may propagate in different strain rate regimes, i.e. the subcritical, quasi-static and dynamic regimes (Schultz 2000). Continued strain under an increased remote displacement loading or a sequence of tectonic episodes further promotes the interactions of multiple fractures (e.g. coalescence, cross-cutting, inhibition, reorientation and arrest) (Price & Cosgrove 1990). Such mechanically-controlled processes result in complex fracture networks with self-organised (i.e. non-random) population statistics, e.g. density, lengths, locations, spacing, intersections, orientations and displacements (Olson 1993; Renshaw & Pollard 1994; Bonnet et al. 2001), which have important consequences on rock engineering applications.

Fractures, along which rupture has caused cohesion loss and mechanical weakness in the rock, often dominate the strength and deformation of geological formations (Hoek 1983). Interconnected fractures can serve as conduits or barriers for fluid and chemical migration in subsurface

(Berkowitz 2002). The characterisation and simulation of the effects of natural fractures on the hydromechanical behaviour of geological media is a challenging issue (Zimmerman & Main 2004), which is relevant to a variety of engineering applications such as hydrocarbon extraction, geothermal production, groundwater remediation and geological disposal of radioactive waste (Rutqvist & Stephansson 2003). Several key issues in hydromechanical modelling of fractured rocks are summarised as follows:

- Characterisation and representation of the geometry of natural fracture systems (Bonnet et al. 2001), and understanding of the underlying mechanisms that create the observed complexities (Pollard & Aydin 1988).
- Development of computational methods for simulating discontinuous phenomena in fractured rocks, including interaction of discrete matrix bodies (Jing 2003), fracturing and fragmentation of intact rocks (Hoek & Martin 2014), opening, shearing and dilation of rough fractures (Barton 2013), fluid flow through fractured porous space (Berkowitz 2002), and coupled hydromechanical processes (Tsang 1991).
- Upscaling of small-scale simulation results for large-scale predictions with the consideration of the scaling nature of natural fracture systems, which may not have any representative elementary volume (REV) (Bonnet et al. 2001).

My research aims to advance the understanding of the geometrical complexity of natural fracture networks (section 2) (Lei & Wang 2016), review the state-of-the-art discrete fracture network modelling (section 3) (Lei et al. 2017a), develop a computational framework for geomechanical modelling of rough rock fractures (section 4) (Lei et al. 2016), investigate the stress effects on fluid flow in 2D (section 5) (Lei et al. 2014) and 3D (sections 7 & 8) (Lei et al. 2015b, 2017c) fracture networks, predict larger-scale fracture network properties by upscaling smaller-scale simulation results (section 6) (Lei et al. 2015a), and further apply the numerical tools to solve engineering problems (section 9) (Lei et al. 2017b). Finally, the implications for rock engineering applications are discussed and a brief summary is given (section 10).

2 GEOMETRY AND TECTONISM OF NATURAL FRACTURE NETWORKS

The growth and interaction of natural fractures result in a hierarchical geometry that may exhibit long-range correlations from macroscale frameworks to microscale fabrics (Barton 1995). The proximity of the connectivity state of natural fracture networks to the percolation threshold remains an unresolved debate. It was argued earlier that natural fracture systems are close to the percolation threshold (Renshaw 1997), because the driving force (tectonic stress or hydraulic pressure) is abruptly released once the system is connected, and a diminished mechanical strength and an enhanced hydraulic conductivity are likely to occur (Chelidze 1982; Madden 1983; Gueguen et al. 1991; Renshaw 1996). However, extensive field observations suggested that crustal fractures can be well-connected and significantly above the threshold (Barton 1995). The geometrical scaling of a fracture population provides clues for a better understanding of the geology and physics behind the statistics. The power law model having no characteristic length scale can be a useful tool to interpret the scaling phenomena of natural fracture systems (Bonnet et al. 2001).

The fracture networks studied here are located in the Languedoc region of Southern France. A study of the regional geological evolution indicates that this area has been affected by multiple tectonic events: normal faulting in the Jurassic (Event I), strike-slip faulting (Event II-A) and thrusting faulting (Event II-B) in the Late Cretaceous to Eocene, and normal faulting in the Oligocene (Event III). A series of outcrop patterns exposed at the Earth's surface over various scales is mapped to measure the statistics of the fracture system (Figure 1).

The spatial organisation is characterised by the fractal dimension D ($\approx$ 1.65), which is derived using the two-point correlation function $C_2(r) = N_d(r)/N \sim r^D$ (Bour & Davy 1999), where N is the total number of fracture barycentres and N_d is the number of pairs of barycentres whose distance is smaller than a scale variable r (Figure 2a). The distribution of fracture lengths is characterised by the power law exponent a ($\approx$ 2.65), which is derived from the density distribution $n(l, L)dl = \alpha L^D l^{-a}$, where l is the fracture length, L is the domain size, and α is a density term (Figure 2b) (Bour et al.

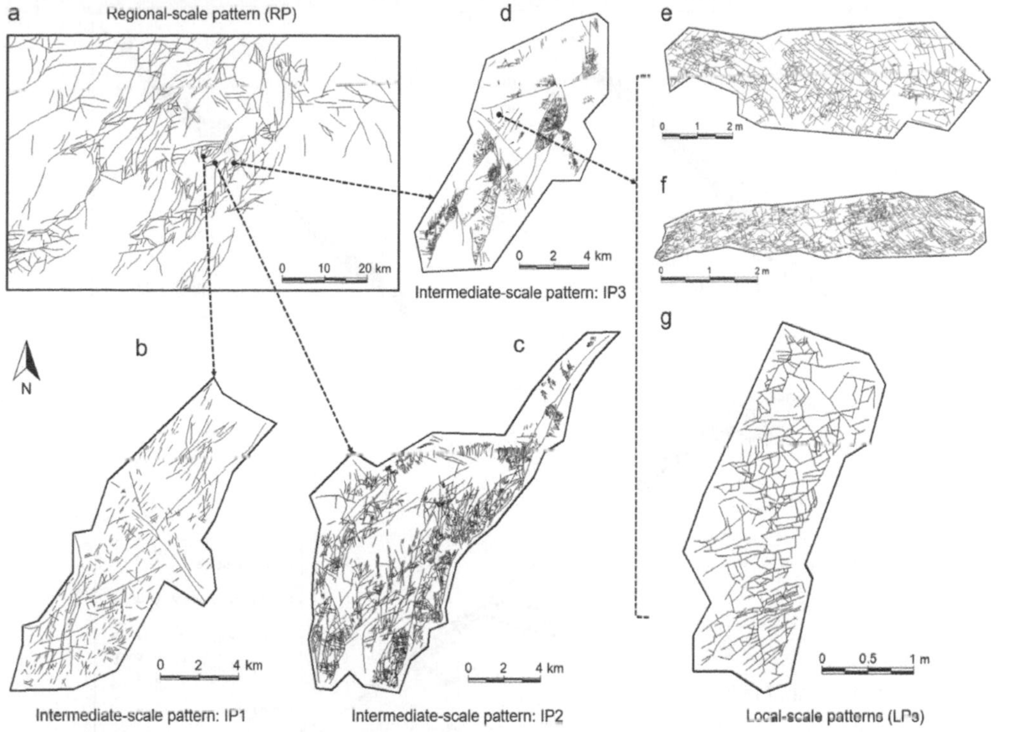

Figure 1. A compilation of multiscale fracture patterns from the Languedoc region, Southern France. (a) A regional-scale lineament pattern generated from the regional structural map, (b)-(d) intermediate-scale fracture patterns obtained from aerial photographs, and (e)-(g) local-scale outcrop patterns derived from geological exposures.

2002). Fractures having a broad-bandwidth power law size distribution are not randomly placed in the geological media, but organised by mechanical interactions that occur during their growth process (Davy et al. 2010). The relationship between a and D of the dataset studied, i.e. $a \approx D+1$, indicates that the multiscale fracture system may be self-similar (Bour et al. 2002). A self-similar fracture pattern emerges under a statistically-valid hierarchical rule that a large fracture inhibits smaller ones from crossing it but not the converse (Davy et al. 2010). The average distance $d(l)$ between the centroid of a fracture having length l and that of the nearest larger neighbour is theoretically correlated with l as $d(l)\alpha l^{x}$ (Bour & Davy 1999), where $x = (a-1)/D$. The current data fit to $x = 1.0$ (Figure 2c), suggesting that the distance of a fracture to its nearest larger one is linearly correlated with its size, and that the faults and joints were well developed and had reached quite a dense state controlled by mechanical interactions (Davy et al. 2010). In addition, the fracture patterns on different scales also exhibit quite similar values in the ratio of $d(l)/l$, implying that the fracture interactions may be governed by a similar mechanism over different scales (this may seem surprising given that faulting is a different process to jointing).

The percolation parameter p as a connectivity metric of fracture networks is calculated using the following equation (Berkowitz et al. 2000):

$$p(l,L) = \int_{l_{\min}}^{L} \frac{n(l,L)l^{D}}{L^{D}} \mathrm{d}l + \int_{L}^{l_{\max}} n(l,L)\mathrm{d}l \tag{1}$$

where l_{max} is the maximum fracture length, and l_{min} is the length over which all fractures were correctly sampled (corresponding to the onset of power law length scaling in each network). The connectivity of a fracture network is made up of two parts, as shown in Eq. (1): the first part describes the contribution made by fractures smaller than the system size L and the second represents the contribution from fractures larger than L (Bour & Davy 1997). Mathematically, the connectivity

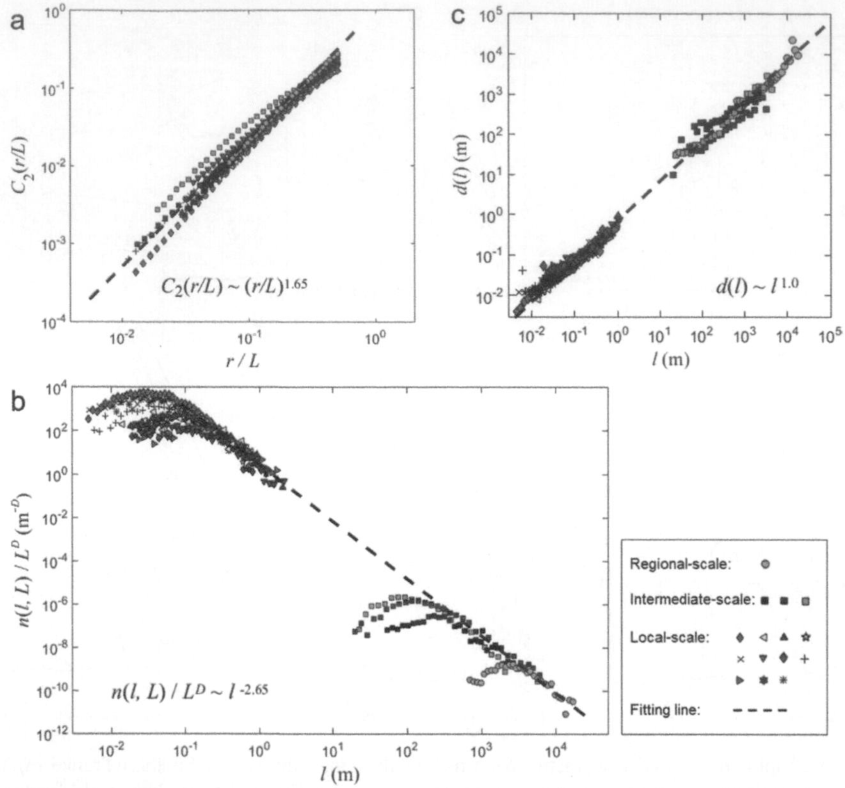

Figure 2. (a) Calculation of the normalised two-point correlation functions $C_2(r/L)$ as a function of r/L. The dashed line represents a power law fitting line with the fractal dimension $D = 1.65$. (b) The normalised density distribution of fracture lengths of the multiscale fracture patterns; the dashed line represents a power law fitting line with an exponent $a = 2.65$ and a density term $\alpha = 3.0$. (c) Scaling of the distance $d(l)$ between the barycentre of a fracture and that of its nearest neighbour having a length larger than l; the dashed line represents a power law fitting line with an exponent $x = 1.0$.

of a self-similar fractal population is scale invariant (Darcel et al. 2003), and the networks are connected at all scales if p is larger than the percolation threshold p_c (i.e. the onset above which a fracture network is, on average, connected from one side of the domain to the other). The range of p_c was determined to be between 5.6 and 6.0 based on 2D random fracture network realisations (Bour & Davy 1997). A correcting factor of $2/\pi$ was suggested to derive a p_c for 3D systems (Lang et al. 2014), which yields $p_c \approx 3.6$-3.8. The p value of the fracture patterns in the study area varies significantly at different scales, ranging from 4.60 to 14.69. The effects of the variation in a and D for different samples and the inconsistency in the ratio of L/l_{min} may not sufficiently explain the high contrast in the calculated p values.

An understanding of the process by which the natural fracture networks evolve might offer an explanation to this. Fracture networks in rock develop over geological time by the superposition of successive fracture sets each linked to a different stress regime and set of crustal conditions. Thus, there is a strong possibility that early fracture sets may become partially or totally cemented as the network evolves and fluids move through it. These sealed or partially sealed early fracture sets may act as barriers to fluid flow and the integrity of the rock has been to some extent recovered. Although the network geometrically remains almost the same, its "effective" connectivity has been reduced well below the percolation threshold. As a result, subsequent stress fields could continue to propagate new fractures until the critical state is re-established. However, if the "apparent" connectivity of trace patterns is measured without considering their internal sealing conditions, it is likely to derive a percolation state significantly above the threshold. In addition, the intrinsic anisotropy of

Table 1. Percolation parameters of the progressively formed fracture networks at the end of each different formation stage.

Pattern	Stage 1 (Event I)	Stage 2 (Event II-A)	Stage 3 (Event II-B & III)
RP	3.87	5.05	7.18
IP1	3.06	4.30	5.30
IP2	8.16	12.62	14.69
IP3	3.62	5.69	6.90
LPs	–	4.38 ± 1.54	6.81 ± 2.17

the fracture network may also permit tectonic energy to accumulate in other directions which have a higher mechanical strength/stiffness and can accommodate more new cracks.

To test this concept, the percolation parameter of the progressively developed fracture networks at the end of each different formation stage is calculated (Table 1), which is achieved simply by re-analysing networks from field data with the appropriate later-staged fractures removed based on the relation between fracture orientations and the tectonic events. Generally, the first stage fracture set exhibits a connectivity state close to the percolation threshold, consistent with the postulation of energy relief at the connecting moment observed in both laboratory experiments (Chelidze 1982) and numerical simulations (Madden 1983; Renshaw 1996; Zhang & Sanderson 1998). However, because of the possibility of early fractures becoming cemented as has been observed in the Languedoc area (Petit & Mattauer 1995; Petit et al. 1999), a fracture network which at the time of its formation was at the percolation threshold may subsequently have an "effective" connectivity considerably lower than p_c. Thus, in response to later tectonic events, further cracking may occur within the network until the system once again becomes connected. The incremental rate of p caused by late-stage fracturing seems to gradually decrease due to the presence of early-stage fractures, since percolation can be reached more easily by reactivating and/or coalescing existing fractures rather than generating new ones. The exceptionally high p in the pattern of IP2 may be attributed to its location very close to one of the regional-scale faults, in the vicinity of which concentrated fracturing paced by active calcite precipitation may occur, i.e. more intensive "crack-seal" cycles may be involved (Petit et al. 1999).

3 STATE-OF-THE-ART DISCRETE FRACTURE NETWORK MODELLING

The recognition of the importance of natural fractures, which can result in heterogeneous stress fields (Pollard & Segall 1987) and channelised fluid flow pathways (Tsang & Neretnieks 1998) in highly disordered geological media, has promoted the development of robust discrete fracture network (DFN) models for numerical simulation of fractured rocks (Herbert 1996). The purpose of this review is to present a summary of various approaches that explicitly mimic natural fracture geometries, and different numerical frameworks that integrate discrete fracture representations for modelling the geomechanical behaviour of fractured rocks as well as further analysis of the impacts on fluid flow.

The geometry of a fracture network model can be generated from geological mapping, stochastic realisation or geomechanical simulation. The geologically-mapped fracture patterns are derived from the exposure of rock outcrops or man-made excavations (Zhang & Sanderson 1996; Belayneh & Cosgrove 2004). The stochastic DFN approach generates randomly distributed fracture networks in which the geometrical properties (e.g. position, size, orientation, aperture) of fractures are treated as independent random variables obeying certain probability functions (Baecher 1983; Dershowitz & Einstein 1988). The geomechanically-based DFN method reproduces natural fracture patterns by applying a geologically-inferred palaeo-stress/strain condition and progressively solving the perturbation of stress fields and capturing the nucleation, propagation and coalescence of discrete fractures (Olson 1993; Renshaw & Pollard 1994; Paluszny & Matthäi 2009). The three types of DFN models have distinct strengths but also suffer from some limitations, as summarised in Table 2.

The geomechanical modelling of fractured rocks can be achieved by continuum or discontinuum approaches, which have important differences in conceptualising geological media and treating displacement compatibility (Jing 2003). The continuum approach treats a rock domain as a continuous

Table 2. Comparison of different numerical models for geometrical and geomechanical modelling of natural fracture networks.

Numerical models	Key inputs	Strengths	Limitations
		Geometrical modelling	
Geological DFNs	Analogue mapping, borehole imaging, aerial photographs, LIDAR scan or seismic survey	◦ Deterministic characterisation of a fracture system ◦ Preservation of geological realisms	◦ Limited feasibility for deep rocks ◦ Difficulty in building 3D structures ◦ Constraints from measurement scale and resolution
Stochastic DFNs	Statistical data of fracture lengths, orientations, locations, shapes and their correlations	◦ Simplicity and convenience ◦ Efficient generation ◦ Applicability for both 2D and 3D ◦ Applicability for various scales	◦ Uncertainties in statistical parameters ◦ Oversimplification of fracture geometries and topologies ◦ Requirement of multiple realisations
Geomechanical DFNs	Palaeostress conditions, rock and fracture mechanical properties	◦ Linking geometry with physical mechanisms ◦ Correlation between different fracture attributes	◦ Uncertainties in input properties and tectonic conditions ◦ Large computational time ◦ Negligence of hydraulic, thermal and chemical processes
		Geomechanical modelling	
Continuum models	Equivalent material properties	◦ Simplicity of geometries ◦ Efficient calculation ◦ Suitability for large-scale industrial applications	◦ No consideration of fracture interaction, block displacement/interlocking/rotation ◦ Complexity in deriving equivalent material parameters and constitutive laws ◦ Valid only if an REV exists
Block-type & particle-based discrete models	Material properties for both fractures and rocks, damping coefficient, bonding strengths	◦ Explicit integration of DFNs ◦ Simple particle/grain bonding logic ◦ Integrated constitutive laws for rocks/fractures ◦ Capturing the interaction of multiple fractures	◦ Limited data on joint stiffness parameters ◦ Calibration of input particle bonding properties ◦ No fracture mechanics principle ◦ Large computational time
Hybrid FEMDEM models	Material properties for both fractures and rocks, fracture energy release rate, damping coefficient	◦ Explicit integration of DFNs ◦ Fracture propagation is based on both the strength criterion and fracture mechanics principles ◦ Integrated constitutive laws for rocks/fractures ◦ Capturing the interaction of multiple fractures	◦ Calibration of fracture energy release rates ◦ Large computational time

medium that can be solved by the finite element method (Oda et al. 1993) or finite difference method (Rutqvist et al. 2013). The discontinuum approaches can explicitly model irregular fracture networks, include complex constitutive laws of rock materials and fractures, and capture the fracturing and fragmentation processes. The commonly used discontinuum models include the distinct element method (Cundall 1971), the discontinuous deformation analysis method (Shi & Goodman 1985), the bonded-particle method (Mas Ivars et al. 2011), and the combined finite-discrete element method (FEMDEM) (Munjiza 2004). The preference for a continuum or discontinuum modelling scheme depends on the scale of the problem and the complexity of the fracture system (Jing & Hudson 2002). A detailed comparison of the strengths and limitations of different types of continuum and discontinuum models can be found in Table 2.

The geomechanical effects on the hydrological properties of fractured rocks has also been investigated with respect to stress-dependent flow pattern, equivalent permeability and mass transport in the literature (Zhang & Sanderson 1996; Min et al. 2004; Baghbanan & Jing 2008; Latham et al. 2013; Rutqvist et al. 2013; Zhao et al. 2013). Several important stress-dependent fluid flow phenomena have been observed such as permeability reduction caused by fracture closure, flow localisation engendered by shear dilation, permeability increase at the critical stress ratio condition, and breakthrough curve shifting with the in-situ stress variation. The stress-dependent fluid flow in fractured rocks demonstrate the importance of using explicit DFN representations and incorporating geomechanical modelling for characterising fluid flow in natural fracture systems. The previous simulation results also show consistency with field measurements/observations, such as only a small portion of fractures are conductive (Tsang & Neretnieks 1998), permeability is less sensitive in deeper formations (Rutqvist & Stephansson 2003), and critically stressed faults tend to have higher hydraulic conductivity (Zoback 2007).

4 A GEOMECHANICAL MODEL FOR SIMULATING ROUGH NATURAL FRACTURES

The geomechanical model for simulating natural fractures is based on the combined finite-discrete element method (FEMDEM) (Munjiza 2004). The FEMDEM method has been extensively developed and applied during the past decades and has proven its strong capability in handling large strain deformation, multi-body interaction, fracturing and fragmentation (Elmo et al. 2013; Latham et al. 2013; Lisjak & Grasselli 2014). However, to model rock fractures associated with intrinsic surface asperities, an extension of the FEMDEM model is needed in order to capture the complex non-linear, scale-dependent strength and deformation behaviour of natural fractures (Bandis et al. 1981, 1983; Barton et al. 1985).

The FEMDEM model represents a fractured rock using a fully discontinuous mesh of triangular (in 2D) or tetrahedral (in 3D) finite elements with their interfaces connected by joint elements. There are two types of joint elements: unbroken joint elements inside the matrix and broken joint elements along existing fractures. The propagation of new fractures is captured as the transition of unbroken joint elements to broken ones in an unstructured grid. The joint constitutive model (JCM) proposed by Bandis et al. (1983) and Barton et al. (1985) based on the empirical parameters of joint roughness coefficient (JRC) and joint compressive strength (JCS) has been implemented into the 2D/3D FEMDEM framework. The coupling between the JCM and FEMDEM fields is achieved with respect to both stress and displacement solutions, such that the aperture calculation accounts for both fracture interaction-induced separations and roughness-controlled openings. The non-linear shear strength of fractures is captured by updating the frictional force computation in the FEMDEM iterations with the mobilised friction coefficient derived from the JCM calculation. The effective fracture sizes (i.e. between fracture intersections) are calculated through a binary (in 2D) or ternary (in 3D) tree search of connected broken joint element chains such that the scale-dependent fracture wall properties (i.e. JRC and JCS) are assigned to individual fractures of different lengths.

The JCM-FEMDEM model is verified by the consistency between the numerical results and empirical solutions for a direct shear test of different sized fracture samples (Figure 3). The rough fractures exhibit significant non-linear shear strength behaviour with the maximum value reached at the peak shear displacement, beyond which the strength gradually decreases to the residual value (Figure 3a). During the shearing process, the fractures exhibit slight contraction in the pre-peak stage and then considerable dilation in the post-peak stage (Figure 3b). As the fracture size increases, a transition from a "brittle" to "plastic" shear failure mode occurs associated with a

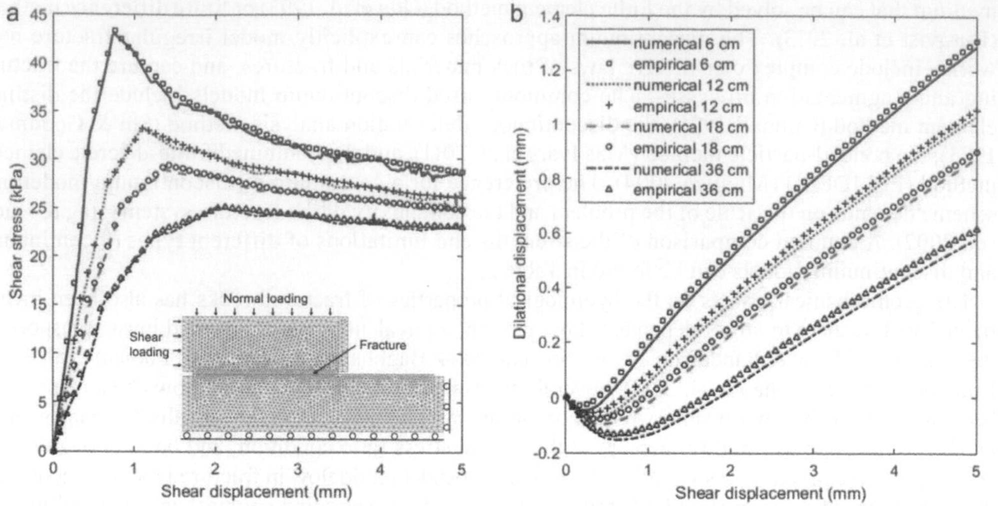

Figure 3. Verification of the numerical model based on a direct shear test of different sized fracture samples under a constant normal stress: (a) shear stress-shear displacement curves and (b) dilational displacement-shear displacement curves obtained from the numerical simulations and empirical solutions.

reduced peak shear strength, an increased peak shear displacement and an enlarged dilational displacement. The JCM-FEMDEM model that simulates the normal and shear displacements of rough fractures with no need to explicitly represent their roughness profiles tends to be advantageous in solving fracture network-scale problems.

5 HYDROMECHANICAL MODELLING OF 2D FRACTURE NETWORKS

The JCM-FEMDEM model is applied to simulate the effects of in-situ stresses on the deformation and fluid flow in 2D fracture networks, which include a 1.5 m × 1.5 m analogue fracture network (AFN) mapped from a limestone outcrop at the Bristol Channel Basin (Belayneh et al. 2009) and its random DFN equivalents. The AFN contains two distinct cross-cut sets of layer-normal joints with generally straight traces and is considered well suited to being represented by stochastic DFNs, whose performance for hydromechanical predictions, however, may need further assessments. Ten DFN realisations are generated based on the measured statistics of the AFN, i.e. fracture orientations, lengths and inhomogeneous spatial distributions.

A series of plane strain numerical experiments is designed with the far-field effective stresses (σ'_1 = 10MPa, σ'_3 = 5MPa) applied at a range of angles (0°, 30°, 60°, 90°, 120° and 150°) to the AFN and DFNs (Figure 4). The models adjust to deformed state under the imposed boundary stresses and exhibit significant stress heterogeneity, which is controlled by both the fracture network geometry and in-situ stress condition. A comprehensive comparison between the AFN and DFNs is made with respect to various geomechanical responses such as shear displacement, crack propagation, hydraulic aperture and network connectivity (Figure 5). The AFN and DFNs exhibit certain similarity in their average response to the change of the in-situ stress orientation, whereas DFNs tend to accommodate larger shear displacement, more crack propagation, wider hydraulic apertures and more intersection nodes. The discrepancy may be attributed to the oversimplification of stochastic DFN geometries in representing natural fractures involving complex curvature, spacing and clustering features.

The equivalent permeability of the stressed 2D fractured rocks is derived from single-phase steady state flow simulations (Geiger et al. 2004). The equivalent permeability is much larger than the assumed matrix permeability (i.e. k_m = 0.1, 1 or 10 mD) (Figure 6), implying that fractures play a dominant role in fluid flow across the fractured rock. The mean permeability of the multiple DFNs shows a reasonably good match with that of the AFN in the x direction associated with an initially good connectivity, whereas a profound discrepancy is observed in the y direction with a

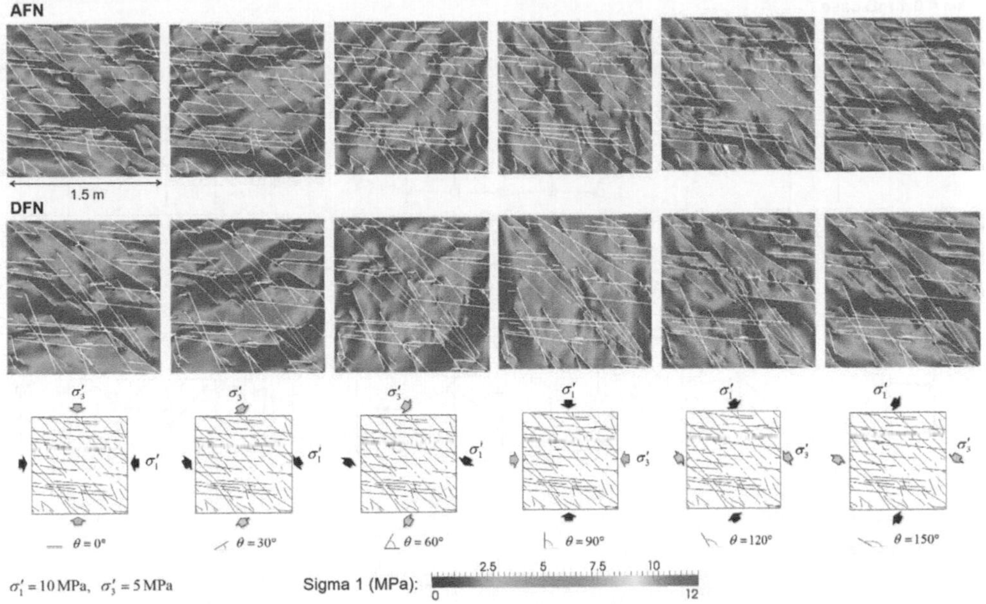

Figure 4. Stress distribution in the AFN and one of its DFN equivalent under in-situ stresses applied at different angles.

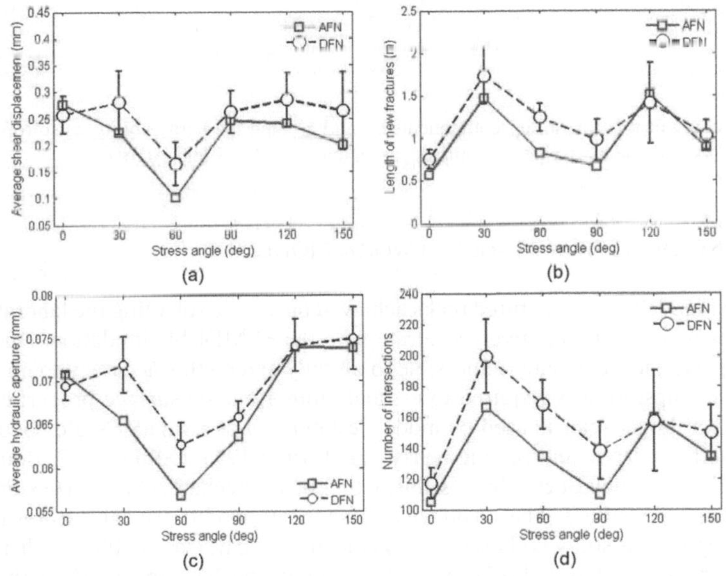

Figure 5. Variation of fracture network properties with the change of the in-situ stress orientation: (a) average shear displacement, (b) length of new fractures (c) average hydraulic aperture, and (d) number of "T" and "X" intersection nodes.

connectivity close to the critical percolating state (more sensitive to geomechanical effects). The high variability in the DFN simulation results (error bar of ± ~30-50%) suggests that large uncertainty tends to exist in random DFNs, especially when the matrix is less permeable such that flow is more dominated by fractures. The quality of DFNs for hydromechanical modelling of fractured rocks is considered to be controlled by the accuracy in representing geologically-formed network geometries and geomechanically-controlled fracture apertures.

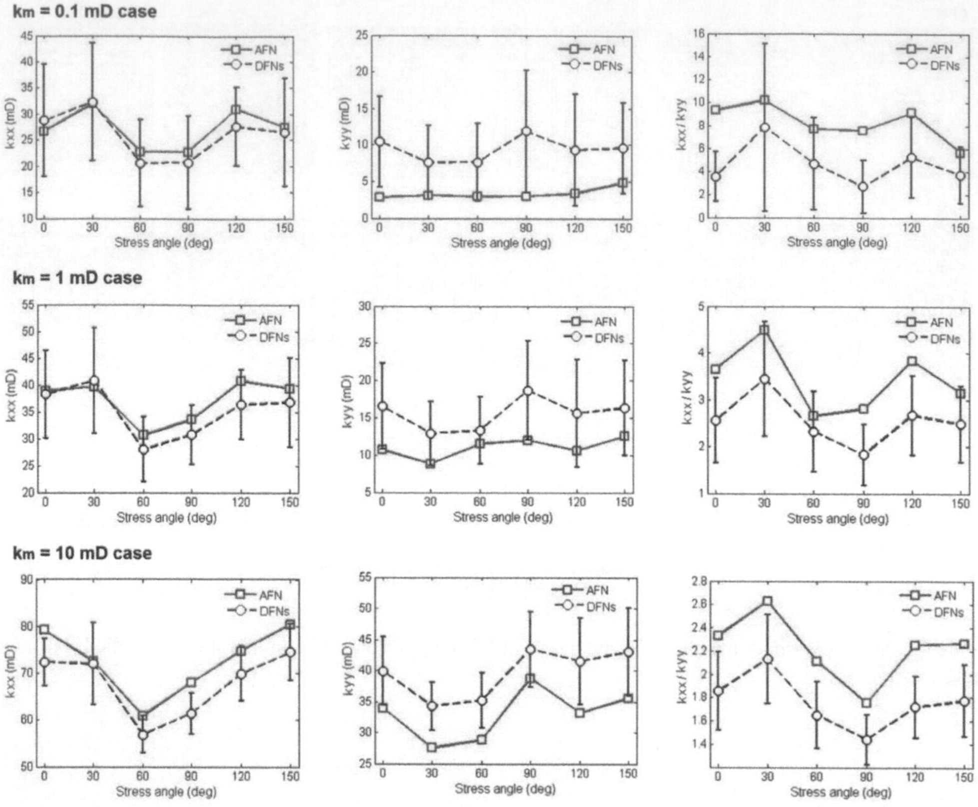

Figure 6. The equivalent permeability components k_{xx} and k_{yy}, and the permeability anisotropy ratio k_{xx}/k_{yy} of the AFN and DFNs with the matrix permeability k_m assumed to be 0.1, 1 and 10 mD.

6 UPSCALING OF 2D FRACTURE NETWORK MODELS

Geomechanical modelling of fractured rocks achieved on a scale spanning the laboratory specimen to a few metres is becoming relatively accurate with the FEMDEM simulation. Due to the limits of processing power, it is currently impossible to directly extend this accuracy to macroscale computations. Hence, upscaling is required to estimate important subsurface properties of naturally fractured rocks at larger scales based on models established at a smaller scale. Geological media often exhibit self-similarity and scaling behaviour (Barton 1995; Odling 1997; Bour et al. 2002), the understanding of which opens the possibility that hydromechanical properties of a macroscale fractured rock may be estimated based on the characterisation of its crucial features from a smaller sample. The scope of this study is chosen to be on the in-situ scale (say, 1-100 m), where flow is often dominated by fractures (Clauser 1992), and to focus on the mechanisms by which the permeability of fractured rocks may vary with the modelling scale over this range.

The scaling properties of a 6 m × 6 m natural fracture system (Figure 7a) are examined in terms of spatial organisation, lengths and connectivity using fractal geometry and power law relations. The fracture pattern is observed to be nonfractal with the fractal dimension $D \approx 2$ (i.e. homogeneous space filling), while its length distribution tends to follow a power law with the exponent $a \approx 2.37$. For this network with $a < D+1$, the critical system size (or the connection length) L_c corresponding to the percolation threshold is calculated to be ~0.80 m, above which the fracture network is expected to be well-connected due to the increased network connectivity with scale (Davy et al. 2006). A smaller domain with size $L = 2$ m (Figure 7b) is used for FEMDEM simulation to generate a realistic distribution of fracture apertures and shear displacements under a hydrostatic or deviatoric stress

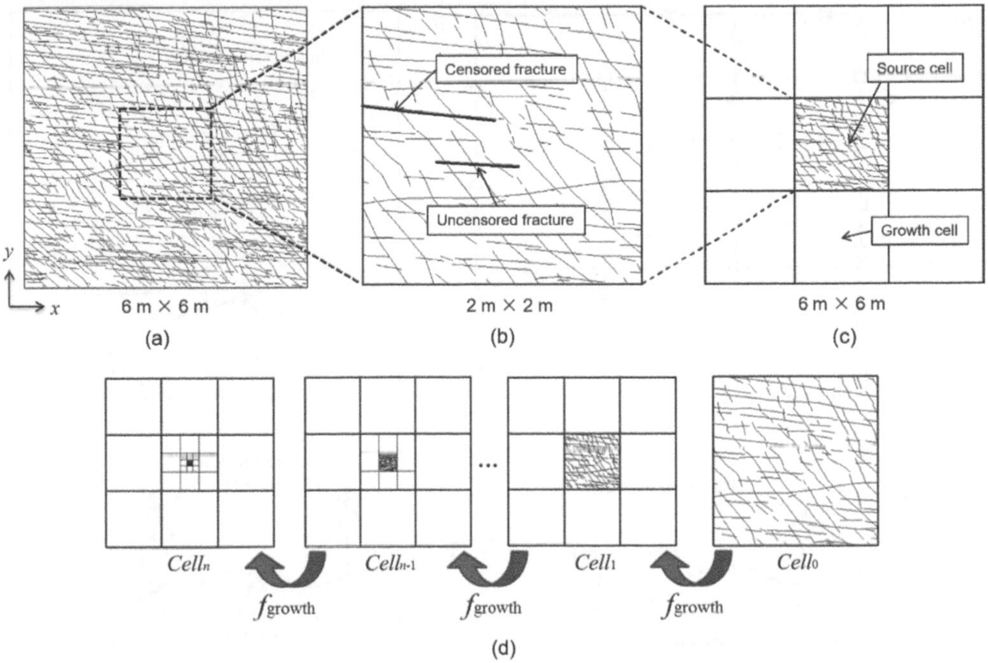

Figure 7. (a) The 6 m × 6 m outcrop pattern for extracting scaling properties, (b) the 2 m × 2 m source cell pattern including censored and uncensored fractures, and (c) the growth lattice for extrapolating fracture networks progressively into larger domains based on (d) a recursive scheme.

condition. This smaller pattern with $L = 2$ m also serves as the source for network extrapolation with its validity examined through a comparison with the original larger pattern of $L = 6$ m.

By assuming the fracture pattern repeats itself in progressively larger and larger Euclidean space, a novel upscaling scheme is developed to extrapolate the geologically-mapped fracture geometry together with its stress-dependent, spatially-variable displacement attributes (i.e. fracture aperture and shear displacement) into larger scales using a recursive growth lattice (Figure 7c & d). There are two types of cells in a growth lattice: the source cell that is the reference for network repetition, and the growth cell that is a clone of the source cell sharing common geostatistics. The fractures are classified into censored (partially sampled) and uncensored (completely observed) types (Figure 7b). Methods of statistics are applied to the source pattern to interpret its topological complexity in a quantitative way. The locations of censored fractures are measured based on the distribution of censoring nodes which truncate the partially sampled fractures at the source cell boundary. The spatial organisation of uncensored fractures is characterised by the distribution of their barycentres as well as the exclusion radius and spacing parameters. The characteristics of individual fractures are also statistically quantified with respect to orientation, length, segmentation, curvature, shear displacement and aperture. The growth of censored and uncensored fractures in a growth cell is implemented in different ways due to their distinct geostatistical features. A censored fracture in a growth cell evolves from a nucleus located on the lattice edge (Figure 8a) and propagates following a random walker (Figure 8b), while an uncensored crack hatches from the barycentre randomly seeded inside the cell (Figure 8c) and propagates as two synchronised walkers jogging towards opposite directions (Figure 8d).

To examine the validity of the growth network for representing larger fracture geometries, a comparison is made at a system scale of $L = 6$ m between the original analogue fracture network (AFN) from outcrop mapping (Figure 9a), ten realisations of growth fracture network (GFN) extrapolated from the central $L = 2$ m source pattern (Figure 9b), and ten realisations of purely random Poisson DFNs (Figure 9c). The advantages of the proposed growth networks are highlighted through a qualitative visual comparison (Figure 9) and a quantitative measurement of the fracture spacing distribution (Figure 10).

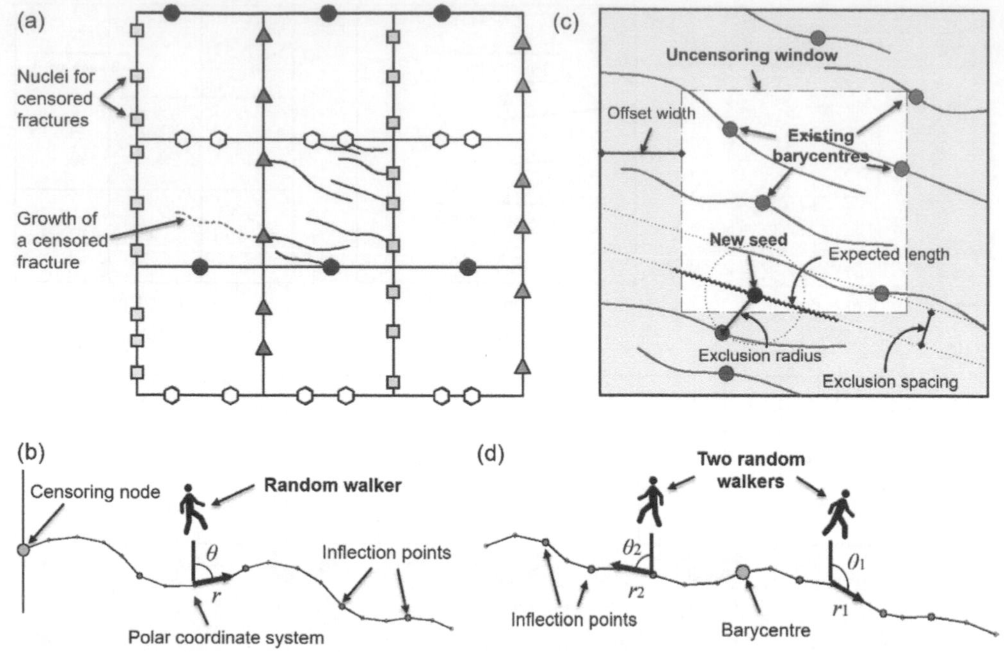

Figure 8. (a) Nucleation of censored fractures by seeding censoring nodes along the edges of a growth lattice, (b) propagation of a censored fracture from a censoring node traced by a random walker, (c) nucleation of uncensored fractures by a point packing process, and (d) propagation of an uncensored fracture from its barycentre captured by two synchronised random walkers.

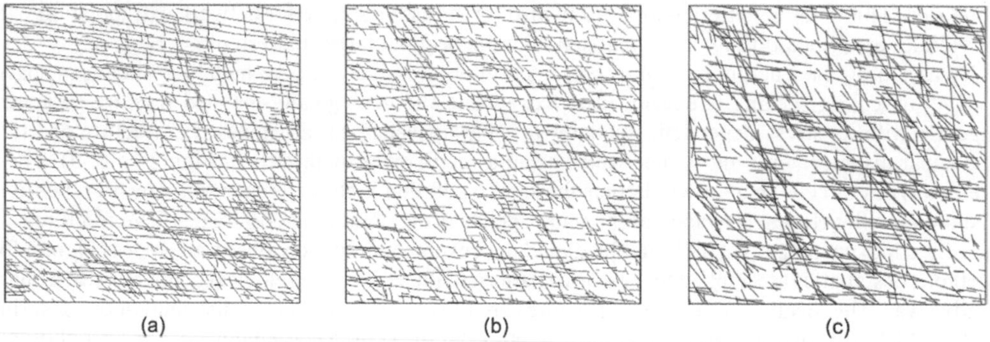

Figure 9. Fracture patterns (domain size $L = 6$ m) of (a) the analogue fracture network (AFN), (b) one of the ten growth fracture network (GFN) realisations, and (c) one of the ten Poisson discrete fracture network (DFN) realisations.

Multiscale growth networks with stress- and scale-dependent apertures are constructed using the recursive growth scheme with the important natural fracture characteristics (e.g. non-planarity, segmentation, local clustering and length scaling) preserved (Figure 11). The equivalent permeability of the growth networks is derived from single-phase flow simulations with the matrix permeability assumed to be 1×10^{-15} m^2. As shown in Figure 12, with the increase of the scale, the permeability of the fractured rock in the deviatoric stress scenario displays an upward trend at the small and intermediate scales (<10-20 m) and a continued downward trend at larger scales (>20 m), whereas the permeability in the hydrostatic case mainly shows a downward trend except a slight increase in the y direction at the small scale (<10 m). Fracture networks under the deviatoric stress condition appear to be more permeable than those under the hydrostatic condition due to the effects of shear dilations in response to differential stresses.

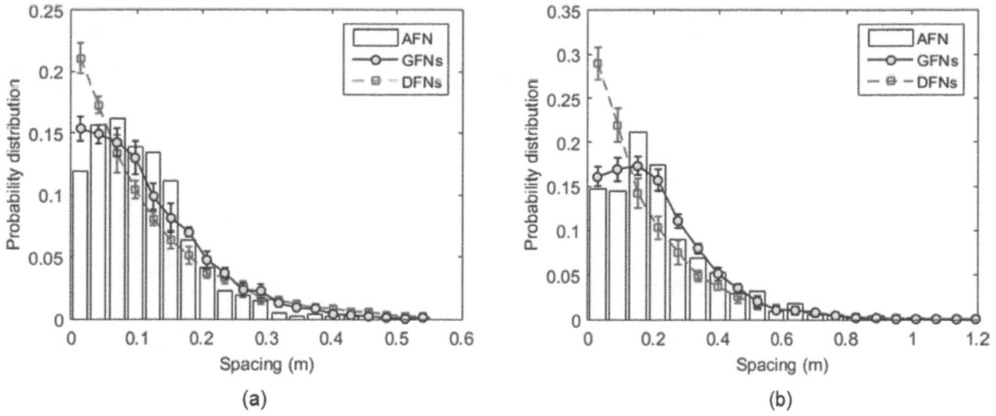

Figure 10. Spacing distribution of the analogue fracture network (AFN), growth fracture networks (GFNs), and Poisson discrete fracture networks (DFNs), measured by twenty scanlines along (a) the y direction and (b) the x direction, respectively.

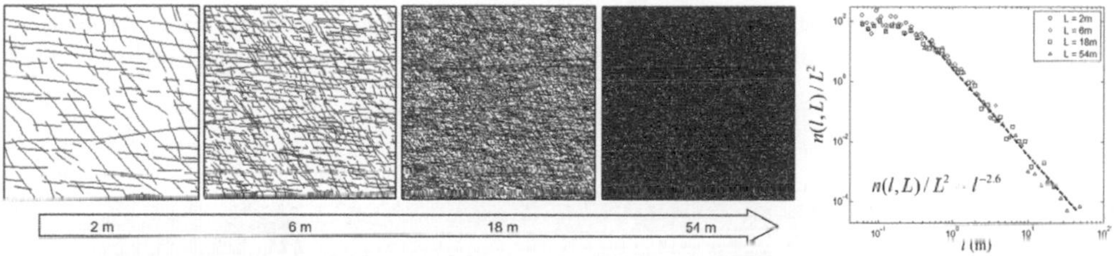

Figure 11. Multiscale growth realisations achieved by the recursive cell culture scheme.

Two factors may dominate the permeability scaling trend: (i) the length exponent a that governs the connectivity scaling of a fracture population (Berkowitz et al. 2000; Darcel et al. 2003), and (ii) the scaling exponents of fracture apertures and shear displacements which control the transmissivity scaling of each individual fracture (de Dreuzy et al. 2002). For the studied case of $2 < a < D+1$, with the increase of domain size L, the number of fractures larger than L (i.e. traversing fractures) increases as $\sim L^{-a+D+1}$ (Davy et al. 2006), whereas the relative percentage of such fractures decreases as $\sim L^{-a+D-1}$. Thus, a global downward trend may be expected for rock permeability at large scales (Renshaw 1998; Klimczak et al. 2010). The flow behaviour is also significantly affected by the distribution of variable apertures, which leads to various fluid flow structures (de Dreuzy et al. 2001a) and permeability scaling trends (Klimczak et al. 2010). Under a higher in-situ stress ratio, longer fractures play a more important role for fluid migration due to their lower resistance (Tsang & Neretnieks 1998) in association with wider apertures that are correlated with fracture length. Hence, at smaller scales, an increased permeability occurs in the deviatoric stress case attributed to the considerable contribution from long fractures. However, a global decreasing trend is inevitable due to the decreasing proportion of traversing fractures at larger scales, where shorter fractures tend to take a heavier role in fluid flow. In the hydrostatic stress case, the equivalent permeability mainly declines with the increased scale, because the slightly scaled apertures with no shear-induced dilation do not endow long fractures with highly conductive capability compared to the decreased relative frequency of long fractures whose length follows the power law.

The trend of rock permeability with scale may be further explained by the flow structure transition zone between the connecting scale and the channelling scale (de Dreuzy et al. 2001b, a; Davy et al. 2006). The connecting scale L_c (or the connection length) is where the fracture network shifts from disconnected to connected, while the channelling scale ξ (i.e. the correlation length in the percolation theory) is where the flow structure transforms from extremely channelled to distributed. For growth networks in the deviatoric stress case (Figure 13), the connection length L_c seems to be

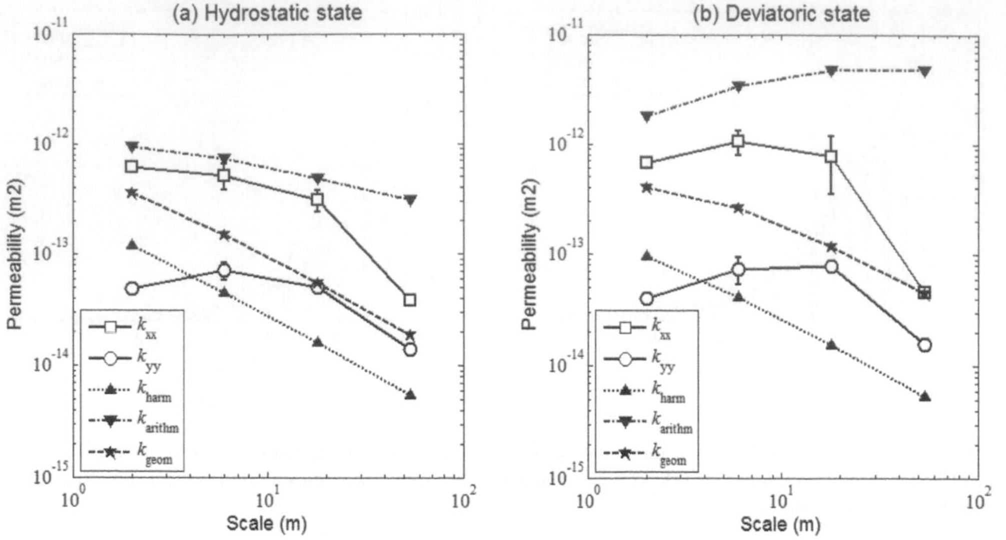

Figure 12. Equivalent permeability k_{xx} and k_{yy} derived from flow simulations and analytical permeability k_{harm}, k_{arithm}, k_{geom} calculated using the harmonic, arithmetic and geometric mean apertures under the (a) hydrostatic and (b) deviatoric stress conditions.

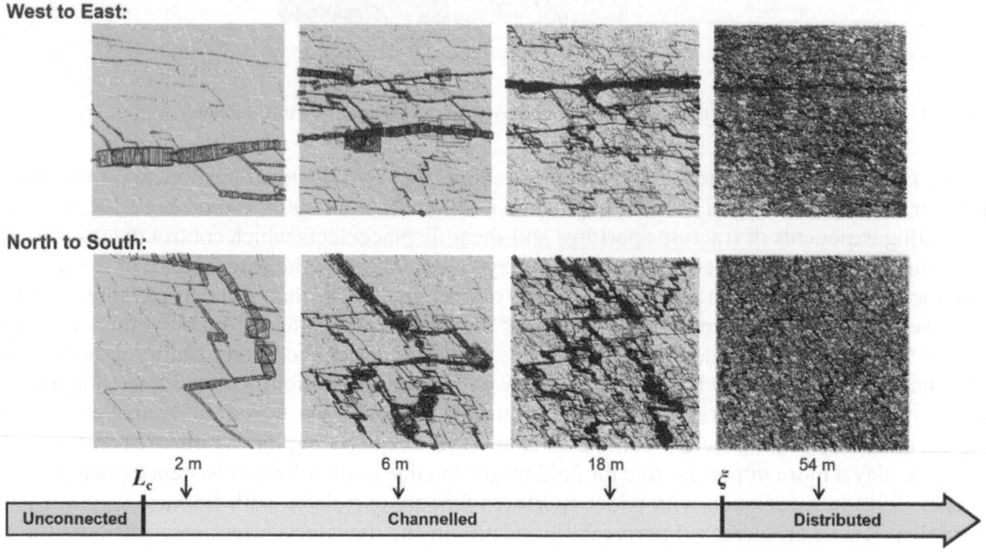

Figure 13. Flow structure transition from extremely channelled to distributed in multiscale growth networks under the deviatoric in-situ stress condition (boxes illustrate the main pathways of the flow structure).

at a scale <2 m, which is consistent with the predicted value of ~0.80 m before, and the channelling scale ξ is at 20-50 m. Within the transition zone (i.e. the system size is between L_c and ξ), the flow structure is made up of a number of quite independent, multi-path, multi-segment channels (Tsang & Neretnieks 1998), under the preference of fluid to flow in least resistance paths through the disordered network of finite-sized, curved fractures. This tortuosity feature has significant impact on the flow properties (Ronayne & Gorelick 2006) and may become even more crucial when the considered rock volume exceeds the channelling scale ξ, beyond which the percentage of domain-sized fractures decreases and flow begins to exhibit dispersive behaviour (like in a homogeneous porous medium) (de Dreuzy et al. 2001b; Davy et al. 2006). A comparison with the analytical solution of the equivalent permeability k_{harm}, k_{arithm}, k_{geom} based on the measured harmonic, arithmetic or geometric mean apertures, respectively, reveals that the numerically derived permeability is well bounded by the harmonic and arithmetic values, while the median trend is better tracked by the geometric one (Figure 12).

Highly conductive fractures with long lengths and wide apertures capable of transmitting fluid across long distances seem to behave more like an "in parallel" connected network (Leung & Zimmerman 2012), so k_{xx} is better captured by the upper arithmetic bound at smaller scales, where the channels formed by very long fractures dominate the flow. However, at larger scales, fluid has to migrate through less conductive branches to reach the opposite boundary due to the proportional reduction of longer fractures, which makes the fracture population act more like an "in series" connected network and k_{xx} tends to approach the lower harmonic bound. The equivalent permeability in the y direction k_{yy} mainly exhibits closer values to the lower limit due to the inherent zigzag feature of the flow structure. Indeed, the mechanism of network alteration from "parallel" to "series" is equivalent to the essence of flow structure transition from "channelled" to "distributed". The permeability magnitudes under the prescribed hydrostatic and deviatoric stresses tend to converge at larger scales but with the intrinsic anisotropy retained. At even larger scales (e.g. >100 m), the fractured rock may behave like a porous medium (Long et al. 1982) with a lower REV permeability conjectured. However, the repetition assumption may not be valid at that scale since many complex larger-scale factors (e.g. seismically visible faults) will be involved (Clauser 1992), which is beyond the current scope.

7 HYDROMECHANICAL MODELLING OF AN IDEALISED 3D PERSISTENT FRACTURE NETWORK

The stress effects on the permeability of fractured rocks have been widely investigated based on 2D fracture network models in the past decades (Zhang & Sanderson 1996; Min et al. 2004; Baghbanan & Jing 2008; Latham et al. 2013). However, the 3D nature of fluid flow in fractured rocks under polyaxial (i.e. true-triaxial) stress conditions remains poorly understood. In this section, the 3D JCM-FEMDEM model is applied to investigate the flow heterogeneity in an idealised 3D persistent fracture network caused by both the fracture-scale roughness effect and the network-scale fracture interactions under polyaxial in-situ stresses.

The discontinuity system involves three orthogonal sets of persistent fractures with one horizontal set of bedding planes and two vertical sets oblique at 45° to the lateral boundaries, on which the far-field horizontal stresses are imposed. All fractures are assumed through-going, tending to provide an upper limit for rock deformability and permeability. The dispersion of fracture orientation is omitted to avoid the numerical difficulty in treating high aspect ratio elements caused by intersection between sub-parallel fractures from the same set. This idealised persistent fracture network might be representative of some special scenarios of highly fractured "non-strata bound" sedimentary rocks. The rock sample (0.5 m × 0.5 m × 0.5 m) is designed to be surrounded by a hollow-box shaped buffer zone with a reduced Young's modulus to provide a semi-free displacement boundary constraint for accommodating potential large slipping in such a persistent system.

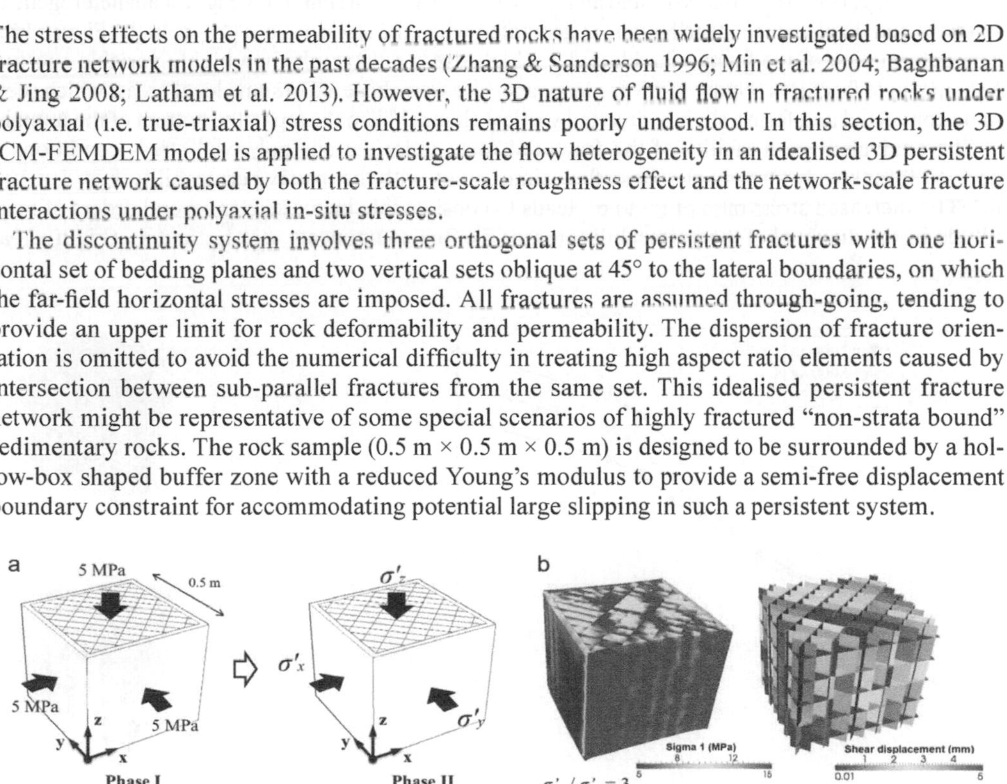

Figure 14. (a) The fractured rock is loaded by two consecutive phases of polyaxial stress conditions and (b) exhibits strong heterogeneity in stress and displacement distributions.

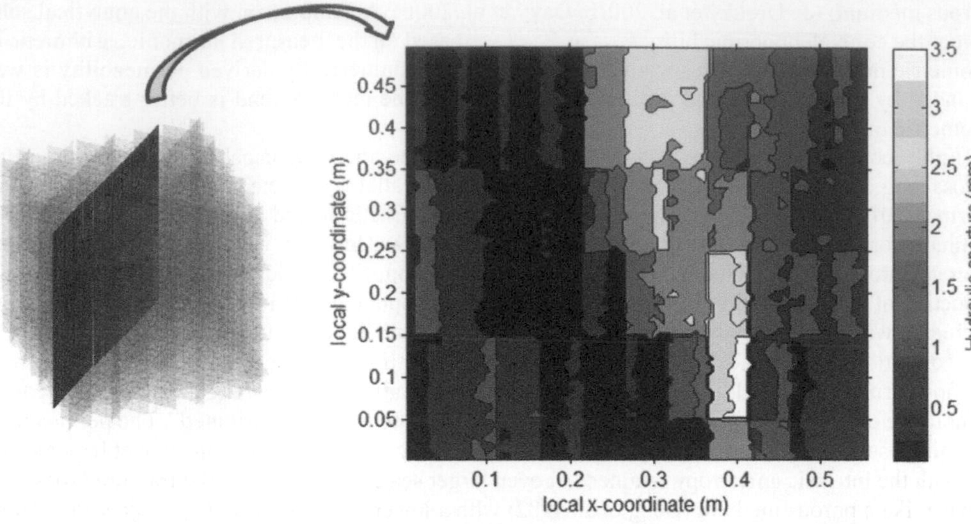

Figure 15. Distribution of hydraulic apertures within a single fracture of the fracture network under a polyaxial stress condition of $\sigma'_x = 5$ MPa, $\sigma'_y = 15$ MPa and $\sigma'_z = 10$ MPa.

The fractured rock is loaded in two consecutive phases (Figure 14a): an isotropic stress field with $\sigma'_x = \sigma'_y = \sigma'_z = 5$ MPa for consolidation (Phase I), and a series of deviatoric stress conditions with a fixed $\sigma'_x = 5$ MPa, various $\sigma'_y = 5$-20 MPa and an increased $\sigma'_z = 10$ MPa (Phase II). Significant heterogeneity in stress and displacement distributions is generated when the stress ratio σ'_y/σ'_x is high (Figure 14b). The deformation of the fractured rock under a high stress ratio results in a highly variable aperture field in single fractures (Figure 15). Very large apertures are clustered in some local areas, which seem to be connected and form a slightly diverted vertical channel from the top to the bottom of the domain.

The equivalent permeability of the fractured rock under various polyaxial stress conditions is derived from steady state single-phase flow simulations with the matrix permeability $k_m = 1 \times 10^{-15}$ m^2. The increased stress ratio of σ'_y to σ'_x leads to considerable increase over several orders of magnitude in the diagonal of the permeability tensor, i.e. the components, k_{xx}, k_{yy}, and k_{zz} (Figure 16a).

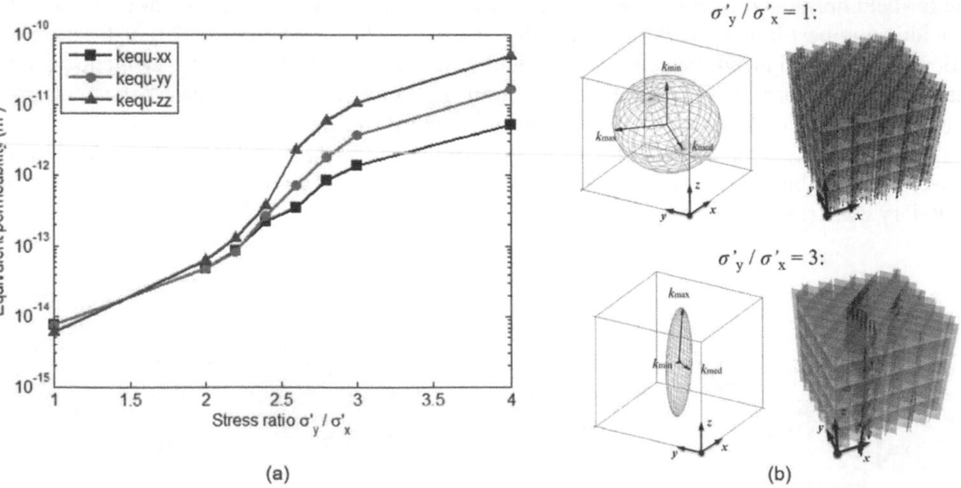

Figure 16. (a) Equivalent permeability of the fractured rock under various polyaxial in-situ stress conditions. (b) Permeability tensor and flow pathways under the stress conditions with $\sigma'_y/\sigma'_x = 1$ and 3.

A transition regime with steep permeability increase occurs when the stress ratio approaches the critical threshold, i.e. 3.1 given that the friction coefficient equals to 0.6 (Zoback 2007). The permeability tensor also changes from isotropic to highly anisotropic with the increase of σ'_y/σ'_x due to the deviatoric stress acting with respect to the favourably oriented vertical fractures, resulting in zig-zag-shaped localised pathways and very high permeability in the subvertical direction (Figure 16b).

8 HYDROMECHANICAL MODELLING OF A REALISTIC 3D FRACTURE NETWORK

The 3D JCM-FEMDEM model is further used to investigate the geomechanical behaviour of a fractured limestone layer embedded with realistic joint sets involving curvature, intersection, abutment and termination features (Figure 17). The 3D fracture system is constructed by extruding a 2D outcrop pattern (2 m × 2 m) of a limestone bed by the layer thickness of 0.1 m. This fracture network exhibits a ladder structure consisting of a "through-going" joint set abutted by later-stage short fractures. A series of in-situ stress conditions is designed to explore the following horizontal stress ratios: σ'_x/σ'_y = 1/3, 1/2, 1, 2, and 3 (Figure 17).

The fractured rocks arrived at equilibrium and exhibit distinct stress distribution patterns under different polyaxial stress conditions (Figure 18, upper panel). The distribution of local maximum principal stresses under an isotropic horizontal stress condition (i.e. $\sigma'_x = \sigma'_y = 5$ MPa) is quite uniform and dominated by the overburden stress (i.e. $\sigma'_z = 10$ MPa). With the increase of the stress ratio (either σ'_y/σ'_x or σ'_x/σ'_y), stress heterogeneity begins to emerge and escalate, with the high stress zones aligning the direction of the far-field maximum horizontal stress. Furthermore, the increased horizontal stress ratio also results in new crack propagation and aperture variability (Figure 18, middle panel). Thus, the vertical flow structure changes from uniformly distributed to highly localised as the horizontal stress ratio increases (Figure 18, lower panel). The magnitude and anisotropy of the equivalent permeability also varies significantly with respect to the change of the in-situ stresses (Figure 19). The distinct stress-dependent variation of the equivalent permeability in the x and y directions (more sensitive to an increased stress ratio of σ'_x/σ'_y than to an increased ratio of σ'_y/σ'_x) is attributed to the inherent anisotropy of the joint network geometries. The bed-normal permeability k_{zz} is much more sensitive to the change of stress loading than k_{xx} and k_{yy}. The results demonstrate that both the magnitude and orientation of the far-field stresses have strong impacts on the permeability of the fractured layer embedded with an anisotropic joint network.

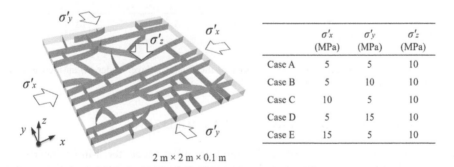

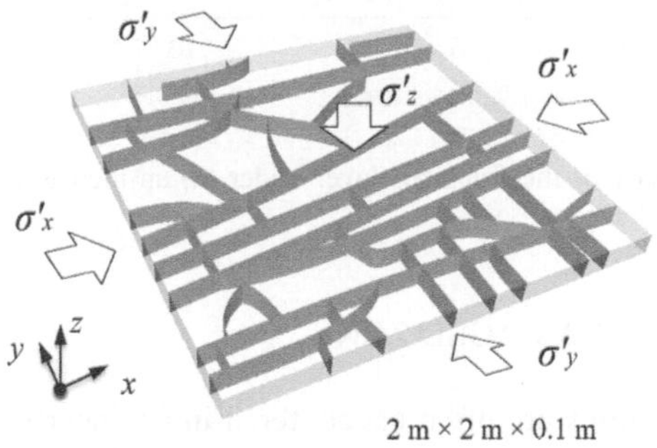

	σ'x (MPa)	σ'y (MPa)	σ'z (MPa)
Case A	5	5	10
Case B	5	10	10
Case C	10	5	10
Case D	5	15	10
Case E	15	5	10

Figure 17. A 3D fractured limestone layer is loaded by various polyaxial stress conditions.

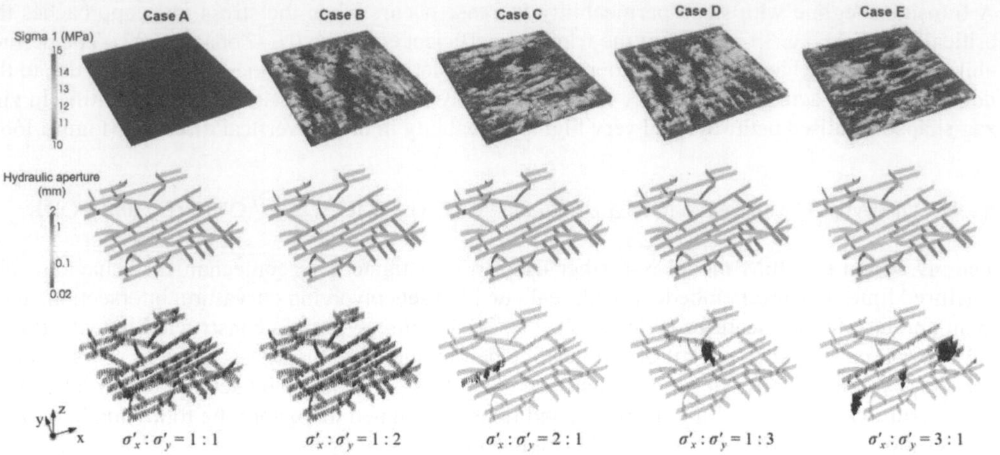

Figure 18. Distribution of maximum principal stresses, hydraulic apertures and vertical flow pathways in the fractured layer under different polyaxial stress conditions (note the flow arrow sizes indicating local flux magnitudes in Case C-E are scaled down by a factor 10 times the one in Case A & B).

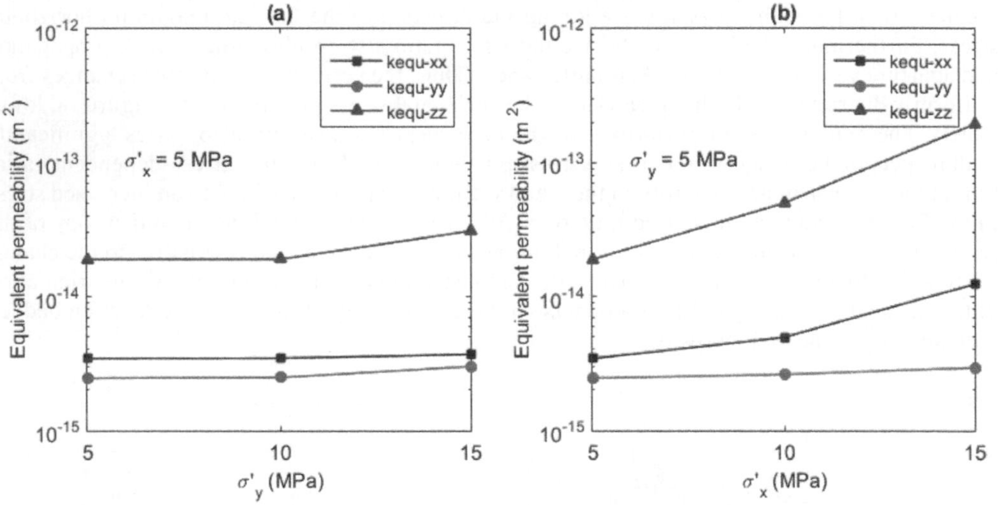

Figure 19. Variations of the equivalent permeability of the fractured layer under (a) an increased σ'_y while $\sigma'_x = 5$ MPa, or (b) an increased σ'_x while $\sigma'_y = 5$ MPa.

9 APPLICATION STUDY OF EXCAVATION DAMAGED ZONE

Subsurface rocks embedded with natural fractures are often encountered in engineering excavations for tunnel and cavern construction, hydrocarbon extraction, mining operations and geological disposal of radioactive waste. Underground excavations that perturb the rock mass from an originally equilibrated state can engender stress redistribution and trigger the formation of excavation damaged zone (EDZ) (Tsang et al. 2005). Previous EDZ numerical studies mainly focused on fracturing in intact rocks, while only a few attempts have been made to address the effects of pre-existing discontinuities. The JCM-FEMDEM model that can capture both the reactivation of pre-existing fractures and the propagation of new cracks is used to simulate the EDZ evolution around a tunnel excavation in a crystalline formation.

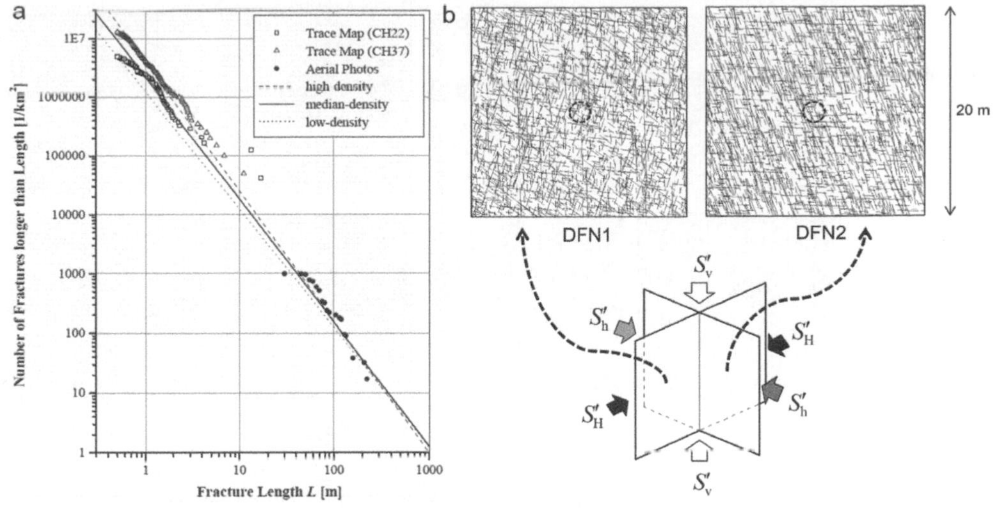

Figure 20. (a) Distribution of fracture lengths mapped at Sellafield that can be fitted by a power law cumulative distribution (Blum et al. 2005), and (b) the 20 m × 20 m DFNs generated in the cross-section plane oriented at 340° (DFN1) and 250° (DFN2) from the North (dashed circles represent the tunnels advancing in two different directions).

The numerical experiment of a hypothetical repository is based on the site characterisation of a volcaniclastic rock at the Sellafield area, Cumbria, UK (Nirex 1997a, b). Four sets of fractures were observed in the field and the fracture lengths tend to follow a power law distribution (Figure 20a). A DFN network is generated through a Poisson process with the four fracture sets assumed to have an equal density. Two 20 m × 20 m 2D cross-sections (i.e. DFN1 oriented at 340° and DFN2 oriented at 250°, from the North) are chosen corresponding to the plane defined by either the maximum (S''_H) or minimum (S'_h) horizontal stress with the vertical stress (S'_v) (Figure 20b). The circular tunnel has a diameter of 2 m and is placed at the centre of the domain. The response of the fractured rock to in-situ stresses and excavation perturbations is simulated for various tunnel depth scenarios (i.e. 250 m, 500 m and 1000 m) through multiple sequential deformation-solving phases: (i) force equilibration under the geological in-situ stress condition, (ii) central core relaxation during the excavation, (iii) physical removal of rocks inside the tunnel after the excavation, and (iv) damage evolution around the unsupported opening.

For DFN1 at the 1000 m depth (Figure 21a), the fractured rock exhibits a homogeneous stress distribution under the initial in-situ stress condition which is close to isotropic ($S'_H/S'_v = 1.17$). With the relaxation of core rocks, stress concentrations begin to appear in the fictitiously softening materials as well as the rocks surrounding the tunnel. After the removal of the rocks inside the tunnel, the model continues to solve for the consequent EDZ evolution around the man-made opening, i.e. the zone where irreversible deformation involving new crack propagation has developed. An interior low stress zone (stress loosening zone) is formed surrounding the tunnel boundary, where intensive rock mass failure develops as a result of structurally-controlled kinematic instability (e.g. key blocks) and stress-driven brittle fracturing (e.g. wing cracks). The stress loosing zone seems to have a long axis along the direction of the in-plane minimum principal stress (i.e. S'_v). A self-organised exterior high stress zone (stress arching zone) is promoted at a certain distance to the tunnel periphery, where compression arches seem to evolve along the direction of the in-plane maximum principal stress (i.e. S'_H). In contrast, the DFN2 model at the 1000 m depth exhibits significant heterogeneity under the initial in-situ stresses (Figure 21b). After the removal of rocks in the tunnel, a stress loosing zone is created around the excavation and also tends to follow the direction of the in-plane minimum principal stress (i.e. S'_h). An exterior stress arching zone is vertically formed along the in-plane maximum principal stress (i.e. S'_v), especially at the right hand side of the tunnel (the marked asymmetry). More interestingly, the high-stress contours of these arching zones seem to be microscopically constrained by the structures of pre-existing fractures.

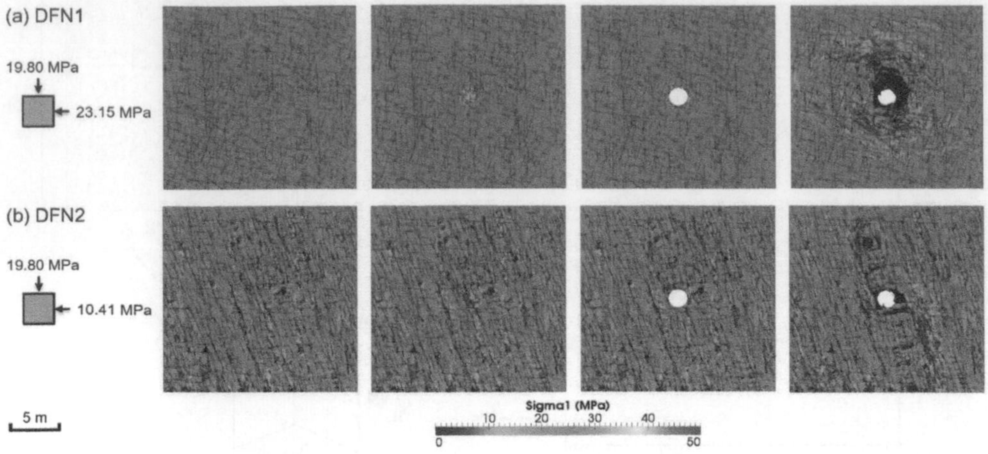

Figure 21. Damage evolution around the tunnel excavation in the 20 m × 20 m fractured rock embedded with the (a) DFN1 or (b) DFN2 at the depth of 1000 m.

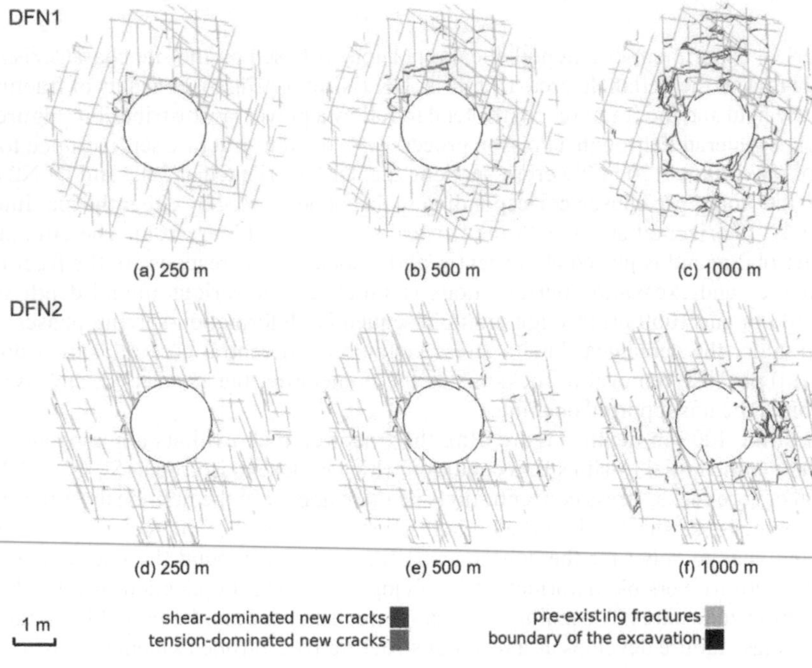

Figure 22. Fracture development at the near field to the excavation boundary in the fractured rocks at different depths, i.e. 250 m, 500 m, and 1000 m.

Figure 22 shows the near-field fracture development around the tunnel excavation in the two DFN models at various depths. At the 250 m depth, quite few new cracks emerge in both networks and the rock can almost remain stable except slight structurally-controlled falling of rock pieces. At the 500 m depth, slightly more new cracks are generated in both DFN networks. However, for the scenario of 1000 m depth, extensive tension-dominated new cracks accompanied by a few shear-dominated ones are created. The propagation of these new cracks tends to follow the direction of the in-plane maximum principal stress in each DFN model, i.e. horizontally in DFN1 and vertically in DFN2. The new cracks in DFN1 are concentrated in the rock above the tunnel top or under the invert (Figure 22c), whereas the fracturing in DFN2 mainly occurs in the lateral space (Figure 22f). It is found that excavation in the condition with a higher far-field maximum principal stress tends to generate more irreversible damage in the host rock. Based on the collapsed state of the rock mass, the shape and depth of the failure zone can be determined, which are useful for the support system design (e.g. lining thickness, bolt length and position). The results of this study have important implications for designing stable underground openings for nuclear waste repositories and other engineering facilities which are intended to generate minimal damage in host media.

10 IMPLICATIONS AND SUMMARY

10.1 *Implications for rock engineering*

(i) A tectonic interpretation was presented to explain the connectivity evolution of a multiscale natural fracture system (section 2). This work proposed an answer to the open question—Are natural fracture networks well or poorly connected? The discussion on the "effective" and "apparent" connectivity has important implications for various rock engineering problems concerned with the percolation state of natural fracture systems such as oil/gas recovery, rock mass stability, and geological disposal of radioactive waste.

(ii) A critical review of the state-of-the-art discrete fracture network modelling of hydromechanical processes in fractured rocks is presented (section 3). This review can serve as a useful guide for practitioners to choose appropriate methods for computer-based engineering design and analysis.

(iii) A general-purpose geomechanical simulator was developed to simulate the realistic fracture behaviour under stress loading including non-linear normal deformation, roughness-controlled shear strength and dilatancy, and their non-trivial size effects (section 4). This numerical method, which has been validated against well-established empirical solutions, permits rock engineers to more accurately simulate the geomechanical behaviour of natural fractures and fractured rocks in response to in-situ stresses and/or engineering perturbations.

(iv) The stochastic Poisson DFN model, which has been commonly used in the rock mechanics community, was thoroughly examined for their validity and quality in representing natural fractured systems (section 5). The results provide useful references for rock engineers about the uncertainty of stochastic DFN modelling.

(v) A novel upscaling approach was proposed for simulating larger-scale properties of fractured rocks including their fracture geometry, aperture variability, and equivalent permeability (section 6). This new method can be used by rock engineers to predict scale- and stress-dependent behaviour of geological formations in hydrocarbon, geothermal or geodisposal systems.

(vi) A workflow for integrating geological information, simulating stress perturbation, deriving aperture distribution, and calculating equivalent permeability of fractured rocks was developed for both 2D and 3D problems (sections 5, 7 & 8). Such a workflow provides a very useful practical tool for rock engineers to solve real problems.

(vii) The technique of modelling EDZ evolution in fractured formations permits rock engineers to characterise the complex interactions between pre-existing fractures and new propagating cracks under excavation-induced perturbations (section 9). The results have important implications for designing underground openings for radioactive waste repositories, mining operations, and public transportations.

10.2 *Summary*

To sum up, this thesis presented a systematic study of the geometry, geomechanics and fluid flow properties of natural fracture networks. The complexity of natural fractures with respect to hierarchical topologies and underlying mechanisms was investigated through a study of the statistics and tectonism of a multiscale fracture system. To simulate the complex geomechanical behaviour of natural fractures associated with intrinsic surface asperities, a joint constitutive model was implemented into the framework of the finite-discrete element method. The numerical model can calculate the stress/strain fields of intact rocks, capture the mechanical interactions of multiple blocks, characterise the non-linear deformation of rough fractures and mimic the propagation of new cracks. This numerical model has been applied to simulate the geomechanical behaviour of various 2D and 3D fracture networks at metre to decametre scales with the consequences on their equivalent permeability further analysed. To estimate the hydromechanical properties of a natural fracture network at larger scales, a novel upscaling approach employing discrete-time random walks in a recursive self-referencing lattice was developed to extrapolate fractures together with their stress- and scale-dependent apertures into larger domains. Distinct permeability scaling behaviour was observed for fracture networks under different in-situ stress conditions. The capability of the model was further demonstrated through an example of modelling the damage evolution around an excavation in a geological formation embedded with pre-existing natural fractures. The research findings of this thesis illustrate the importance of realistic fracture network representation and systematic geomechanical simulation for modelling the hydromechanical behaviour of naturally fractured rocks.

ACKNOWLEDGEMENT

I would like to express my sincerest gratitude to my PhD supervisor, Dr. John-Paul Latham at Imperial College London, for his continuous help, encouragement, guidance and support during my PhD studies. I acknowledge Prof. Chin-Fu Tsang for his insightful comments and advice on my PhD research. Another special thank is given to Prof. John Cosgrove, who guided me to the world of structural geology. I appreciate the help from my former colleagues at Imperial College London including Dr. Jiansheng Xiang, Dr. Xiaoguang Wang, Dr. Philipp Lang and Dr. Liwei Guo for their contributions to some of my publications during my PhD period. I thank Prof. Peter King and Prof. John Hudson at Imperial College London, and Prof. David Sanderson at University of Southampton for their constructive review and comments on my PhD thesis. In addition, I am very grateful for the financial support from the itf-ISF industrial consortium project "Improved simulation of faulted and fractured reservoirs, itf-ISF Phase 3: gravity assisted recovery processes" funded by Total, Statoil, ConocoPhillips, OMV, ExxonMobil and Armaco, and the Janet Watson Scholarship awarded by the Department of Earth Science and Engineering, Imperial College London. Finally, I would like to acknowledge the ISRM committee for financially supporting my attendance to the 2019 International Congress on Rock Mechanics and Rock Engineering at Foz do Iguassu, Brazil.

REFERENCES

Baecher, G.B. 1983. Statistical analysis of rock mass fracturing. *Mathematical Geology* 15(2): 329–348.

Baghbanan, A. & Jing, L. 2008. Stress effects on permeability in a fractured rock mass with correlated fracture length and aperture. *International Journal of Rock Mechanics and Mining Sciences* 45: 1320–1334.

Bandis, S., Lumsden, A.C. & Barton, N.R. 1981. Experimental studies of scale effects on the shear behaviour of rock joints. *International Journal of Rock Mechanics and Mining Sciences & Geomechanics Abstracts* 18: 1–21.

Bandis, S.C., Lumsden, A.C. & Barton, N.R. 1983. Fundamentals of rock joint deformation. *International Journal of Rock Mechanics and Mining Sciences & Geomechanics Abstracts* 20: 249–268.

Barton, C.C. 1995. Fractal analysis of scaling and spatial clustering of fractures. In C.C. Barton & P.R. La Pointe (eds), *Fractals in the Earth Sciences*: 141–178. New York: Plenum Press.

Barton, N. 2013. Shear strength criteria for rock, rock joints, rockfill and rock masses: Problems and some solutions. *Journal of Rock Mechanics and Geotechnical Engineering* 5(4): 249–261.

Barton, N., Bandis, S. & Bakhtar, K. 1985. Strength, deformation and conductivity coupling of rock joints. *International Journal of Rock Mechanics and Mining Sciences & Geomechanics Abstracts* 22: 121–140.

Belayneh, M. & Cosgrove, J.W. 2004. Fracture-pattern variations around a major fold and their implications regarding fracture prediction using limited data: an example from the Bristol Channel Basin. In J.W. Cosgrove & T. Engelder (eds), *The Initiation, Propagation, and Arrest of Joints and Other Fractures*: 89–102. London: Geological Society London, Special Publications.

Belayneh, M.W., Matthai, S.K., Blunt, M.J. & Rogers, S.F. 2009. Comparison of deterministic with stochastic fracture models in water-flooding numerical simulations. *American Association of Petroleum Geologists Bulletin* 93: 1633–1648.

Berkowitz, B. 2002. Characterizing flow and transport in fractured geological media: A review. *Advances in Water Resources* 25: 861–884.

Berkowitz, B., Bour, O., Davy, P. & Odling, N. 2000. Scaling of fracture connectivity in geological formations. *Geophysical Research Letters* 27: 2061–2064.

Blum, P., Mackay, R., Riley, M.S. & Knight, J.L. 2005. Performance assessment of a nuclear waste repository: Upscaling coupled hydro-mechanical properties for far-field transport analysis. *International Journal of Rock Mechanics and Mining Sciences* 42: 781–792.

Bonnet, E., Bour, O., Odling, N.E., Davy, P., Main, I., Cowie, P. & Berkowitz, B. 2001. Scaling of fracture systems in geological media. *Review of Geophysics* 39: 347–383.

Bour, O. & Davy, P. 1999. Clustering and size distributions of fault patterns: theory and measurements. *Geophysical Research Letters* 26:2001–2004.

Bour, O. & Davy, P. 1997. Connectivity of random fault networks following a power law fault length distribution. *Water Resources Research* 33: 1567–1583.

Bour, O., Davy, P., Darcel, C. & Odling, N. 2002. A statistical scaling model for fracture network geometry, with validation on a multiscale mapping of a joint network (Hornelen Basin, Norway). *Journal of Geophysical Research* 107: ETG4-1-ETG4–12.

Chelidze, T.L. 1982. Percolation and fracture. *Physics of the Earth and Planetary Interiors* 28: 93–101.

Clauser, C. 1992. Permeability of crystalline rocks. *Eos, Transactions American Geophysical Union* 73: 233–233.

Cundall, P.A. 1971. A computer model for simulating progressive, large-scale movements in blocky rock systems. In *Proc. Intern. Symp. Rock Mech., Nancy, France*: 129–136.

Darcel, C., Bour, O., Davy, P. & de Dreuzy, J.-R. 2003. Connectivity properties of two-dimensional fracture networks with stochastic fractal correlation. *Water Resources Research* 39: 1272.

Davy, P., Bour, O., de Dreuzy, J.-R. & Darcel, C. 2006. Flow in multiscale fractal fracture networks. In G. Cello & B.D. Malamud (eds), *Fractal Analysis for Natural Hazards*: 31–45. London: Geological Society London, Special Publications.

Davy, P., Le Goc, R., Darcel, C., Bour, O., de Dreuzy, J.-R. & Munier, R. 2010. A likely universal model of fracture scaling and its consequence for crustal hydromechanics. *Journal of Geophysical Research* 115: B1041.

de Dreuzy, J.-R., Davy, P. & Bour, O. 2002. Hydraulic properties of two-dimensional random fracture networks following power law distributions of length and aperture. *Water Resources Research* 38: 1276.

de Dreuzy, J.-R., Davy, P. & Bour, O. 2001a. Hydraulic properties of two-dimensional random fracture networks following a power law length distribution: 2. Permeability of networks based on lognormal distribution of apertures. *Water Resources Research* 37: 2079–2095.

de Dreuzy, J.-R., Davy, P. & Bour, O. 2001b. Hydraulic properties of two-dimensional random fracture networks following a power law length distribution: 1. Effective connectivity. *Water Resources Research* 37: 2065–2078.

Dershowitz, W.S. & Einstein, H.H. 1988. Characterizing rock joint geometry with joint system models. *Rock Mechanics and Rock Engineering* 21: 21–51.

Elmo, D., Stead, D., Eberhardt, E. & Vyazmensky, A. 2013. Applications of finite/discrete element modeling to rock engineering problems. *International Journal of Geomechanics* 13: 565–581.

Geiger, S., Roberts, S., Matthäi, S.K., Zoppou, C. & Burri, A. 2004. Combining finite element and finite volume methods for efficient multiphase flow simulations in highly heterogeneous and structurally complex geologic media. *Geofluids* 4: 284–299.

Gueguen, Y., David, C. & Gavrilenko, P. 1991. Percolation networks and fluid transport in the crust. *Geophysical Research Letters* 18: 931–934.

Herbert, A.W. 1996. Modelling approaches for discrete fracture network flow analysis. In O. Stephansson, L. Jing & C.-F. Tsang (eds), *Coupled Thermo-Hydro-Mechanical Processes of Fractured Media*: 213–229. Amsterdam: Elsevier.

Hoek, E. 1983. Strength of jointed rock masses. *Géotechnique* 23: 187–223.

Hoek, E. & Martin, C.D. 2014. Fracture initiation and propagation in intact rock – A review. *Journal of Rock Mechanics and Geotechnical Engineering* 6(4): 287–300.

Jing, L. 2003. A review of techniques, advances and outstanding issues in numerical modelling for rock mechanics and rock engineering. *International Journal of Rock Mechanics and Mining Sciences* 40: 283–353.

Jing, L. & Hudson, J.A. 2002. Numerical methods in rock mechanics. *International Journal of Rock Mechanics and Mining Sciences* 39: 409–427.

Klimczak, C., Schultz, R., Parashar, R. & Reeves, D. 2010. Cubic law with aperture-length correlation: implications for network scale fluid flow. *Hydrogeology Journal* 18: 851–862.

Lang, P.S., Paluszny, A. & Zimmerman, R.W. 2014. Permeability tensor of three-dimensional fractured porous rock and a comparison to trace map predictions. *Journal of Geophysical Research: Solid Earth* 119: 6288–6307.

Latham, J.-P., Xiang, J., Belayneh, M., Nick, H.M., Tsang, C.-F. & Blunt, M.J. 2013. Modelling stress-dependent permeability in fractured rock including effects of propagating and bending fractures. *International Journal of Rock Mechanics and Mining Sciences* 57: 100–112.

Lei, Q., Latham, J.-P. & Tsang, C.-F. 2017a. The use of discrete fracture networks for modelling coupled geomechanical and hydrological behaviour of fractured rocks. *Computers and Geotechnics* 85: 151–176.

Lei, Q., Latham, J.-P., Tsang, C.-F., Xiang, J. & Lang, P. 2015a. A new approach to upscaling fracture network models while preserving geostatistical and geomechanical characteristics. *Journal of Geophysical Research: Solid Earth* 120: 4784–4807.

Lei, Q., Latham, J.-P. & Xiang, J. 2016. Implementation of an empirical joint constitutive model into finite-discrete element analysis of the geomechanical behaviour of fractured rocks. *Rock Mechanics and Rock Engineering* 49: 4799–4816.

Lei, Q., Latham, J.-P., Xiang, J., Tsang, C.-F., Lang, P. & Guo, L. 2014. Effects of geomechanical changes on the validity of a discrete fracture network representation of a realistic two-dimensional fractured rock. *International Journal of Rock Mechanics and Mining Sciences* 70: 507–523.

Lei, Q., Latham, J.-P., Xiang, J. & Tsang, C.-F. 2017b. Role of natural fractures in damage evolution around tunnel excavation in fractured rocks. *Engineering Geology* 231: 100–113.

Lei, Q., Latham, J., Xiang, J. & Tsang, C.-F. 2015b. Polyaxial stress-induced variable aperture model for persistent 3D fracture networks. *Geomechanics for Energy and the Environment* 1: 34–47.

Lei, Q. & Wang, X. 2016. Tectonic interpretation of the connectivity of a multiscale fracture system in limestone. *Geophysical Research Letters* 43: 1551–1558.

Lei, Q., Wang, X., Xiang, J. & Latham, J.-P. 2017c. Polyaxial stress-dependent permeability of a three-dimensional fractured rock layer. *Hydrogeology Journal* 25: 2251–2262.

Leung, C.T.O. & Zimmerman, R.W. 2012. Estimating the hydraulic conductivity of two-dimensional fracture networks using network geometric properties. *Transport in Porous Media* 93: 777–797.

Lisjak, A. & Grasselli, G. 2014. A review of discrete modeling techniques for fracturing processes in discontinuous rock masses. *Journal of Rock Mechanics and Geotechnical Engineering* 6: 301–314.

Long, J.C.S., Remer, J.S., Wilson, C.R. & Witherspoon, P.A. 1982. Porous media equivalents for networks of discontinuous fractures. *Water Resources Research* 18: 645–658.

Madden, T.R. 1983. Microcrack connectivity in rocks: A renormalization group approach to the critical phenomena of conduction and failure in crystalline rocks. *Journal of Geophysical Research* 88: 585–592.

Mas Ivars, D., Pierce, M.E., Darcel, C., Reyes-Montes, J., Potyondy, D.O., Young, R.P. & Cundall, P.A. 2011. The synthetic rock mass approach for jointed rock mass modelling. *International Journal of Rock Mechanics and Mining Sciences* 48: 219–244.

Min, K.-B., Rutqvist, J., Tsang, C.-F. & Jing, L. 2004. Stress-dependent permeability of fractured rock masses: a numerical study. *International Journal of Rock Mechanics and Mining Sciences* 41: 1191–1210.

Munjiza, A. 2004. *The Combined Finite-Discrete Element Method*. London: Wiley.

Nirex. 1997a. *Assessment of the in situ stress field at Sellafield (Nirex Report)*. UK: Harwell.

Nirex. 1997b. *Evaluation of heterogeneity and scaling of fractures in the Borrowdale Volcanic Group in the Sellafield Area (Nirex Report)*. UK: Harwell.

Oda, M., Yamabe, T., Ishizuka Y, Kumasaka, H., Tada, H. & Kimura, K. 1993. Elastic stress and strain in jointed rock masses by means of crack tensor analysis. *Rock Mechanics and Rock Engineering* 26: 89–112.

Odling, N.E. 1997. Scaling and connectivity of joint systems in sandstones from western Norway. *Journal of Structural Geology* 19: 1257–1271.

Olson, J.E. 1993. Joint pattern development: Effects of subcritical crack growth and mechanical crack interaction. *Journal of Geophysical Research* 98: 12251–12265.

Paluszny, A. & Matthäi, S.K. 2009. Numerical modeling of discrete multi-crack growth applied to pattern formation in geological brittle media. *International Journal of Solids and Structures* 46: 3383–3397.

Petit, J.-P. & Mattauer, M. 1995. Palaeostress superimposition deduced from mesoscale structures in limestone: the Matelles exposure, Languedoc, France. *Journal of Structural Geology* 17: 245–256.

Petit, J.-P., Wibberley C.A.J. & Ruiz, G. 1999. "Crack-seal", slip: A new fault valve mechanism? *Journal of Structural Geology* 21: 1199–1207.

Pollard, D.D. & Aydin, A. 1988. Progress in understanding jointing over the past century. *Geological Society of America Bulletin* 100: 1181–1204.

Pollard, D.D. & Segall, P. 1987. Theoretical displacements and stresses near fractures in rock: With applications to faults, joints, veins, dikes, and solution surfaces. In B.K. Atkinson (ed), *Fracture Mechanics of Rock*: 277–349. San Diego: Academic Press.

Price, N.J. & Cosgrove, J.W. 1990. *Analysis of Geological Structures*. New York: Cambridge University Press.

Renshaw, C.E. 1997. Mechanical controls on the spatial density of opening-mode fracture networks. *Geology* 25: 923–926.

Renshaw, C.E. 1996. Influence of subcritical fracture growth on the connectivity of fracture networks. *Water Resources Research* 32: 1519–1530.
Renshaw, C.E. 1998. Sample bias and the scaling of hydraulic conductivity in fractured rock. *Geophysical Research Letters* 25: 121–124.
Renshaw, C.E. & Pollard, D.D. 1994. Numerical simulation of fracture set formation: A fracture mechanics model consistent with experimental observations. *Journal of Geophysical Research* 99: 9359–9372.
Ronayne, M.J. & Gorelick, S.M. 2006. Effective permeability of porous media containing branching channel networks. *Physical Review E* 73: 1–10.
Rutqvist, J., Leung, C., Hoch, A., Wang, Y. & Wang, Z. 2013. Linked multicontinuum and crack tensor approach for modeling of coupled geomechanics, fluid flow and transport in fractured rock. *Journal of Rock Mechanics and Geotechnical Engineering* 5: 18–31.
Rutqvist, J. & Stephansson, O. 2003. The role of hydromechanical coupling in fractured rock engineering. *Hydrogeology Journal* 11: 7–40.
Schultz, R. 2000. Growth of geologic fractures into large-strain populations: review of nomenclature, subcritical crack growth, and some implications for rock engineering. *International Journal of Rock Mechanics and Mining Sciences* 37: 403–411.
Shi, G.-H. & Goodman, R.E. 1985. Two dimensional discontinuous deformation analysis. *International Journal for Numerical and Analytical Methods in Geomechanics* 9: 541–556.
Tsang, C.-F. 1991. Coupled hydromechanical-thermochemical processes in rock fractures. *Review of Geophysics* 29(4): 537–551.
Tsang, C.-F., Bernier, F. & Davies, C. 2005. Geohydromechanical processes in the Excavation Damaged Zone in crystalline rock, rock salt, and indurated and plastic clays—in the context of radioactive waste disposal. *International Journal of Rock Mechanics and Mining Sciences* 42: 109–125.
Tsang, C.-F. & Neretnieks, I. 1998. Flow channeling in heterogeneous fractured rocks. *Review of Geophysics* 36: 275–298.
Zhang, X. & Sanderson, D.J. 1998. Numerical study of critical behaviour of deformation and permeability of fractured rock masses. *Marine and Petroleum Geology* 15: 535–548.
Zhang, X. & Sanderson, D.J. 1996. Effects of stress on the two-dimensional permeability tensor of natural fracture networks. *Geophysical Journal International* 125: 912–924.
Zhao, Z., Rutqvist, J., Leung, C., Hokr, M., Liu, Q., Neretnieks, I., Hoch, A., Havlíček, J., Wang, Y., Wang, Z., Wu, Y. & Zimmerman, R.W. 2013. Impact of stress on solute transport in a fracture network: A comparison study. *Journal of Rock Mechanics and Geotechnical Engineering* 5: 110–123.
Zimmerman, R.W. & Main, I. 2004. Hydromechanical behavior of fractured rocks. In Y. Gueguen & M. Bouteca (eds), *Mechanics of Fluid-Saturated Rocks*: 363–421. London: Elsevier.
Zoback, M.D. 2007. *Reservoir Geomechanics*. Cambridge: Cambridge University Press.

Author index

Proceedings in Earth and geosciences series

The *Proceedings in Earth and geosciences series* contains proceedings of peer-reviewed international conferences dealing in earth and geosciences. The main topics covered by the series include: geotechnical engineering, underground construction, mining, rock mechanics, soil mechanics and hydrogeology.

ISSN: 2639-7749
eISSN: 2639-7757

1. Tunnels and Underground Cities: Engineering and Innovation meet Archaeology, Architecture and Art
   Edited by Daniele Peila, Giulia Viggiani & Tarcisio Celestino
   ISBN: 978-1-138-38865-9 (Hbk + USB)
   ISBN: 978-0-429-42444-1 (eBook)

2. Geotechnics Fundamentals and Applications in Construction
   Edited by Rashid Mangushev, Askar Zhussupbekov, Yoshinori Iwasaki & Igor Sakharov
   ISBN: 978-0-367-17983-0 (Hbk)
   ISBN: 978-0-429-05888-2 (eBook)

3. Mining Goes Digital
   Edited by Christoph Mueller, Winfred Assibey-Bonsu, Ernest Baafi, Christoph Dauber, Chris Doran, Marek Jerzy Jaszczuk & Oleg Nagovitsyn
   ISBN: 978-0-367-33604-2 (Hbk)
   ISBN: 978-0-429-32077-4 (eBook)

4. Earthquake Geotechnical Engineering for Protection and Development of Environment and Constructions
   Edited by Francesco Silvestri &Nicola Moraci
   ISBN: 978-0-367-14328-2 (Set of Hbk and Multimedia)
   ISBN: 978-0-429-03127-4 (eBook)

5. Rock Mechanics for Natural Resources and Infrastructure Development. Invited Lectures
   Edited by Sergio A.B. da Fontoura, Ricardo José Rocca & José Félix Pavón Mendoza
   ISBN: 978-0-367-42285-1 (Hbk)
   ISBN: 978-0-367-82318-4 (eBook)

6. Rock Mechanics for Natural Resources and Infrastructure Development. Full Papers
   Edited by Sergio A.B. da Fontoura, Ricardo José Rocca & José Félix Pavón Mendoza
   ISBN: 978-0-367-42284-4 (Hbk)
   ISBN: 978-0-367-82317-7 (eBook)